Bioresource Utilization and Bioprocess

Sadhan Kumar Ghosh · Ramakrishna Sen ·
H. N. Chanakya · Agamuthu Pariatamby
Editors

Bioresource Utilization and Bioprocess

Springer

Editors
Sadhan Kumar Ghosh
Department of Mechanical Engineering
Jadavpur University
Kolkata, West Bengal, India

H. N. Chanakya
Centre for Sustainable Technology
Indian Institute of Science Bangalore
Bengaluru, India

Ramakrishna Sen
Indian Institute of Technology Kharagpur
Kharagpur, West Bengal, India

Agamuthu Pariatamby
Jeffrey Sachs Center on Sustainable Development
Sunway University
Kuala Lumpur, Malaysia

High Level End Foreign Expert
Ministry of Science and Technology
Beijing, China

ISBN 978-981-15-1606-1 ISBN 978-981-15-1607-8 (eBook)
https://doi.org/10.1007/978-981-15-1607-8

This Springer imprint is published by the registered company Springer Nature Singapore Pte Ltd.
The registered company address is: 152 Beach Road, #21-01/04 Gateway East, Singapore 189721, Singapore

Preface

Extensive use of fossil fuels, rampant industrialization with unabated GHG emissions and mindless deforestation coupled with unscientific disposal of municipal and industrial wastes have caused unceasing environmental pollution leading to global warming-driven climate change. In order to reverse the deleterious effects of global environmental pollution and consequent climate change, the developed and developing countries have mandated to promote and support sustainable approaches towards bioresource development, processing and utilization through well-designed and optimized bioprocesses. Utilization of bioresources via thermo-chemical and biochemical processes has attracted researchers from all over the world. It is a wide area which has given birth to concepts like biorefinery as well as a new stream known as biotechnology. The bioprocesses open up a new dimension of utilization of abundant biomass, agricultural waste, fruit peel refuse and even organic fraction of municipal solid waste (OFMSW). While there are several challenges which hinder the successful implementations, there is immense potential for these technologies for scale-up. Bioethanol and biodiesel are no longer a myth today. This inspires more research, discussion and policy implementation for successful conversion of waste to bioethanol, waste to biobutanol, algae to bio-diesel, algae to value-added products, etc. from myth to reality.

Undoubtedly, it requires discussions and fine-tuning of the relevant factors along with effective supply chain network design and effective business models for a sustainable future. This book focuses on the utilization of bioresources and their conversion pathways for a sustainable future.

The 8th IconSWM 2018 received 380 abstracts and 320 full papers from 30 countries. A total of 300 accepted full papers have been presented as oral and poster presentations in November at ANU, Guntur, AP, India. The chapters finally selected by the board have been thoroughly reviewed by the experts for the book "*Bioresource Utilization and Bioprocesses*" dealing with the study of food waste, resource recovery from organic fraction of MSW, anaerobic digestion and hydrothermal carbonization, biosorption, algal biodiesel and petro-diesel, production of bioethanol, Jatropha biodiesel blendings, nano-additive using D.I. diesel engine, metal extraction from discarded PCB using *Aspergillus tubingensis*, leaf litter biogas, dye adsorption in water treatment, bioconversion of mango industrial waste into vermi-compost, co-composting of cattle dung and pigeon pea stalk, effective recycling of flower waste, biodiesel production, enzyme activity of lipolytic bacteria isolated from degrading oil cakes, etc.

The IconSWM movement was initiated for better waste management, resource circulation and environmental protection since the year 2009 through generating awareness and bringing all the stakeholders together from all over the world under the aegis of the International Society of Waste Management, Air and Water (ISWMAW). It established a few research projects across the country including the CST at the Indian Institute of Science, Jadavpur University, KIIT and Calcutta University. Consortium of Researchers in International Collaboration (CRIC) and many other organizations across the world are helping the IconSWM-CE movement. IconSWM has become significantly one of the biggest platforms in India for knowledge sharing, awareness generation and encouraging the urban local bodies, government departments, researchers, industries, NGOs, communities and other stakeholders in the area of waste management. The primary agenda of this conference is to reduce the waste generation by encouraging the implementation of 3R (Reduce, Reuse and Recycle) and circular economy concept and management of the generated waste ensuring resource circulation. The conference will show a paradigm and provide holistic pathways to waste management and resource circulation conforming to urban mining and circular economy.

The success of the 8th IconSWM is the result of a significant contribution of many organizations and individuals, specifically the government of Andhra Pradesh, several industry associations, chamber of commerce and industries, the AP higher education council, Swachh Andhra Mission and various organizations in India and in different countries as our partners including the UNEP, UNIDO and UNCRD. The 8th IconSWM 2018 was attended by nearly 823 delegates from 22 countries. The 9th IconSWM-CE 2019 was held at KIIT, Bhubaneswar, Odisha, during 27–30 November 2019, participated by 21 countries. Shri Venkateswara University (SVU) has expressed their willingness to organize the 10th IconSWM-CE at SVU in the temple city Tirupati, Andhra Pradesh, tentatively during 02–05 December 2020. This book will be helpful for the researchers, educational and research institutes, policy makers, government, implementers, ULBs and NGOs. Hope to see you all in the 10th IconSWM-CE 2020 at Tirupati.

Jadavpur University, Kolkata, India
March 2020

Prof. Sadhan Kumar Ghosh
Prof. Agamuthu Pariatamby
Dr. H. N. Chanakya
Prof. Ramakrishna Sen

Acknowledgements

The Hon'ble Chief Minister of the state of Andhra Pradesh and Hon'ble Minister of MA&UD, Andhra Pradesh for taking personal interest in this conference.

We are indebted to Shri. R. Valavan Karikal, IAS, Dr. C. L. Venkata Rao, Shri. B. S. S. Prasad, IFS (Retd.), Prof. S. Vijaya Raju and Prof. A. Rajendra Prasad, VC, ANU, for their unconditional support and guidance for preparing the platform for successful 8th IconSWM at Guntur, Vijayawada, AP.

We must express my gratitude to Mr. Vinod Kumar Jindal, ICoAS; Shri. D. Muralidhar Reddy, IAS; Shri. K. Kanna Babu, IAS; Mr. Vivek Jadav, IAS; Mr. Anjum Parwez, IAS; Mr. Bala Kishore; Prof. S. Varadarajan; Mr. K. Vinayakam; Prof. Shinichi Sakai, Kyoto University, JSMCWM; Prof. Y. C. Seo and Prof. S. W. Rhee of KSWM; Shri. C. R. C. Mohanty of UNCRD; members of Industry Associations in Andhra Pradesh; Prof. Agamuthu Pariatamby, WM&R; Prof. M. Nelles, Rostock University; Dr. Rene Van Berkel of UNIDO; and Ms. Kakuko Nagatani-Yoshida and Mr. Atul Bagai of UNEP and UN Delegation to India for their active support.

IconSWM-ISWMAW Committee acknowledges the contribution and interest of all the sponsors, industry partners, industries, co-organizers, organizing partners around the world; the government of Andhra Pradesh; Swachh Andhra Corporation as the principal collaborator, the vice chancellor and all the professors and academic community at Acharya Nagarjuna University (ANU); the chairman, vice chairman, secretary and other officers of AP State Council of Higher Education for involving all the universities in the state; the chairman, member secretary and the officers of the AP Pollution Control Board, the director of factories, the director of boilers, director of mines and officers of different ports in Andhra Pradesh; and the delegates and service providers for making successful 8th IconSWM.

We must specially mention the support and guidance by each of the members of the International Scientific Committee, CRIC members, the core group members and the local organizing committee members of 8th IconSWM who are the pillars for the success of the programme. The editorial board members including the reviewers, authors and speakers and Mr. Aninda Bose and Ms. Kamiya Khatter of

M/s. Springer India Pvt. Ltd. deserve thanks who were very enthusiastic in giving me inputs to bring this book.

We must mention the active participation of all the team members in IconSWM Secretariat across the country with special mention of Prof. H. N. Chanakya and his team in IISc Bangalore, Ms. Sheetal Singh and Dr. Sandhya Jaykumar and their team in CMAK and BBMP, Mr. Saikesh Paruchuri, Mr. Anjaneyulu, Ms. Senophiah Mary, Mr. Rahul Baidya, Ms. Ipsita Saha, Mr. Suresh Mondal, Mr. Bisweswar Ghosh, Mr. Gobinda Debnath and the research team members in Mechanical Engineering Department and ISWMAW, Kolkata HQ, for various activities for the success of the 8th IconSWM 2018. Special thanks to Sannidhya Kumar Ghosh, being the governing body member of ISWMAW supported the activities from the USA.

I am indebted to Mrs. Pranati Ghosh who gave me guidance and moral support in achieving the success of the event. Once again, the IconSWM and ISWMAW express gratitude to all the stakeholders, delegates and speakers who are the part of the success of 8th IconSWM 2018.

Prof. Sadhan Kumar Ghosh

Contents

About the Editors

Prof. Sadhan Kumar Ghosh, Ph.D. (Engg.), Dean, Faculty of Engineering & Technology & Professor, Jadavpur University, India, is an internationally renowned figure in the fields of waste management, supply chain management, circular economy, green manufacturing, low-carbon technologies, and ISO standards, has authored 210 publications, and edited 30 books/proceedings. He has participated in research collaborations in 34 countries, and his projects have been funded by EU Horizon 2020 (2018), the Royal Academy of Engineering, Government of India, etc. He was an International Expert/Consultant for the UN DESA/UNCRD in 2016 and APO in 2014. He holds 2 Indian patents and 1 in Bangladesh.

Dr. Ramakrishna Sen, Ph.D. (Eng.), is a Professor of Biotechnology, IIT Kharagpur. He served on a Fulbright Visiting Professor Fellowship at Columbia University and has pursued R&D activities in Healthcare, Energy, Environment & Water, focusing on Green Processes & Product Development for anticancer biomolecules; marine bacterial green surfactants; microbial and microalgal biorefineries for biofuels, and value-added products with concomitant CO2 sequestration and waste valorization. He has authored 200 publications, holds 1 patent, and wrote the book 'Biosurfactants.'

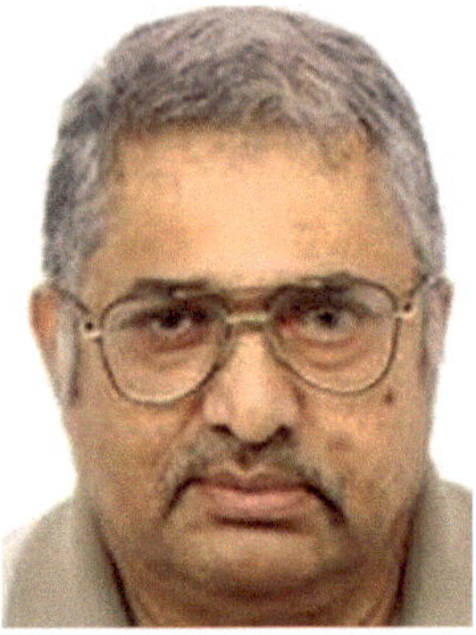

H. N. Chanakya, Ph.D. (Microbiology), is Chief Scientist at the CST (formerly ASTRA Center), Associate Faculty at CiSTUP, CCS & the National Coordinator of the Indo-UK project on hybrid energy RHEES at the Indian Institute of Science, India. His R&D efforts concern the following broad areas: Biomethanation, AD and W2E, Wastewater Purification, Decentralized/Sustainable Rural Energy, Participatory Technology Development & Appropriate Technologies. He has authored 250 publications.

Dr. Agamuthu Pariatamby, is a Senior Professor in the Jeffrey Sachs center on Sustainable Development at Sunway University. Prior to this he was attached to University of Malaya for 44 years. He is a Fellow of the Academy of Sciences, Malaysia. He is the Founder Head of the Center for Research in Waste Management. He is appointed as the High-Level Foreign Expert for the Ministry of Science and Technology, China and is also a Visiting Professor for Zhejiang University of Technology in Hangzhou, China. Current responsibilities include Senior Editor in Chief of Waste Management and Research, and Vice-President of the Society of Solid Waste Management Experts in Asia and Pacific Islands (SWAPI) and the Founder President of the Malaysian Society of Waste Management and Environment. He has authored over 460 peer-reviewed articles, proceedings and invited papers. He has done over 75 consultancy projects and supervised over 200 Master's Degree students and 25 doctoral students. He has international cooperation in several countries such as UK, China, Austria, Japan, India to name a few.

A Comprehensive Study of Food Waste to Biogas Plant: Paths to Improve the Performance—A Case Study

N. Dhinesh, J. Metilda Annamary, K. Iswarya, K. Vadivel Murugan and V. Kirubakaran

Abstract In India, currently 49 lakhs of small- and medium-sized biogas plant setups are available and working with average efficiency of 40–60%. Normally, most of the biogas plants are fed with any one variant of biodegradable material. A 25 m^3 biogas plant has been set up in The Gandhigram Rural Institute—Deemed University with food waste as feed stock. The biogas has been used for cooking application in the ladies hostel. This paper analyzes the daily feed rate vs gas generated. The composition of biogas generated has been analyzed using online biogas analyzer. From the analysis, a chart has been prepared to add inoculum for improvement of gas quality as well as quantity. The burning efficiency has also been estimated and reported in this paper. The feasibility study of partial replacement of fuel with biogas for the boiler is also presented in the paper.

Keywords Biogas · Food waste · Inoculum · Boiler

1 Introduction

Biogas is a green energy which produced from the decomposition of organic matter. It is emitted from the anaerobic digestion of biodegradable material and comprises mainly methane. Those gases are predominantly used in cooking. In the world, mostly natural gas is used for cooking purpose, which is obtained from fossil fuels. Fossil fuels are non-renewable energy which is exhaustible in nature. These resources are depleted easily and produce large amount of greenhouse gas. To avoid the over depletion of natural gas, biogas acts as the alternate; hence, the substitution for natural gas is biogas. Nowadays, biogas plants are very popular in India, large-scale biogas plant and some small-scale biogas plants are also installed for domestic purpose. But there is a lack of performance maintaining in our country. In India, currently 49 lakhs of biogas plants are constructed and working in the average efficiency.

N. Dhinesh · J. Metilda Annamary · K. Iswarya · K. Vadivel Murugan · V. Kirubakaran (✉)
Centre for Rural Energy, The Gandhigram Rural Institute—Deemed to be University, Dindigul, Tamil Nadu 624302, India
e-mail: kirbakaran@yahoo.com

S. K. Ghosh et al. (eds.), *Bioresource Utilization and Bioprocess*,
https://doi.org/10.1007/978-981-15-1607-8_1

The generation of biogas from decomposition of biodegradable matter is carried out in different temperature conditions. In India, there is a generation of 960 million tons of solid wastes per annum. In these wastes, up to 50% contain biodegradable matter and rest of the 50% are non-biodegradable matter. The biodegradable matter is prior material for production of biogas. The vegetable and food wastes are collected and fed into the biogas plant continuously; then, it produces biogas in nature. But the performance of the biogas plant is very poor, because of the improper feed. Hence, this plant moves to inoculum addition process in different ratio. The purpose of adding inoculums is to accelerate the decomposition of material by microorganisms. Inoculums contain the required microorganisms to degrade the organic matter efficiently.

If inoculums added in biogas plant, then wait for hydraulic retention period, i.e., for 7 days or if the plant continuously feeds digester, then it produces the biogas in 2–3 days. The burning efficiency is noted, and it is improved in the following days after adding inoculums in it. Hence, the partial replacement of natural gas with biogas for the boiler is also studied.

2 Literature Survey

Ofoefule et al. (2010) reported that biogas production was increased by blending of paper waste and cow dung. This blending was subject to anaerobic digestion under a 45 days retention period in a mesophilic temperature 26–43 °C. From observing these conditions on the feedstock flammability increased due to blending paper waste with animal wastes. Paper waste in the environment creating nuisance are prevented by using these method. The biogas yield of waste paper is lower than compared with a blending. The performance of biogas plant is increased by variation in the thermal characteristics of the microbes (Chamarthi et al. 2013). Temperature is prime factor in the production of biogas. Because the microbes in the feedstock is working based the temperature maintenance in the digester. Experiment carried out to modify the existing mild steel dome into transparent dome to capture greenhouse effect. The greenhouse gas will increase the temperature inside of the dome. The transparent dome will penetrate the light and trapped the heat inside of the dome. Food wastes are collected from the kitchen and feed into digester to produce biogas (Ziauddin and Rajesh 2015). In continuously fed digester requires addition of sodium hydroxide to maintain the alkalinity and pH to neutral. In this batch reactor the inoculums is added along with the kitchen waste and evaluate the performance of producing gas. The performance of biogas is improved by method of water scrubbing method (Tira et al. 2015). The circulated water absorption method is effective method for purification of biogas. This method is to increase the level of methane. This method is concentrated on decreasing the level of carbon dioxide and hydrogen sulphide. These compounds are flowed outward from the scrubbing unit which increases the quantity of methane. Biogas is produced from the decomposition of organic wastes. In leather industry the wastes are cleaved as inorganic wastes which are also used as a feedstock for the production of biogas (Apruzzese et al. 2017). In the pulp and paper industry, the removal of hemicelluloses from cellulose by means of enzymatic treatment is bioleaching. The leather wastes are difficult to manage hence the decomposition is undertaken by mixing of paper in it and through the process of anaerobic digestion.

3 Biogas Plant

The plant which is helpful to produce biogas under anaerobic condition is called as biogas plant. There are various dimensions are existing to construct biogas plant, and also different types of biogas plants are available in the market, based upon the drum and yield in the plant to produce biogas. Based on the drum, they are differentiated as fixed drum and floating drum type. The biogas reactors are also classified based upon their yield; they are low-yield reactors and high-yield reactors. The high-yield reactors are having very efficient design, and the cost of the plant is also varying, due to the design of the plant. The low-yield reactors are mostly used in market, because of the low cost. The most important design of the plant depends on the pressure and volume. In this, two different types of design are available, i.e., constant pressure and constant volume. In constant pressure type, the plant should be in floating drum design; only the drum will move up and down depending on the volume of gas. Due to constant pressure, this type of design needs suction pump to take out the gas. In constant volume type, the plant should be in fixed drum design so that the pressure will vary depending upon the gas level. Hence, there is no need of suction pump here to take out the gas.

A. *Layout of Biogas Plant*

Figures 1 and 2 show the top view and section view of biogas plant which is installed in The Gandhigram Rural Institute. In this plant, the constant pressure model is designed so it is floating drum method. Basically, all the biogas plants contain four stages, such as crusher machine, pre-digester, main digester, and slurry collection tank drum. Because of the constant pressure method, the plant needs the support of suction pump to take out the gas for cooking purpose.

Digester and gas holder are made up of fiber glass-reinforced plastic (FRP) because of the prevention of corrosion and maintenance cost. Table 1 shows the dimension of the existing biogas plant.

B. *Electrical Energy Consumption of Biogas Plant*

The electrical energy is utilized for the equipments which are used to prepare feedstock in biogas plant. The equipments are crusher machine, sludge pump, and biogas blower. The crusher machine is used for the purpose of sizing of feedstock. Sludge pump is handling the slurry disposed from the biogas plant. A blower is a machine used for moving volumes of a gas with moderate increase of pressure (Table 2).

C. *Snapshots of Existing Biogas Plant*

See Figs. 3, 4, 5, 6, 7 and 8.

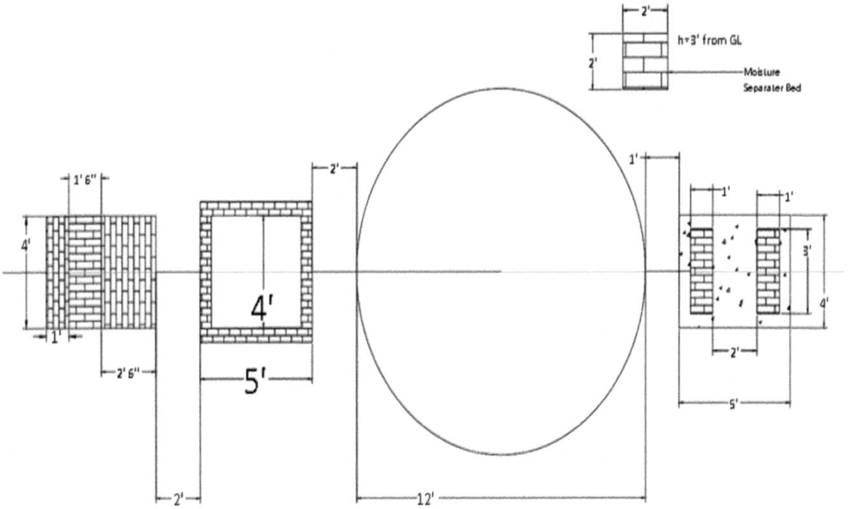

Fig. 1 Top view of biogas plant

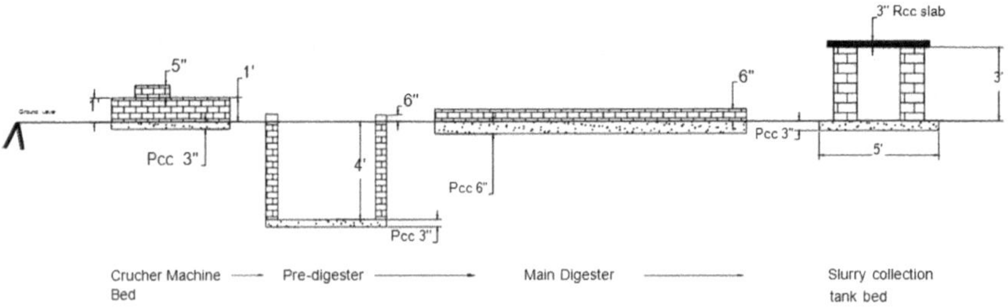

Fig. 2 Section view of plant

Table 1 Details of existing biogas plant

S. no.	Particulars	Digester	Gas Holder
1	Made by	FRP	FRP
2	Height	10′	4′ 6″
3	Diameter	10′ 8″	10′
4	Volume (m^3)	25	10
5	Gas line	1. Braided hose and uPVC pipe are used for installation work	

Table 2 Electrical usage of biogas plant

S. no.	Particulars	Phase	H.P	Current (amps)	K.W
1	Crusher machine	3	2	3.5	1.5
2	Sludge pump	3	1	2.5	0.75
3	Biogas blower	Single	0.25	1.7	0.2

Fig. 3 Full view of existing plant

4 Production of Biogas

For the production of biogas, there are different types of steps to be processed in the biogas plant. Those different processes are as follows:

1. Food waste preparation
2. Digestion (fermentation), consisting of hydrolysis, acetogenesis, acidogenesis, and methanogenesis
3. Conversion of the biogas to renewable electricity and useful heat with cogeneration/combined heat and power
4. Post-treatment of the digester.

Initially, the food waste to the digesters is received in a liquid storage tank. The food waste should be properly sized to feed into the digester. Food waste is mixed with same ratio of water to solubilize which is the hydrolysis process. From here, it is laid into the digester. Anaerobic digestion process is carried out under absence of oxygen in the digester. The biogas is produced by the biological process involved inside the digestion tank. It then undertakes the transitional steps of acidogenesis and acetogenesis. This creates the predecessor molecule to methanogenesis. Methanogens produce methane as a cellular waste product. That methane is called biogas, and it is produced inside the gas holder. The gas holder rises above the digester and indicates the production of biogas. From this way, it is possible to convert the waste disposed

Fig. 4 Surge collection point

from the kitchen into biogas. This biogas can be able to convert thermal energy form and also electricity. Biogas constituent is mainly methane which is flammable. So it is alternative for liquid petroleum gas which is used for cooking purposes in kitchen. It is adoptable way for the biodegradable waste to decompose without affecting the environment. Because biodegradable waste is mostly disposed in the storm drains, it will cause the water and land pollution. By this way, the biogas can be produced, and it also gave a solution for the disposal of food waste in proper manner.

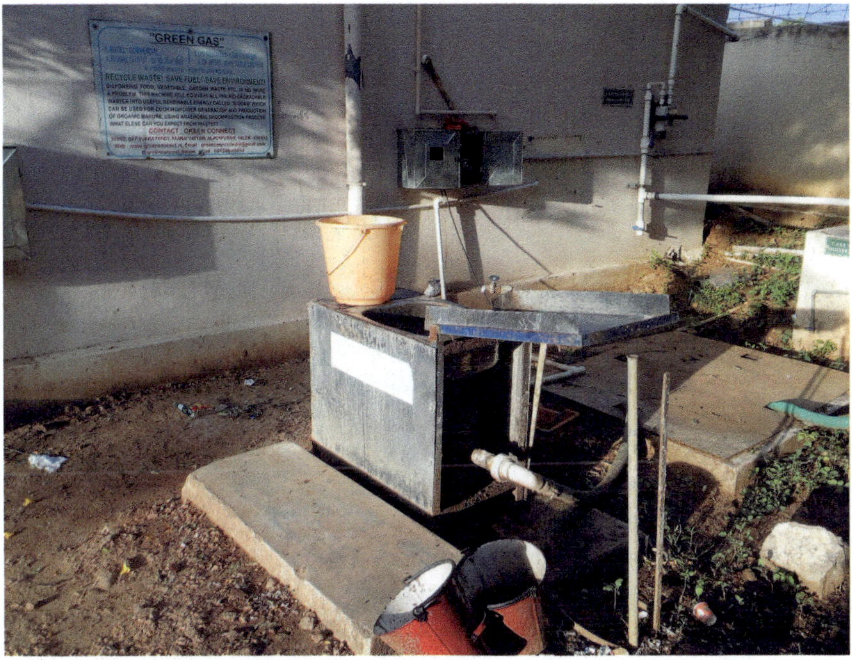

Fig. 5 Feed grinding machine

5 Performance Analysis of Existing Plant

This paper evaluates the functioning of existing biogas plant with capacity of 200 kg food waste. Here the food waste is taken from the Gandhigram Rural Institute women's hostel. In previous years, the biogas plant was working with only food waste and the biogas used for cooking purpose and water heating. The time taken for cooking in LPG gas and biogas varies slightly; there is no immense different between those two gases. But the level of methane percentage in this biogas will vary due to the different input food waste. This fluctuation in the methane level will cause the possibility of low flame and also in sometime flickers will occur while cooking. This reaction will reduce the plant efficiency and disuse of biogas. To avoid this problem, the permanent inoculum is introduced in this plant with the feed.

Fig. 6 Pre-digester

6 Performance Analysis of Plant with Inoculum

Biogas plant will decrease its efficiency, when there is a lack of maintenance and usage. In vacation period, institute affects the incoming feed to the biogas plant. If there is any change in continuous feeding of the biogas plant, it leads to affect the pH level of the digester in the biogas plant. If there is a lack of feeding for some months will degrade the pH level. The decrease of pH level affects the production of biogas. The calorific value and the quantity of the biogas will decrease because of the pH imbalance. The produced biogas is also having low methane and high CO_2 level in it. Our aim here is to increase the level of methane and also the quantity of biogas. To increase the amount of methane, it is revised to add inoculum in the digester. The addition of inoculum is necessary to improve the methanogenic bacteria; these

Fig. 7 Biogas flow meter

bacteria help in the production of biogas. Cow dung is affluent of methanogenic bacteria, and it is used as an inoculum in this plant. So adding cow dung as an inoculum will increase the amount of biogas production. Here the cow dung also increased the methane level from 7.6 to 28% and decreases the CO_2 level from 70.6 to 40%. This percentage variation in the methane and CO_2 level will help the biogas to burn effectively, and the flickers is also reduced. The calorific value of the biogas is also increased by the addition of inoculum.

Graph 1 shows the percentage of methane, CO_2, and also H_2S in PPM variation in the biogas while taken before and after the addition of inoculum in the biogas plant.

Fig. 8 Online biogas analyzer

7 Future Work

This work extends up to maintaining the permanent inoculum inside the digester. In digester, the inoculum is perpetuated inside, and it continuously helps the bacteria growth to digest the waste. Because the inoculum enriched with the methanogenic bacteria will elevate the reaction. In this, biogas plant feed is the food waste from hostel; this food waste mainly contains the biodegradable matter. Biodegradable matter is mainly required for the production of biogas. But here the food waste feed inside the digester is varying because of the different types of recipes in the hostel. This variation in the feed leads to less quantity of biogas production. By adding this inoculum with the food waste helps to increase the production. But here the concept of permanent inoculum is introduced to reduce the separate inoculum addition process

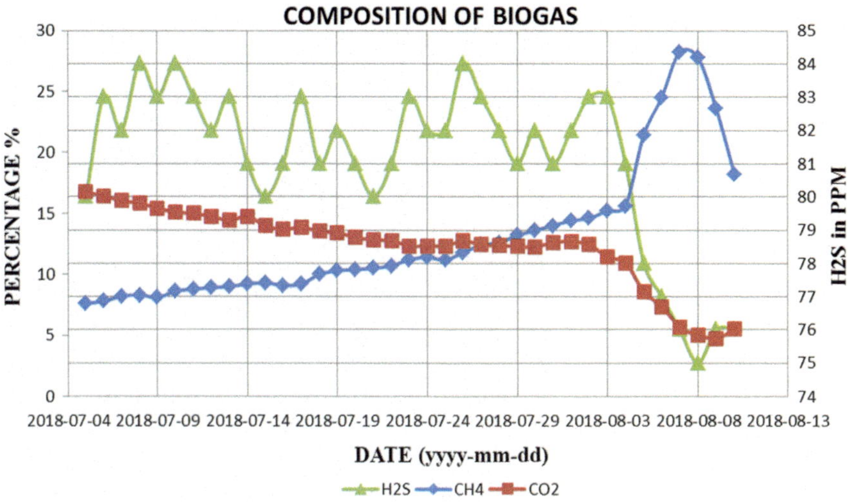

Graph 1 Composition of biogas plant

with feed in the biogas plant. The permanent inoculum will enhance the digestion of food waste. Figure 9 shows the modification of existing biogas plant.

8 Conclusion

Several biodegradable solid wastes are not in use and dumped in waste land or open wiring. Instead of wasting those solid wastes, we can use it for our energy purpose. Those wastes can be converted into biogas; from the biogas, we can able to produce electricity. In the hostel, they are using different types of recipes, and it leads to produce different types of food waste. This food waste will yield low percentage of methane in the biogas. This paper works with the addition of inoculum with the existing biogas plant; it gives new result as improving the level of methane and reduces the carbon dioxide. This leads the biogas plant to increase the calorific value of biogas, and it avoids flicker in the flame. On account of increasing the efficiency, the permanent inoculum concept is also performed in this paper. Hence, various uses of biodegradable waste will help to reduce the solid waste generation in the country.

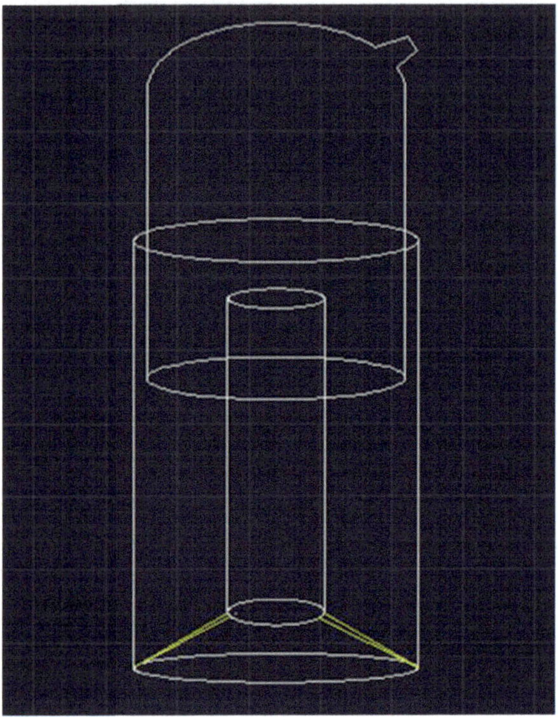

Fig. 9 Sectional view of modified biogas plant

References

Apruzzese, I., Bhardwaj, V., Degrassi, G., & Bhardwaj, R. K. (2017, July–August). Biogas production from organic wastes of paper and leather industries. *IOSR Journal of Biotechnology and Biochemistry, 3*(4), 8–14.

Chamarthi, S., Srinivasa Reddy, N., & Ravi Chandra, G. (2013, May–June). Experimental investigation on performance improvement of biogas plant. *International Journal of Engineering Research and Applications, 3*(3), 1218–1223.

Ofoefule, A. U., Nwankwo, J. I , & Ibeto, C. N. (2010). Biogas production from paper waste and its blend with cow dung. *Advances in Applied Science Research, 1*(2), 1–8.

Tira, H. S., Padang, Y. A., Mirmanto, M., & Hendriono, H. (2015). Improving biogas quality through circulated water scrubbing method. *Applied Mechanics and Materials, 776*, 443–448.

Ziauddin, Z., & Rajesh, P. (2015). Production and analysis of biogas from kitchen waste. *International Research Journal of Engineering and Technology, 2*, 2395–0056.

Resource Recovery from Organic Fraction of Municipal Solid Waste Using Anaerobic Digestion and Hydrothermal Carbonization

Hari Bhakta Sharma, Sagarika Panigrahi, Brajesh K. Dubey and Satyanarayan Narra

Abstract The possible sustainable technologies for waste-to-energy production are anaerobic digestion and hydrothermal carbonization. In this paper, aforementioned technologies were explored on yard waste. Yard waste was thermally pretreated to overcome inherent recalcitrant nature during anaerobic digestion to enhance biogas production. In addition, hydrothermal carbonization (HTC) of yard waste was also conducted for bio-solid fuel production. After pretreatment, the biogas production was improved from 311 ± 5 to 361 ± 11 mL/g VS. HTC of yard waste at different treatment conditions yields a higher calorific value up to 24.59 MJ/kg as compared to 15.37 MJ/kg for raw. In addition, the structural and chemical changes in biomass after pretreatment and hydrothermal carbonization were also studied. The HCT of yard waste at 200 °C for 24 h converted it into the dark lignite like coal with enhanced energy.

Graphical Abstract

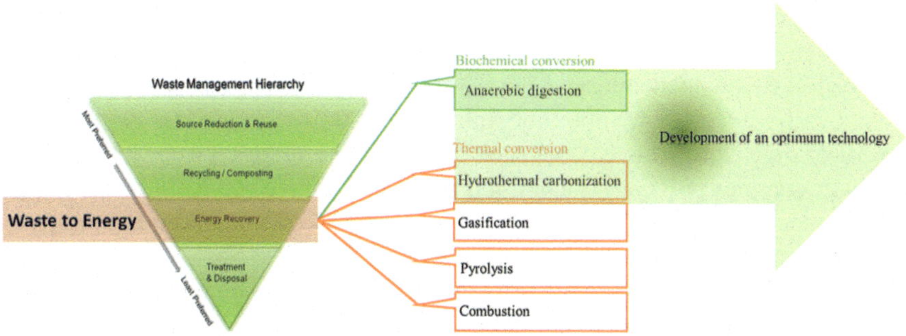

H. B. Sharma · S. Panigrahi · B. K. Dubey (✉)
Department of Civil Engineering, Indian Institute of Technology Kharagpur, Kharagpur, West Bengal 721302, India
e-mail: bkdubey@civil.iitkgp.ac.in

S. Narra
Waste and Resource Management, Faculty of Agricultural and Environmental Sciences, University of Rostock, Justus v. LiebigWeg 6, Rostock, Germany

S. K. Ghosh et al. (eds.), *Bioresource Utilization and Bioprocess*,
https://doi.org/10.1007/978-981-15-1607-8_2

Keywords Hydrothermal carbonization · Anaerobic digestion · Pretreatment · Biogas · Yard waste

1 Introduction

Industrialization and economic growth attract more and more people to a cities, and as a result, rate of urbanization is going to manifold in coming years which will also increase municipal solid waste (MSW) generation rate. Many of the Urban Local Bodies (ULBs) in India are struggling to manage MSW. This struggle is mainly due to inadequate waste managing facility, expertise and also due to lack of funds. India, a country with diverse religious groups, cultures and traditions, produces a waste composition which is unique to different cities. However, organic waste is a major fraction of MSW in ULBs which ranges between 54 and 64% (Hoornweg and Bhada-Tata 2012). Additionally, due to poor source segregation practices by urban population, wet and dry waste gets mixed which increases the overall moisture content. Thermal treatment of high moisture and low calorific waste is not suitable as wet waste has to be predried. Predrying itself is an energy-consuming process. Due to mixed nature of waste owing to poor segregation and also due to poor demand, composting is not feasible technically as well as economically.

Energy harvesting from waste is the most feasible option to handle the society transition for sustainable SWM along with the implementation of promising renewable technologies. Pertinent to Indian scenario and waste composition, anaerobic digestion (AD) and thermochemical treatment process like hydrothermal carbonization or wet torrefaction which processes a wet waste could be a reliable option. However, organic waste due to its recalcitrance nature needs to be pretreated before using it for an anaerobic digestion. Recalcitrance nature of organic waste is mainly due to lignocellulosic component like lignin, cellulose and hemicellulose. These lignocellulosic components are a complex polymer of a sugar and phenols which are not easily degraded by micro-organism and therefore yields less biogas during anaerobic digestion. To overcome such defiant nature of an organic waste for AD, due to its complex polymeric nature, it is pretreated, where polymer is broken down into simpler individual sugar or phenolic component to which micro-organism acts and yields more biogas.

Hydrothermal carbonization (HTC) is another relatively newer thermochemical treatment technique which treats wet waste or dry waste in the presence of water (Benavente et al. 2015). Waste is heated in a closed reactor at the subcritical temperature range of water generally between 200 and 260 °C for 30 min to 3 h in a auto-generated pressure. HTC takes advantage of special property of water that manifests during high temperature and pressure. During high temperature, in a subcritical range, water behaves both as acid as well as base (Funke and Ziegler 2010). Ionization of water increases during high temperature which promotes bond cleavage of complex polymeric chain of lignocellulosic component (Funke and Ziegler 2010; Kruse and Dahmen 2015). The broken glucose or phenolic molecule of lignin, cellulose and hemicellulose gets converted into coal-like material owing to series of chemical reaction like dehydration, decarboxylation, demethanation, aromatization and polymerization (Berge et al. 2011). The final solid char formed is called as a hydrochar whose value-added application is reported in a wide range of area in a literature including fuel, adsorbent, soil amender and as a precursor to a noble carbon materials.

This paper provides an insight in an AD and HTC of organic waste from MSW as a means for resource recovery, thereby contributing to a sustainable SWM transition.

2 Materials and Methods

2.1 Yard Waste Collection, Preparation and Characterization

YW [consisting of mainly dry leaves (65%), grass (33%) and fallen sticks (2%)] was collected in November 2017 from Indian Institute of Technology Kharagpur campus in West Bengal, India. Then, the feedstock was dried in an ambient temperature to reduce the moisture content, and then, it was grinded with a domestic mixture grinder (Havells Marathon, 2200 W, 230 V) to reduce its particle size and referred as prepared biomass. The prepared biomass was stored in an airtight bag.

The effluent from liquid anaerobic digester handing wastewater of Indian Institute of Technology Kharagpur was used as the inoculum. Prior to use, the collected inoculum was centrifuged to increase its total solid (TS) content. The key initial characteristics of YW and inoculum are presented in Table 1 (Fig. 1).

2.2 Experimental Procedure for Pretreatment and Batch Anaerobic Digestion

Thermal pretreatment of the prepared biomass was carried out by a hot air oven (600 W, 2450 MHz). Pretreatment temperature and time were selected according to the literature reported by Ennouri et al. (2016), González-Fernández et al. (2013),

Table 1 Initial characteristics of yard waste and inoculum (With permission from the Elsevier with the License Number 4703421429126 (Panigrahi et al., 2019))

Parameter	Yard waste	Inoculum
TS (%)	99.45 ± 6.12	51.83 ± 0.09
VS (%)	84.75 ± 4.37	98.10 ± 6.11
pH	6.15 ± 0.2	7.23 ± 0.15
VFA (mg/L)	363.75 ± 20.16	157.45 ± 13.16
sCOD (mg/g TS)	189.35 ± 5.45	216.30 ± 4.92
C/N	49.73 ± 4.18	7.89 ± 1.09

Fig. 1 Heap of collected yard waste consisting of dry leaves **O** (65%), grasses and garden trimmings **O** (33%), fallen sticks and wild flower petals **O** (2%) (With permission from Elsevier with the license number 4703420251430 (Sharma et al., 2019))

Agbor et al. (2011). All pretreatments were carried out in 250 mL glass conical flask bottles contained 5 mg YW and 150 mL deionized water. After each pretreatment, the conical flasks were kept in a water bath to bring its temperature to normal temperature. One control sample was kept without any pretreatment named as untreated. Whole experiment and analysis were carried out in triplicate.

Batch AD of pretreated and untreated samples was carried out to examine the effect of pretreatment on biogas production. The batch AD process was carried out in a 1 L glass bottle under mesophilic conditions. Centrifuged inoculum was used. Food to micro-organism ratio of 2:1 was maintained in the reactor. Complete anaerobic condition inside the reactor was ensured by purging nitrogen gas into the reactor. The batch biochemical methane potential was measured by water displacement method. The batch BMP test was conducted for an approximately 30 days until biogas production was stopped.

2.3 Experimental Procedure for Hydrothermal Carbonization

The HTC was carried out in a 50 mL Teflon lined autoclave. For each experiment, 4 g of YW was mixed with 40 mL of distilled water. The autoclave was closed and placed in an electric furnace and heated to the desired temperature at a heating ramp of 13 ± 4 °C/min. The HTC was conducted as a two set of experiments. In the first set, only temperature was increased (160, 180 and 200 °C) keeping time constant (2 h). In the second set of experiment, time was increased (4, 8, 12 and 24 h) keeping temperature constant (200 °C). Calorific value of hydrochar produced at 200 °C in the first experiment was higher, so we decided to increase the temperature at it. After the desired reaction time, autoclave was immersed in a cold water bath to avoid any reaction beyond the desired residence time. Then, the dark slurry (hydrochar) was vacuum filtered and washed with distilled water for multiple times and stored in a zip lock bag for further analysis. The each sample was designated as per the reaction temperature and the time. Hydrochar prepared at the temperature and time of 160 °C and 2 h, respectively, was coded H-160-2. Similarly, rest were coded as H-180-2, H-200-2, H-200-4, H-200-8, H-200-12, H-200-24 for the hydrochar prepared at temperature–time of 180 °C–2 h, 200 °C–2 h, 200 °C–4 h, 200 °C–8 h, 200 °C–12 h and 200 °C–24 h, respectively.

2.4 *Instrumentation*

Physico-chemical parameters TS, volatile solid (VS), pH, alkalinity and soluble COD (sCOD) of the samples were measured using standard methods (APHA 2005). Volatile fatty acids (VFAs) were measured by the procedure described by (Lahav et al. 2002). Ultimate analysis in terms of carbon (C), hydrogen (H) and nitrogen (N) contents of YW was carried out using standard procedure of CHNS analysis by using EUROEA, fully automatic elemental analyser.

2.5 *Structural Characterization*

The surface structure properties of untreated and pretreated YW were analysed by SEM (Carl Zeiss SMT, Germany). The effect of thermal pretreatment on chemical composition of biomass was analysed by FTIR (NEXUS-870, Thermo Nicolet Corporation, USA).

3 Result and Discussion

3.1 *Thermal Pretreatment and Anaerobic Digestion*

3.1.1 Effect of Thermal Pretreatment on Organic Matter Solubilization and Subsequent Anaerobic Digestion

Thermal pretreatment of YW by hot air oven was conducted for a temperature range of 70–200 °C for a treatment duration of 45 min (Fig. 2a). A highest treatment temperature of 200 °C was selected to avoid pyrolysis reaction and formation of inhibitors. The sCOD of untreated YW was 189.35 ± 5.45 mg/L. There was an increase in sCOD to 208.6 ± 4.88 was found with increase in treatment temperature to 70 °C, with further increase in temperature to 120 °C led to decrease in sCOD to 175.6 ± 5.02 mg/L. Then, an increase in organic matter solubilization was found with increase in treatment temperature from 120 to 170 °C. The increase in sCOD with increase in temperature up to 70 °C is due to solubilization of extra polymeric substances (González-Fernández et al. 2013). However, the decrease in sCOD with further increase in temperature up to 120 °C was due to vaporization of solubilized compounds. The increase in organic matter solubilization at 170 °C was due to solubilization of structural components of YW. Hot air oven pretreatment for a treatment temperature of 200 °C led to decrease in organic matter solubilization was due to the formation of melanoidin, which was produced by the millard reactions among sugars and amino acids (Paudel et al. 2017).

There was a decrease in pH was found with increase in temperature which was due to the formation of short-chain fatty acids. The pH of the liquid hydrolysate obtained after pretreatment was found to decrease from 6.15 to 5.4 with increase

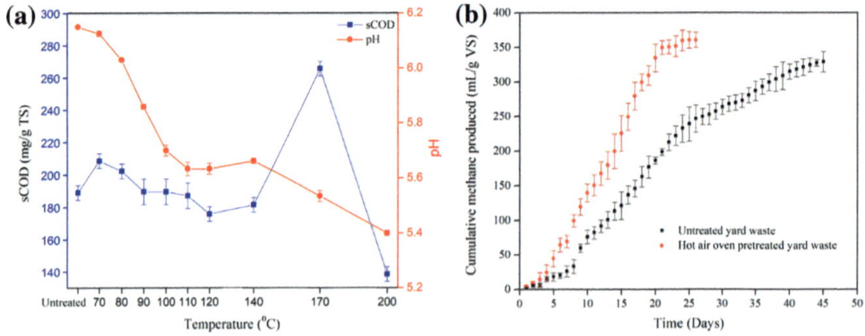

Fig. 2 **a** Variation in sCOD and pH of the effluent obtained after hot air oven pretreatment at different temperature for a duration of 45 min, **b** Cumulative methane produced from untreated and hot air oven pretreated yard waste (With permission from the Elsevier with the license number 4703421429126 (Panigrahi et al., 2019))

in temperature from 70 to 200 °C, which is due to the increase in formation of short-chain fatty acids such as VFA.

Thermal pretreatment improves the sCOD by solubilizing some part of structural constituents like hemicellulose and lignin. However, the solubilization of lignin is undesired in AD process. Solubilization of lignin may produce inhibitors like phenolicfurfural and hydroxyl methyl furfural, which affects the activity of methanogenic bacteria. In order to address the effect of pretreatment on biogas production, batch AD of untreated and hot air oven pretreated (170 °C/45 min) YW was conducted for 45 days. The cumulative biogas produced by untreated and hot air oven pretreated YW is presented in Fig. 2b. Higher cumulative biogas was produced in pretreated sample. In pretreated YW, biogas was produced after five days, whereas in untreated YW, it was nine days. The cumulative biogas produced in untreated and pretreated YW was 361 ± 11 mL/g VS and 311 ± 5 mL/g VS, respectively.

3.1.2 Effect of Thermal Pretreatment on Surface Structure of Biomass

The effect of hot air oven pretreatment (170 °C/45 min) on structural properties of biomass was mapped through SEM analysis (Fig. 3). Well ordered and smooth surface was seen in SEM micrograph of untreated biomass. After thermal pretreatment, the biomass SEM micrographs revealed cracks and void which mostly indicated the destructed cell wall of the biomass.

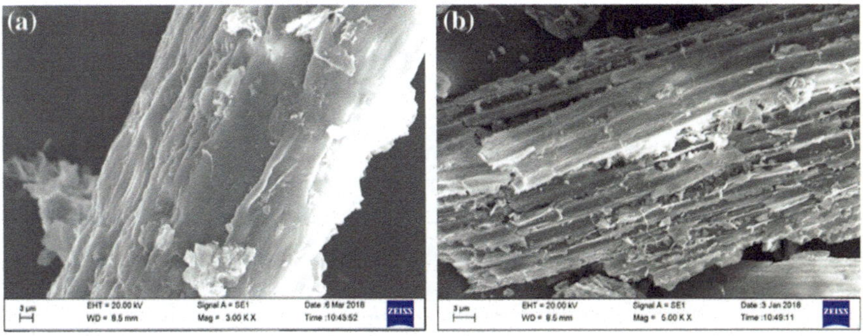

Fig. 3 SEM images of yard waste. **a** Untreated; **b** hot air oven pretreated (With permission from the Elsevier with the license number 4703421429126 (Panigrahi et al., 2019))

Fig. 4 HTC of municipal yard waste for solid biofuels production

3.2 Hydrothermal Carbonization

3.2.1 Effect of Treatment Temperature and Time on Hydrochar

The proximate, ultimate analysis of raw YW and hydrochar along with its energetic properties are presented in Table 1. Biomass has high volatile matter content, attributable to that it burns quickly (flash burn), and so, it is not appropriate for co-firing with coal because it has comparatively less volatile matter and takes time to burn (Fig. 4).

The volatile matter content in raw YW was 84.75% that got reduced to 68.99% when treatment temperature at 2 h time was increased from 160 to 200 °C. Volatile matter further decreased from 68.99 to 63.24% when treatment time was increased from 2 h to 24 h at 200 °C which makes us conclude that the treatment temperature plays major role during HTC of the YW. However, overall increase in ash content (Table 1) was observed with the increase in the reaction time and temperature. The HTC of YW improved overall carbon content, which varied between 45.70 and 60% and decreased overall oxygen content which varied between 49.9 and 32.48%. The main reaction mechanisms that lower oxygen content and increased the carbon content are dehydration and decarboxylation reaction as shown in the Van Krevelen diagram in Fig. 5. Van Krevelen diagram also helps in predicting reaction pathways during the HTC process, and it also helps in assessing quality of fuel. Basically, Van Krevelen diagram depicts coalification process and is plot of O/C and H/C atomic ratios. The yield decreased from 78.5% for the sample H-160-2 to 45.6% for the sample H-200-24 as shown in Table 2. This significant reduction in the yield was mainly due to high volatilization of organic matter in the sample during the HTC process. The energetic properties like fuel ration, calorific value, energy yield also improved after HTC. The caloric value for raw YW was 15.37 MJ/kg and that of hydrothermally treated YW at 200 °C for 24 h was 24.59 MJ/kg.

Dehydration and decarboxylation are the major reaction that takes place during the HTC of YW. The hydrochar H-200-24 shows the fuel properties similar to that of lignite coal, meaning HTC is effective in treating YW to produce solid coal-like fuel. The fuel ration which is a ration of fixed carbon to volatile matter also increased during HTC. Energy yield up to 82% was obtained. With HTC reaction, severity

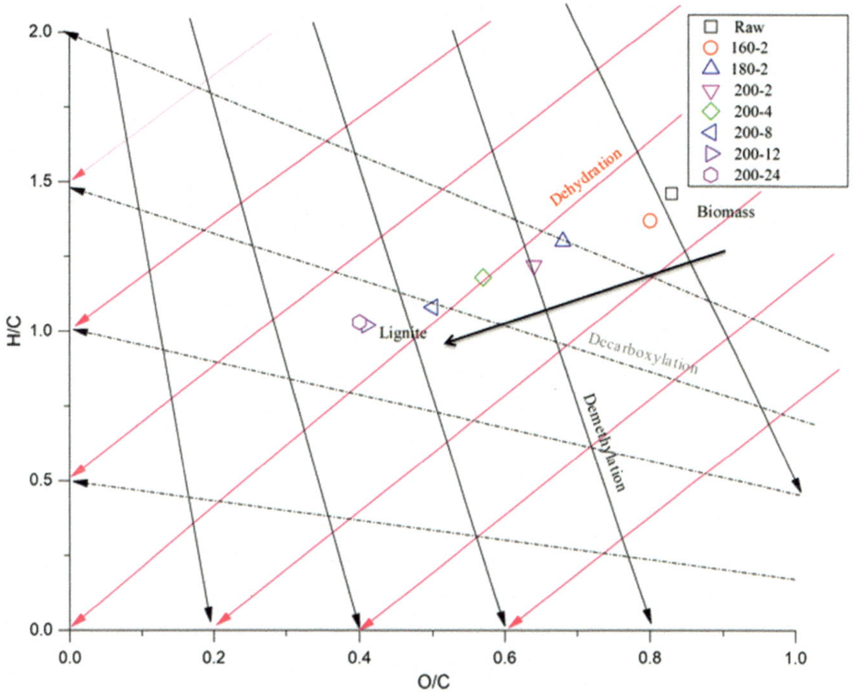

Fig. 5 Van Krevelen diagram showing delineation of reaction pathway during hydrothermal carbonization process for raw and hydrochar sample prepared at different reaction severity (With permission from the Elsevier with the license number 4703420251430 (Sharma et al., 2019))

hydrochar yield decreased. The improvement in fuel quality is mainly attributed to increase in carbon content. The improvement in the carbon content is influenced both by reaction time and temperature.

4 Effect of Hydrothermal Carbonization on Surface Structure of Biomass

The HTC of YW also helps in rupture of physical structure mainly due to weakening of cell wall promoted by decomposition of lignocellulosic components in YW. The decomposed physical structure can be seen in Fig. 6(2) in comparison with smooth well-defined structure as seen in Fig. 6(1). The destruction of physical structure due to HTC promotes friability of hydrochar which makes it easy to be pelletized due to improved surface area which promotes more mechanical interlocking and liquid bridge during compression (Liu et al. 2014).

Table 2 Proximate, ultimate and energetic analysis of raw and hydrothermally treated yard waste (With permission from the Elsevier with the license number 4703420251430 (Sharma et al., 2019))

Sample	Temp (h)	[b]RT (h)	Proximate analysis (%)			Ultimate analysis (wt%)				Energy yield (%)	HHV[a] (MJ/kg)	Yield (wt%)
			FC	A	V	C	H	O	N			
Raw	–	–	6.98	8.30	84.75	44.90	6.38	49.9	0.89	–	15.37	–
H-160-2	160	2	11.28	5.11	83.01	45.70	5.98	49.3	1.06	78.5	15.72	78.5
H-180-2	180	2	17.18	5.32	77.50	48.20	6.60	44.14	1.03	82.67	17.83	75.20
H-200-2	200	2	24.95	6.06	68.99	50.39	5.73	43.02	0.84	80.37	19.60	61.80
H-200-4	200	4	26.83	6.11	67.06	53.79	6.15	41.18	0.92	73.12	19.63	56.30
H-200-8	200	8	28.48	6.62	64.90	56.09	5.56	37.68	0.66	66.98	20.16	51.50
H-200-12	200	12	28.45	7.74	64.01	59.99	6.24	32.70	1.05	75.75	23.33	50.50
H-200-24	200	24	27.80	8.96	63.24	60.00	7.52	32.48	1.03	72.96	24.59	45.60

[a]HHV (Higher heating value) was calculated using Dulong's formula (HHV = 0.3383 * %C + 1.422 * (%H – %O/8)) FC = Fixed carbon; A = Ash; V = Volatile content; C = Carbon; H = Hydrogen; O = Oxygen; N = Nitrogen content. [b]RT = Residence time or reaction time

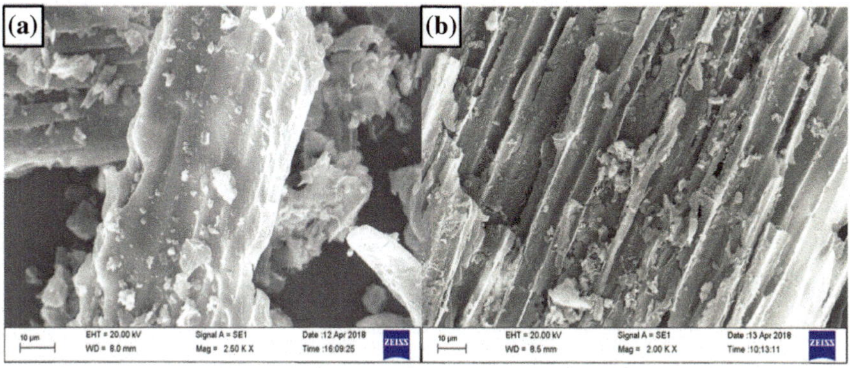

Fig. 6 SEM image of raw yard waste (**a**) and Hydrochar (**b**) (With permission from the Elsevier with the license number 4703420251430 (Sharma et al., 2019))

5 Conclusion and Future Scope

A positive effect of thermal pretreatment on organic matter solubilization and subsequent AD was found. Thermal pretreatment for duration of 45 min at a temperature of 170 °C was best for organic matter solubilization. After thermal pretreatment, the biogas production was improved from 311 ± 5 to 361 ± 11 mL/g VS. The result of this study shows that the HTC lignocellulosic fraction of municipal solid provides us a newer technique to produce energy-dense lignite like solid fuel with a calorific value up to 24.59 MJ/kg. The hydrochar produced had higher carbon content, improved calorific value and reduced volatile matter content.

Since AD of organic waste will produce large solid residue and HTC of waste will produce organic rich water, therefore, we propose a combined HTC-AD or AD-HTC process to be more beneficial. The solid residue from AD plant can be used as a feedstock for HTC to produce bio-solid fuel or process water, rich in organic from HTC plant can be anaerobically digested to produce biogas. The hypothetical framework HTC-AD or AD-HTC is presented in a Fig. 7.

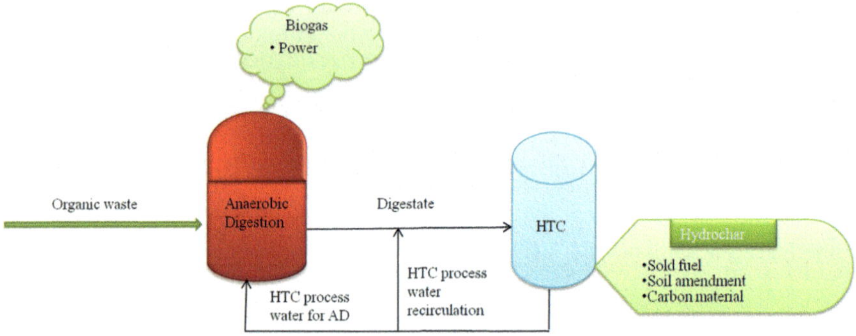

Fig. 7 Hypothetical design for better utilization of integrated HTC-AD or AD-HTC

References

Agbor, V. B., Cicek, N., Sparling, R., Berlin, A., & Levin, D. B. (2011). Biomass pretreatment: Fundamentals toward application. *Biotechnology Advances, 29*(6), 675–685.

Apha, A. (2005). WPCF. 1989. Standard methods for the examination of water and wastewater, vol. 17, pp. 1–5.

Benavente, V., Calabuig, E., & Fullana, A. (2015). Upgrading of moist agro-industrial wastes by hydrothermal carbonization. *Journal of Analytical and Applied Pyrolysis, 113,* 89–98.

Berge, N. D., Ro, K. S., Mao, J., Flora, J. R., Chappell, M. A., & Bae, S. (2011). Hydrothermal carbonization of municipal waste streams. *Environmental Science and Technology, 45*(13), 5696–5703.

Ennouri, H., Miladi, B., Diaz, S. Z., Güelfo, L. A. F., Solera, R., Hamdi, M., et al. (2016). Effect of thermal pretreatment on the biogas production and microbial communities balance during anaerobic digestion of urban and industrial waste activated sludge. *Bioresource Technology, 214,* 184–191.

Funke, Axel, & Ziegler, Felix. (2010). Hydrothermal carbonization of biomass: a summary and discussion of chemical mechanisms for process engineering. *Biofuels, Bioproducts and Biorefining, 4*(2), 160–177.

González-Fernández, C., Sialve, B., Bernet, N., & Steyer, J. P. (2013). Effect of organic loading rate on anaerobic digestion of thermally pretreated Scenedesmus sp. biomass. *Bioresource Technology, 129,* 219–223.

Hoornweg, D., & Bhada-Tata, P. (2012). *What a waste: a global review of solid waste management.*

Kruse, A., & Dahmen, N. (2015). Water–A magic solvent for biomass conversion. *The Journal of Supercritical Fluids, 96,* 36–45.

Lahav, O., Morgan, B. E., & Loewenthal, R. E. (2002). Rapid, simple, and accurate method for measurement of VFA and carbonate alkalinity in anaerobic reactors. *Environmental Science and Technology, 36*(12), 2736–2741.

Liu, Z., Quek, A., & Balasubramanian, R. (2014). Preparation and characterization of fuel pellets from woody biomass, agro-residues and their corresponding hydrochars. *Applied Energy, 113,* 1315–1322.

Panigrahi, S., Sharma, H. B., & Dubey, B. K. (2019). Overcoming yard waste recalcitrance through four different liquid hot water pretreatment techniques–Structural evolution, biogas production and energy balance. *Biomass and Bioenergy, 127,* 105268.

Paudel, S. R., Banjara, S. P., Choi, O. K., Park, K. Y., Kim, Y. M., & Lee, J. W. (2017). Pretreatment of agricultural biomass for anaerobic digestion: Current state and challenges. *Bioresource Technology, 245,* 1194–1205.

Sharma, H. B., Panigrahi, S., & Dubey, B. K. (2019). Hydrothermal carbonization of yard waste for solid bio-fuel production: Study on combustion kinetic, energy properties, grindability and flowability of hydrochar. *Waste Management*, *91*, 108–119.

Effect of Inoculation on Anaerobic Digestion of Food Waste

N. Anand, V. V. Chinnumole and P. Sankar Ganesh

Abstract Anaerobic digestion of food waste and concomitant production of methane-rich biogas will vary majorly depending on the quality of the inoculum, and due to the presence or absence of inhibitory substances. This study focuses on the effect of two types of inocula on anaerobic digestion of food waste. For this study, two laboratory-scale digesters, with working volume of 40 L, were used. While one of the digesters was inoculated with cow dung (digester 1), the other was inoculated with sludge obtained from an UASB reactor treating municipal sewage (digester 2). Both the digesters were maintained at mesophilic conditions (38 °C), and the initial pH of the digester mixture was adjusted to 7. Feeding was done daily, based on Ripley's ratio, and the average mass of the food waste fed daily to the digesters was 150 g. The digesters were operated for a period of 112 days. As the digesters started stabilizing, Ripley's ratio was in the range of 0.085–0.5 in digester 1, and in digester 2, it was 0.157–0.6. The VFA concentration was in the range of 1000–3500 ppm and 1300–4000 ppm in digesters 1 and 2, respectively. The cumulative biogas production was 6531 L in digester 1 and 6761 L in digester 2. From the results, it was clear that there was only marginal difference in the biogas production between the digesters. The effect of inocula in the anaerobic digestion of food waste was minimal; however in terms of conversion rate of VFAs to biogas, digester 2 was operating better at mesophilic conditions.

Keywords Anaerobic digestion · Food waste · Inoculum · Ripley's ratio · VFA

N. Anand · V. V. Chinnumole · P. Sankar Ganesh (✉)
Environmental Biotechnology Laboratory, Department of Biological Science, BITS Pilani, Hyderabad Campus, Jawahar Nagar, Kapra Mandal, Medchal District, Hyderabad, Telangana 500078, India
e-mail: sangan@hyderabad.bits-pilani.ac.in

N. Anand
e-mail: h20161290006@hyderabad.bits-pilani.ac.in

V. V. Chinnumole
e-mail: chinnumolevv@gmail.com

S. K. Ghosh et al. (eds.), *Bioresource Utilization and Bioprocess*,
https://doi.org/10.1007/978-981-15-1607-8_3

1 Introduction

According to the survey done United Nation's Food and Agricultural Organization (2018) approximately 1.3 billion tons of food is being wasted every year around the globe. Industrialized and developing countries account for 670 and 630 million tons of food waste per year, respectively. India being a developing country accounts for 67 million tons of food waste annually.

Food waste is characterized by high moisture (70–90%) and organic matter (more than 80% of VS/TS) contents. Since food waste has high organic matter content, it can be used as a substrate for anaerobic digestion (Zhang et al. 2014). There are many methods available for the disposal of food waste such as incineration, landfilling and composting processes, but anaerobic digestion (AD) is considered to be an appropriate technology for utilizing food waste since it generates revenue in the form of biogas and compost. AD involves a series of biochemical reactions converting organic wastes (carbohydrates, lipids and proteins) to biogas. Biogas mainly consists of methane, carbon dioxide and water vapour, with trace amounts of hydrogen and hydrogen sulphide (Rasi et al. 2007).

In any anaerobic digestion process, the production of biogas might vary depending on two important things: the origin of inoculum and the type of organic waste. Eventually, the activity of microbes in the inoculum affects the production of biogas (Nyholm et al. 1984). The source of inoculum determines the adaptation of the microbial community to the substrate and to the ambience of the digester. To cite two examples, (i) using inoculum from digester treating agricultural waste to a different digester treating sewage or landfill leachate will vary in its characteristics and efficiency; (ii) when inoculum from a digester operating at thermophilic conditions is used in a digester operating in mesophilic conditions, production of biogas may vary) (Thouand and Block 1993; Barkay and Pitchard 1988; Thouand et al. 1995). Hence, this work mainly focuses on investigating the effect of two different inocula—fresh cow dung and sludge from UASB reactor that were used during anaerobic digestion of food waste and their influence on the production of biogas.

2 Materials and Methods

2.1 Substrate and Inoculum

The substrate (food waste) was collected from student's hostel mess located at Birla Institute of Technology and Science, Pilani, Hyderabad Campus, India. For this study, two types of inocula were used—(i) fresh cow dung slurry and (ii) sludge obtained from UASB reactor treating municipal sewage.

2.2 *Digester Design and Operation*

This study was performed using duplicate anaerobic digesters having 40 L as working volume. The digesters were inoculated with 13.5 L of cow dung (digester 1) and UASB sludge (digester 2) and were operated for a period of 112 days at 38 °C, and pH of both the digesters was adjusted to 7. Food waste was collected from student's hostel mess, ground using a mixer and fed to the digesters. VFA concentration and production of biogas were measured daily. Feeding of the digesters started with 500 g of food waste on day 1. Starting from day 2, feeding was done based on Ripley's ratio. When Ripley's ratio was more than 0.5, feeding was halted, and when it was less than 0.5, then the feeding was resumed. The mass of substrate fed to the reactors was increased gradually to 1000 g in a span of 83 days.

2.3 *Analytical Methods*

50 mL sample was collected daily for the analysis of total VFA and Ripley's ratio. The analyses were done using titration methods. The sample was centrifuged at 6500 rpm for 30 min, and the supernatant was collected and used for the analysis. Daily biogas production was measured using wet flow biogas meter.

2.3.1 Ripley's Ratio

Supernatant (25 mL) was taken in 200-mL conical flask. It was titrated with 0.4 N H_2SO_4. pH was measured, and the volume of 0.4 N H_2SO_4 when it reaches the pH values of 5.75 and 4.3 was noted. The values were substituted in the following equation.

$$\text{Ripley's ratio} = \frac{(\text{volume of } 0.4\,\text{N}\,\text{H}_2\text{SO}_4 \text{ till } 4.3 - \text{volume of } 0.4\,\text{N}\,\text{H}_2\text{SO}_4 \text{ till } 5.75)}{\text{volume of } 0.4\,\text{N}\,\text{H}_2\text{SO}_4 \text{ till } 5.75}$$

2.3.2 Total Volatile Fatty Acid

The sample used for determination of Ripley's ratio was also used for total VFA determination. It was digested at 100 °C for 4 h in COD apparatus. The sample was titrated with 0.1 N NaOH till it reaches pH 4. And then the pH of the sample was increased to 7 and the value was recorded. The values were used in the following equation:

$$\text{Total VFA (ppm)} = \frac{\text{volume of } 0.1\text{N NaOH required to bring the pH from 4 to 7}}{(\text{volume of the sample})}$$

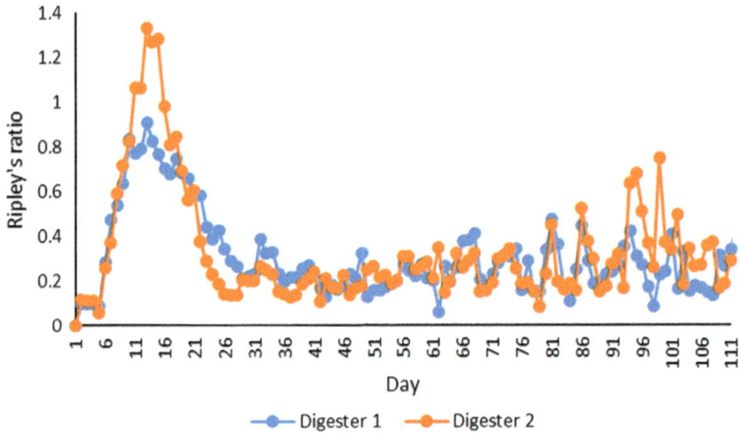

Fig. 1 Ripley's ratio in digesters 1 and 2

3 Results and Discussion

3.1 *Ripley's Ratio*

Ripley's ratio measures the ratio of intermediate alkalinity to partial alkalinity, thereby determining the ratio of VFA::alkalinity. If Ripley's ratio is high, then it indicates that the VFAs that are produced are not getting converted to biogas efficiently.

Ripley's ratio was below 1 during the first 21 days in digester 1, whereas Ripley's ratio of digester 2 reached the highest value of 1.33 on 15th day. The reason for this is that methanogens in digester 2 were taken from an anaerobic reactor treating sewage which contains very less organic matter, whereas the food waste used in this study had high organic content. As the reactor started stabilizing, Ripley's ratio got reduced to 0.5 or lesser. As for digester 1, the inoculum was cow dung and the microbes easily got acclimatized to these conditions at lesser time than the microbes in digester 2. As the digester reached stabilization, the VFA turnover (i.e., production of VFA and its simultaneous conversion into biogas) got higher. Figure 1 shows Ripley's ratio for both the digesters. As the digesters start stabilizing, Ripley's ratio decreased, indicating that the VFAs produced are being converted to biogas.

3.2 *Ripley's Ratio-Based Feeding Regime*

In Figs. 2 and 3, the relation between Ripley's ratio and food waste addition is presented. When Ripley's ratio is high (more than 0.5), the digesters were not fed, and

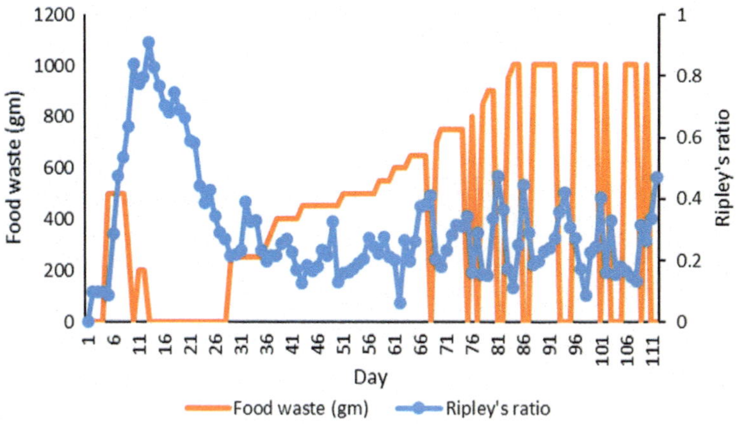

Fig. 2 Ripley's ratio and feeding regime in digester 1

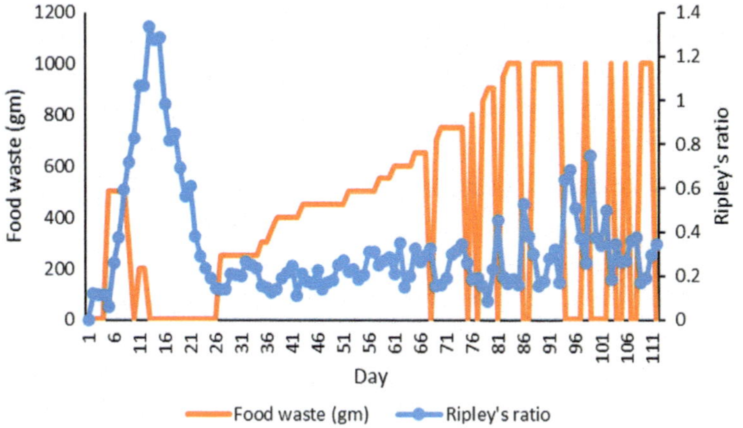

Fig. 3 Ripley's ratio and feeding regime in digester 2

when it was low (less than 0.5), the digesters were fed. During the initial stages, Ripley's ratio was high in both the digesters; hence, the feeding was halted temporarily. On day 28, feeding got resumed in digester 1. In the digester 2, it was restarted on 26th day. After this, for a period of 40 days, Ripley's ratio was less than 0.5 and the feeding was done on a daily basis. And when the digesters were fed with 1000 gm of food waste, there was again a fluctuation in Ripley's ratio.

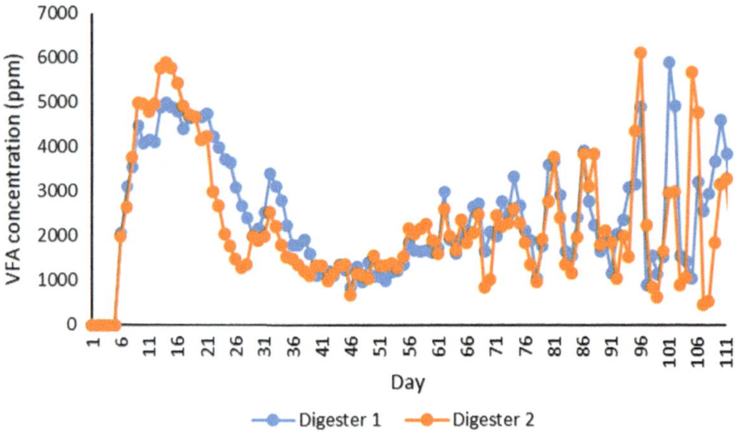

Fig. 4 VFA concentration in digesters 1 and 2

3.3 Total Volatile Fatty Acids

From Fig. 4, it can be seen that total VFA concentration was high in the initial stages. As the digesters attained stabilization from day 26 to day 70, the VFA concentration got stabilized. This is because the conversion rate of VFA to biogas was higher. And when 1000 gm of food waste was fed to the digesters, there was VFA influx, which hindered the process and resulted in the rise of Ripley's ratio by which the feeding regime got altered. This directly affects the microorganisms, since increase in VFA means the pH shifts towards acidic conditions, which affects the activity of methanogens (Chanakya and Malayil 2012).

3.4 Biogas Production

From Fig. 5, it can be noted that the biogas production during the initial stages was low. When the microorganisms started to acclimatize to the digester conditions, biogas production gradually increased. In digester 2, the production of biogas was marginally higher than that of digester 1. This could be linked with the fact that the VFAs which are being produced are converted rapidly to biogas; i.e., VFA turnover is high in digester 2. In the final stage, there is a fluctuation in the biogas production because of the increase in the VFA production. When there was a decrease in the pH to acidic values, the methanogens got affected (Chanakya and Malayil 2012), which affects the conversion of VFA to biogas. The cumulative biogas production in digesters 1 and 2 was 6531 L and 6761 L, respectively.

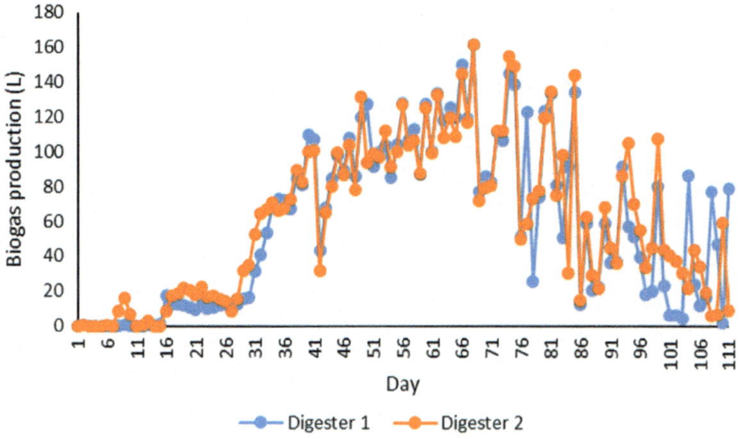

Fig. 5 Daily biogas production in digesters 1 and 2

4 Conclusion

It is expected that digester 2 with sewage sludge from UASB reactor as inoculum should work better than digester 1 which was started with cow dung as inoculum. This is because the inoculum from UASB reactor will have microorganisms which are active since they are already acclimatized to the anaerobic conditions, whereas the inoculum from cow dung requires time for the microorganisms to grow and to acclimatize (Forster-Carneiro et al. 2007). However, the results showed that there was no significant difference in the biogas production. This might be attributed to the fact that the inoculum from UASB reactor was used to treat sewage, which is a heterogenic substrate, and the organic content of sewage is very less as compared to the food waste. When this inoculum is used for treating food waste, which has high organic content, the activity of methanogens decreased initially as compared to the hydrolytic and acidogenic bacteria. The methanogens took sometime to acclimatize and contribute towards the biogas production. One important finding is that after digester 2 got stabilized, the production of VFA and its conversion to biogas was better. And in digester 1, the production of VFA was higher, but the conversion of VFA to biogas was relatively lesser. Hence, it can be concluded that both fresh cow dung slurry and a UASB sludge could serve as a potential inocula for anaerobic digestion of food waste in specific and any organic material in general.

Acknowledgements PSG as principal investigator thanks CoRE WWE Project of BITS Pilani for funding this research work. NA and VVC as junior research fellows thank BITS Pilani, Hyderabad Campus, and BIRAC, respectively, for providing fellowships.

References

Barkay, T., & Pitchard, T. (1988). Adaptation of aquatic microbial communities to pollutant stress. *Microbiological Sciences, 5*(6), 165–169.

Chanakya, H. N., & Malayil, S. (2012). Anaerobic digestion for bioenergy from agro-residues and other solid wastes—an overview of science, technology and sustainability. *Journal of the Indian Institute of Science, 92*(1), 111–144.

Forster-Carneiro, T., Pérez, M., Romero, L. I., & Sales, D. (2007). Dry-thermophilic anaerobic digestion of organic fraction of the municipal solid waste: focusing on the inoculum sources. *Bioresource Technology, 98*(17), 3195–3203.

Nyholm, N., Lindgaard-Jorgensen, P., & Hansen, N. (1984). Biodegradation of 4-nitrophenol in standardized aquatic degradation test. *Ecotoxicology and Environmental Safety, 8*(5), 451–470.

Rasi, S., Veijanen, A., & Rintala, J. (2007). Trace compounds of biogas from different biogas production plants. *Energy, 32,* 1375–1380.

Thouand, G., & Block, J. C. (1993). The use of precultured inocula for biodegradability test. *Environmental Technology, 14*(7), 601–614.

Thouand, G., Friant, P., Bois, F., Cartier, A., Maul, A., & Block, J. K. (1995). Bacterial inoculum density and probability of para-nitrophenol biodegradability test response. *Ecotoxicology and Environmental Safety, 30*(3), 274–282.

United Nation's Food and Agricultural Organization. (2018).

Zhang, C., Su, H., Baeyens, J., & Tan, T. (2014). Reviewing the anaerobic digestion of food waste for biogas production. *Renewable and Sustainable Energy Reviews, 38*, 383–392.

Bio-Sorption of Cr (VI) from Aqueous Solutions by Pericarp of *Pongamia pinnata*

P. V. V. Prasada Rao, R. V. Ramana Murthy and Ch. Durga Prasad

Abstract Chromium (VI) is a major pollutant released during several industrial operations and if not managed properly affects the environment. The present study reports adsorption of Cr (VI) from aqueous solutions by pericarp of *Pongamia pinnata* (ppp), with special reference to different operating variables like pH, adsorbent dose, particle size, agitation speed of the solution, contact time and initial concentration of adsorbate. Adsorption process was found to be highly pH dependent, while the optimum pH for the adsorption of Cr (VI) was 2.0. Removal of Cr (VI) increased with the increase in metal concentration and contact time. The adsorption data have been analyzed and tested for both Langmuir and Freundlich adsorption isotherm models and the first-order kinetic rate was observed with Lagergren. The kinetics of adsorption is found to be first order with intra-particle diffusion as one of the rate determining steps. ppp is found to be effective for the removal of Cr (VI) from aqueous solutions and industrial waste waters.

Keywords Adsorption · Desorption · Isotherms · Kinetics · Pericarp · pH · Pongamia pinnata

1 Introduction

Rapid urbanization and industrialization have led to increased generation and disposal of wastewater, containing a wide spectrum of pollutants affecting the ecology and environment. With the growing demands and decimation of resources, scarcity of freshwaters made it necessary to consider conversion of wastewaters for reuse as an additional resource (AquaFed 2013), which in turn lowers the burden of pollution load. Chromium, a steely gray, lustrous, hard and brittle transition metal is widely used in various industrial operations like leather, chemical manufacture, metal finishing, electroplating and many other industries (Jacobs and Testa 2004). Chromium

P. V. V. Prasada Rao (✉) · Ch. Durga Prasad
Department of Environmental Sciences, Andhra University, Visakhapatnam, India
e-mail: peddineniprasadarao@gmail.com

R. V. Ramana Murthy
Department of Chemistry, Mrs. AVN College, Visakhapatnam, India

S. K. Ghosh et al. (eds.), *Bioresource Utilization and Bioprocess*,
https://doi.org/10.1007/978-981-15-1607-8_4

is present in industrial effluents primarily as trivalent and hexavalent states. Cr (III) is relatively less toxic and less mobile (Oliveira 2012), while Cr (VI) above 50 ppb concentration is toxic, carcinogenic and mutagenic to animals as well as humans (Costa and Klein 2006), while at lower concentrations, it is considered as one of the essential elements for maintaining proper carbohydrate and lipid metabolism (Vincent 2010) and promotes weight loss and lean muscle (Kiefer 2013) when it is present in trace amounts.

Acute exposure to Cr (VI) causes nausea, diarrhea, liver and kidney problems, dermatitis, internal hemorrhaging and respiratory problems (TURI report 2006). The USEPA (2003) drinking water guidelines recommended 100 μg/L of Cr (VI) as the maximum permissible limit in the drinking waters. Various techniques for the reduction and removal of Cr (VI) have been reported from time to time and include, cementation (Lin et al. 1992), chemical precipitation (Zhou et al. 1993), polymer-based filtration and membrane separation (Chakravarti et al. 1995), electrochemical precipitation (Kongsricharoern and Polprasert 1996), ion-exchange (Tiravanti et al. 1997) chemical reduction (Seaman et al. 1999), adsorption (Dahbi et al. 1999), solvent extraction (Pagilla and Canter 1999), electrokinetic remediation (Sawada et al. 2004) and microbial adsorption for the removal of heavy metals (Sen et al. 2005). Nevertheless, many of these approaches are marginally cost-effective, practically difficult and environmentally incompatible to implement, especially in the developing world. Hence, there is a need and necessity for a treatment strategy that is simple, robust, cost-effective and environmentally acceptable.

Adsorption can be an effective and versatile method for removing Cr (VI), particularly when combined with an appropriate regeneration step. This addresses the problem of sludge disposal and renders the system more economically viable, especially if low-cost adsorbents are used. Many adsorbents were reported to remove Cr (VI) from aqueous solutions and wastewaters. Studies on tamarind nut carbon (Srinivasan and Geetamani 2004), mixture of fly ash and peepal bark (Vasanty et al. 2003), soybean hull (Parlayıcı and Pehlivan 2012), corncob (Muthusamy et al. 2008), rice husk (Kumar et al. 2012) and bituminous coal (Nitin et al. 2012) are reported. The present work reports the use of pericarp dust of *Pongamia pinnata* as an adsorbent for the removal of Cr (VI) from aqueous solutions.

2 Materials and Methods

The pericarp of *Pongamia pinnata* was collected from the suburbs of Visakhapatnam, cleaned thoroughly with water and soaked in distilled water for 24 h, washed with distilled water and sun dried. The dried drupes were pulverized, and the material was screened for three particle sizes: 0.430 mm, 0.600 mm and 0.800 mm. All the chemicals used in the experiment were of AR grade. Double distilled water was used throughout the experiment. Solutions of 1 mg/mL of Cr (VI), 1 mg/mL of 1, 5-Diphenylcarbazide in 95% ethyl alcohol and 10% sulfuric acid solution were employed in the experiment.

3 Chromium Adsorption Studies

Batch mode studies: Batch mode adsorption studies were carried out by agitating 50 mL of Cr (VI) solution of desired concentrations at pH 2.0 with 100 mg of pericarp of *Pongamia pinnata* dust at different time intervals. At the end of pre-determined time intervals, the sorbate was separated by centrifugation. The supernatant liquid was analyzed for left-over Cr (VI) by a UV–Visible spectrophotometer using 1, 5-Diphenylcarbazide at 540 nm (APHA 2005). The adsorption of Cr (VI) was studied at different pH, initial Cr (VI) concentration, adsorbent dose and particle size at different agitation times. The pH studies were carried out with 50 mL of 10 mg/L of Cr (VI) solution at different pH values, agitated with 100 mg of adsorbent for 120 min. The supernatant liquid was analyzed for the remaining Cr (VI) (APHA 2005).

4 Desorption Studies

The adsorbent from Cr (VI) adsorption experiment (10 mg/L of Cr (VI) solutions and 100 mg of adsorbent) was separated and gently washed with distilled water to remove any unadsorbed Cr (VI). Several such adsorbent samples were re-suspended in 50 mL of distilled water containing various concentrations of NaOH and agitated for 120 min. Then, the desorbed Cr (VI) was estimated. All the experiments were done in triplicate ($n = 3$), and the results were reproducible within ±5% deviation.

5 Results and Discussion

5.1 Effect of pH

pH of the aqueous solution plays a very important role in the adsorption of metal ions. The adsorption efficacy depends on pH value of the solution. A series of experiments at different pH values were conducted to explore the optimum pH value for conducting adsorption experiments. Figure 1 shows the extent of Cr (VI) adsorption as a function of pH for an initial concentration of 10 mg/L. The percent removal of Cr (VI) was found to be maximum at pH 2.0. This indicates that at pH 2.0, pericarp of *Pongamia pinnata* was effective for Cr (VI) removal as reported earlier (Gupta et al. 1988). This may be due to protonation of adsorbent surface in acidic pH, and hence, the positively charged adsorbent removes higher amounts of Cr (VI) in the anionic forms $HCrO_4^{-1}$, CrO_4^{-2} (Henefield 1982).

With an increase in the pH of the system, the degree of protonation of the surface reduced gradually, and hence, decreased adsorption was noticed (Singh et al. 1992).

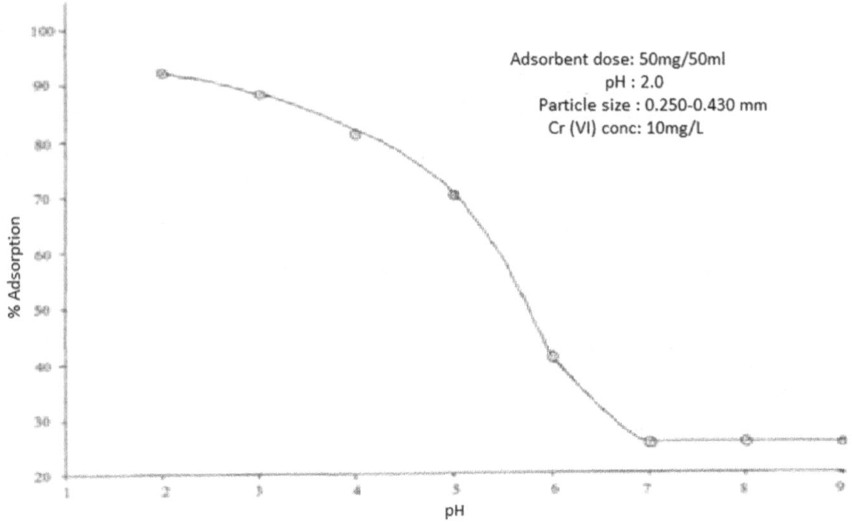

Fig. 1 Effect of pH on Cr (VI) removal

5.2 Effect of Agitation Time on Initial Cr (VI) Concentration

A series of experiments at different Cr (VI) concentrations with varying agitation times were conducted to explore the relation between the two. Figure 2 shows that the percent adsorption of Cr (VI) on the adsorbent increased with an increase in agitation time and attained equilibrium after 120 min. The percent removal was found to be 95.4, 86.5, 77.2 and 67.1% at the initial concentrations of 10, 20, 30 and 40 mg/L of Cr (VI) concentrations, respectively. The equilibrium time was found to be independent of Cr (VI) concentration. The time curves (leading to saturation) in Fig. 2 are single, smooth and continuous suggesting the possibility of formation of monolayer coverage of Cr (VI) on the surface of the adsorbent (Kadirvelu and Namasivayam 2003; Kannan and Balamurugan 2004).

5.3 Effect of Agitation Time on Particle Size

Cr (VI) solutions of 10 mg/L concentration were agitated with 100 mg of adsorbent at different particle sizes, such as 0.43, 0.60 and 0.80 mm for different agitation times. Cr (VI) removal was found to be 95.4, 84.6 and 69.1% at adsorbent particle sizes of 0.43, 0.60 and 0.80 mm, respectively (Fig. 3). This shows that the decrease in particle size increases the adsorption of Cr (VI) on pericarp of *Pongamia pinnata*. This may be due to breaking of large particles that tend to open tiny cracks and channels on the particle surface (Weber and Morris 1963), providing an additional surface area

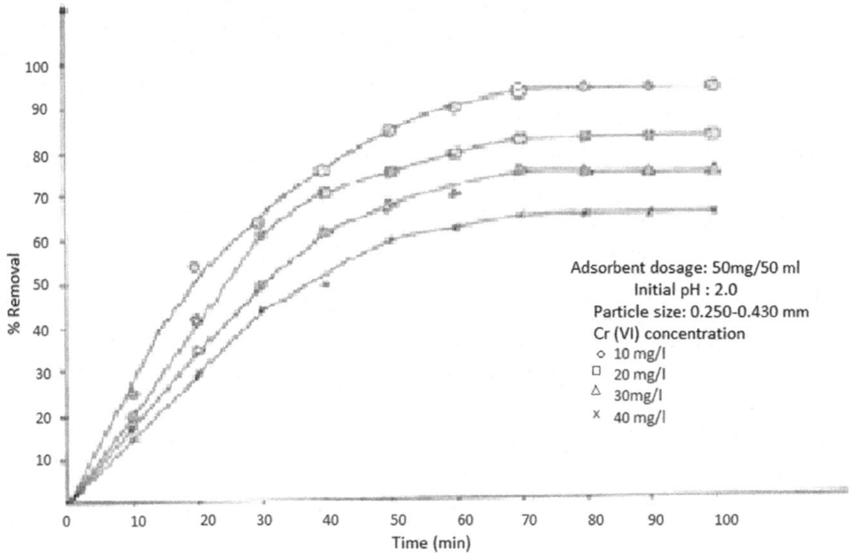

Fig. 2 Effect of agitation time and Cr (VI) concentration on Cr (VI) adsorption

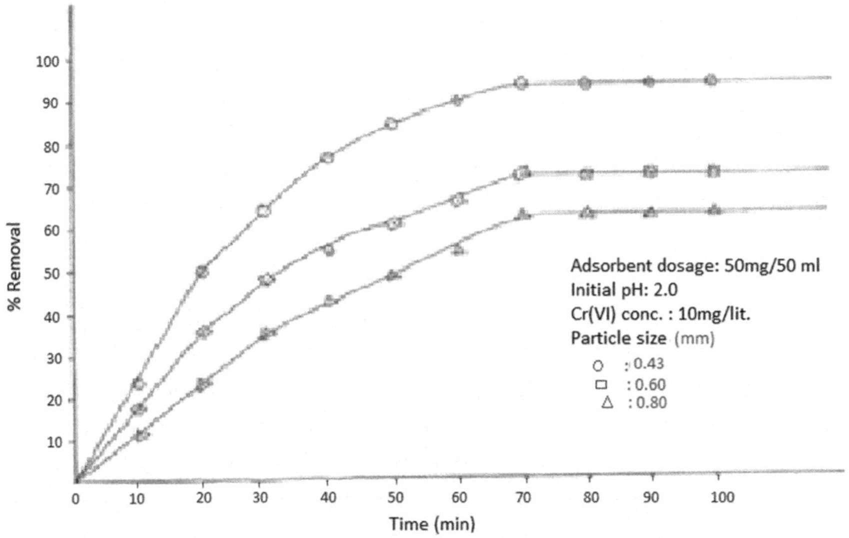

Fig. 3 Effect of particle size on Cr (VI) removal

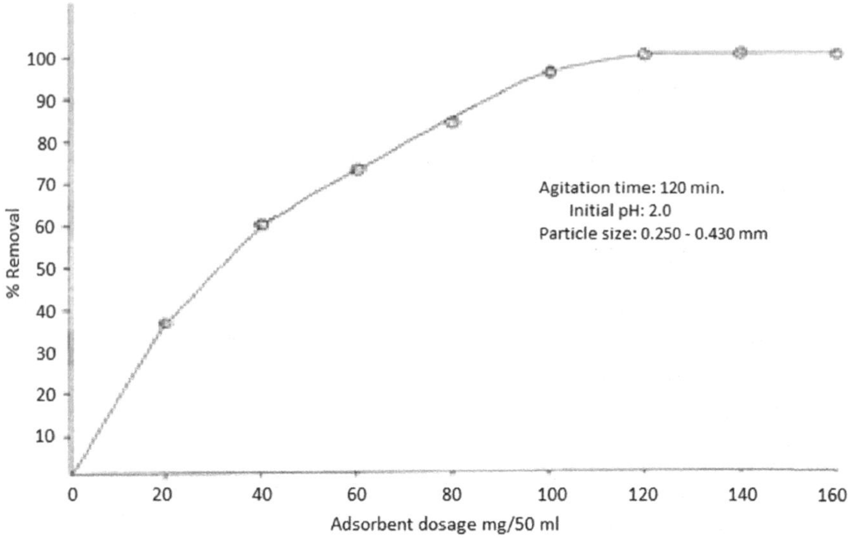

Fig. 4 Effect of adsorbent dosage on Cr (VI) removal

which can be used in the adsorption process. For the larger particles, intra-particle diffusion will be the predominant mechanism (Junyapoon and Weerapong 2006; Venkateswarulu et al. 2008).

5.4 Effect of Adsorbent Dosage

The relation between adsorbent dosage and Cr (VI) adsorption was studied with increasing dosage of adsorbent at pH 2.0 and 10 mg/L of Cr (VI) concentration. Figure 4 shows the relation between dosage of the adsorbent and Cr (VI) adsorption at adsorbent concentrations from 20 to 120 mg/50 mL of Cr (VI) solutions. The results show that quantitative removal of Cr (VI) is possible with 120 mg of adsorbent at 120 min equilibrium time. The increase in percent adsorption with the increase in adsorbent dosage may be due to the availability of more surface area of the adsorbent for adsorption (Nigam and Rama 2002; Joshi et al. 2003).

5.5 Desorption

Desorption studies help to elucidate the nature of adsorption, recovery of metals from wastewater and recycling of adsorbent. In the present study, desorption studies were carried out at different NaOH concentrations. An increase in the concentration of

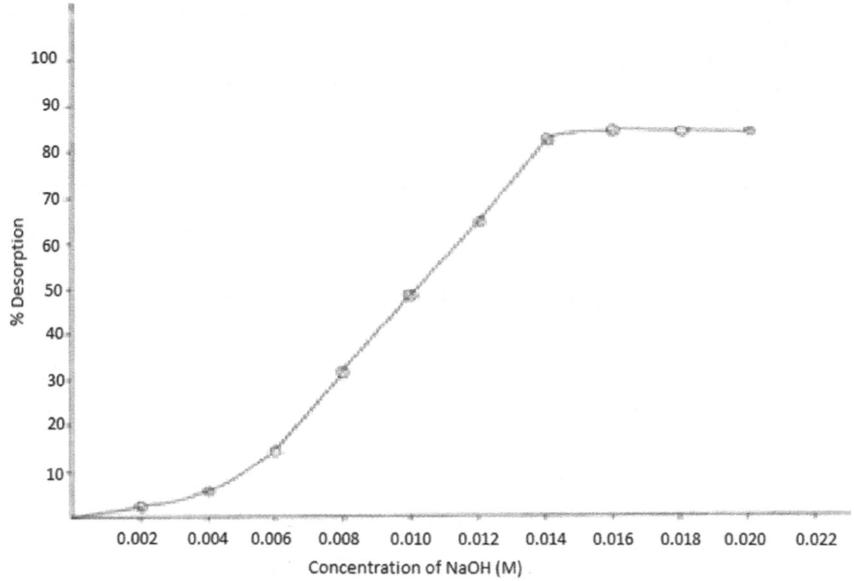

Fig. 5 Effect of NaOH on desorption of Cr (VI)

NaOH has increased the desorption of Cr (VI), and maximum desorption of 81.3% was recorded at 0.014 M NaOH at 120 min agitation time (Fig. 5). Increase in alkali concentration would deprotonate the surface, and hence, the alkaline adsorbent surface would abandon the negatively charged $HCrO_4^{-1}$, CrO_4^{-2} species. The incomplete desorption may be due to the presence of organic constituents which may also act as an adsorbent for chromium (Manonmani 2002; Selvaraj et al. 1998).

5.6 Adsorption Isotherms

Adsorption isotherms are one of the most important tools to understand the mechanism and quantifying the distribution of the adsorbate between liquid phase and solid adsorbent phase at equilibrium during adsorption process. The present study has considered both Langmuir (1916) and Freundlich (1906) isotherms to evaluate the experimental data.

Langmuir isotherm: The Langmuir model takes the form,

$$\frac{C_e}{q_e} = \frac{1}{Q_o} + \frac{C_e}{Q_o} \tag{1}$$

where

C_e—the concentration of metal ion (mg/L) at equilibrium

q_e—the amount of metal ion (mg) adsorbed per 'gm' of adsorbent at equilibrium
Q_o and 'b'—Langmuir constants related to adsorption capacity and energy of adsorption, respectively.

The linear plot between C_e/q_e versus C_e shows that the adsorption obeys Langmuir model (Fig. 6), and the values of Q_o and 'b' are 33.91 and 0.11, respectively (Gupta and Babu 2006; Singh et al. 2003). The essential characteristics of a Langmuir isotherm can be expressed in terms of a dimensionless constant separation factor/equilibrium parameter R_L (Mckay et al. 1982) which takes the form

$$R_L = \frac{1}{1 + bC_e} \tag{2}$$

where
C_e and b represent initial metal ion concentration (mg/L) and Langmuir constant (L/mg), respectively. The R_L values at different concentrations are between 0 and 1 indicating favorable adsorption for the initial concentrations of Cr (VI) studied (McKay et al. 1982).

Freundlich isotherm: The Freundlich isotherm (1906) takes the form

$$\log(x/m) = \log K + \frac{1}{n}\log C_e \tag{3}$$

where C_e—the equilibrium concentration mg/L, x/m—the amount adsorbed (mg/gm).

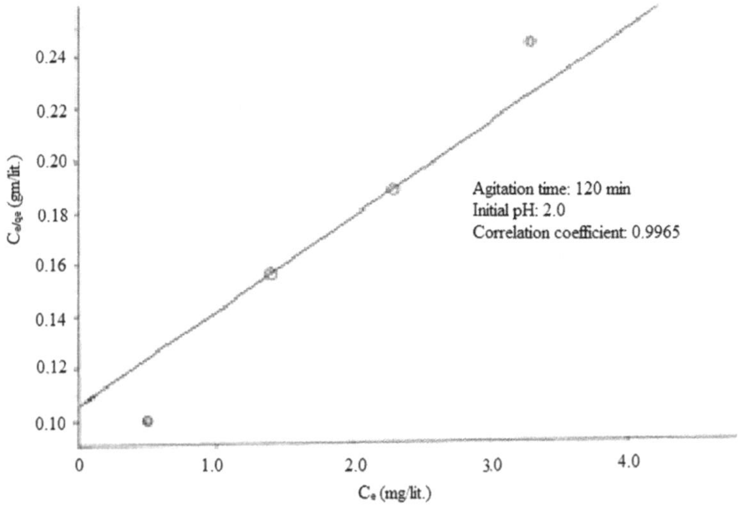

Fig. 6 Langmuir plot for Cr (VI) adsorption

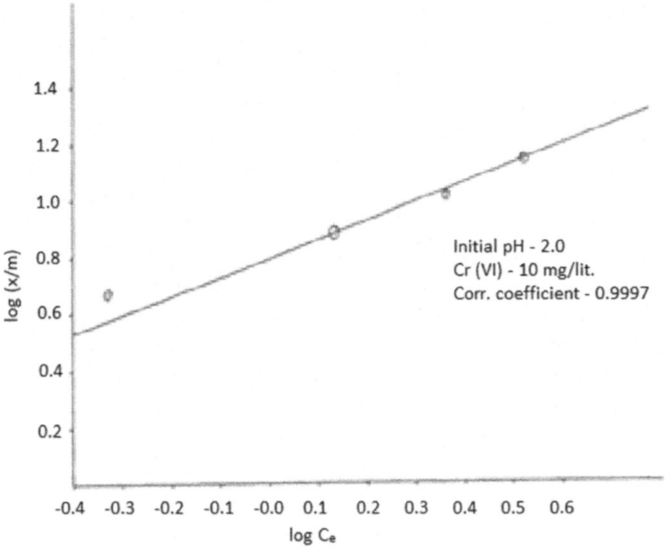

Fig. 7 Freundlich plot for Cr (VI) adsorption

The plot between $\log(x/m)$ versus $\log C_e$ as shown in Fig. 7 is linear which shows that the adsorption follows Freundlich isotherm for Cr (VI). The observed 'n' and K values are 1.59 and 0.53, respectively.

If the 'n' value is between 1 and 10, it indicates good adsorption (McKay et al. 1982), and in the present study, the 'n' value being 1.59 may be considered as good adsorption (Rao et al. 2003).

5.7 Adsorption Kinetics

The kinetics of Cr (VI) adsorption on pericarp of *Pongamia pinnata* follows the first-order rate expression given by Lagergren (Singh et al. 2006).

$$\log(q_e - q) = \log q_e - \frac{Kadt}{2.303} \tag{4}$$

where 'q' is the amount of Cr (VI) adsorbed (mg/g) at time 't' (min), and q_e is the amount adsorbed (mg/g) at equilibrium time. The relation between $\log(q_e - q)$ versus 't' is linear as presented in Fig. 8, indicating the applicability of the above equation.

The K_{ad} values calculated from the slops of plots at 10, 20, 30 and 40 mg/L of Cr (VI) concentrations are 3.32×10^{-2}, 3.36×10^{-2}, 3.12×10^{-2} and 2.24×10^{-2} min^{-1}, respectively. The above data indicate that the metal ion concentration has no

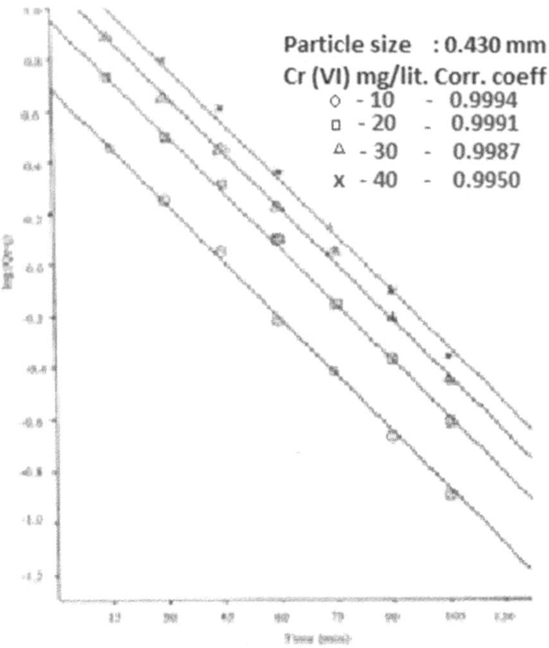

Fig. 8 Relation between Cr (VI) concentrations on Lagergren plot for Cr (VI) adsorption

significant effect on the first-order rate of adsorption. The values are comparable with those reported earlier (Ali and Deo 1992), and the rate constants for the adsorption are also comparable with those reported in the literature (Singh et al. 1992).

5.8 Adsorption Studies on Field Samples

Wastewater samples from an electroplating industry from the suburbs of Visakhapatnam were collected and brought to the laboratory for further analysis. The pH of the samples was adjusted to 2.0, and the samples were subjected to bio-sorption by pericarp of *Pongamia pinnata*. It has been found that the increase in adsorbent dosage increased the percent adsorption (Fig. 9), and 85.3% removal of Cr (VI) was achieved with 1000 mg of adsorbent for 50 mL of wastewater at 120 min agitation time. This indicates the validity of the data obtained from the batch studies, and hence, it may be concluded that the present adsorbent has the potential to remove Cr (VI) from chrome plating industrial wastewater (Manonmani 2002; Selvaraj et al. 1998).

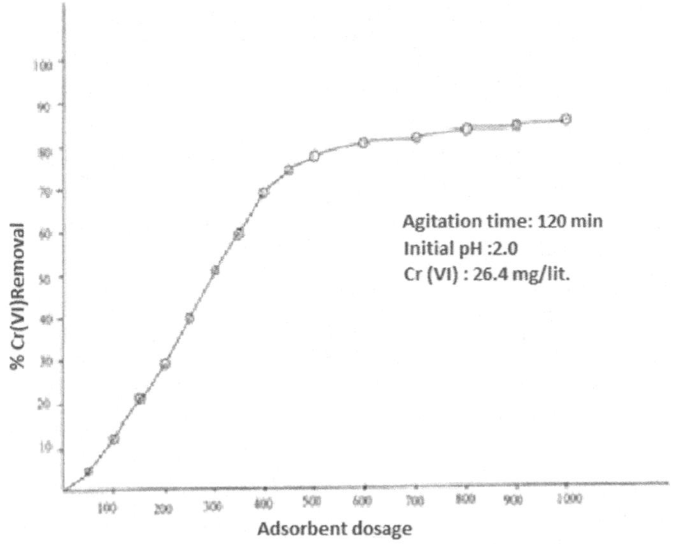

Fig. 9 Effect of adsorbent dosage on Cr (VI) removal from electroplating industry wastes

5.9 Practical Significance

Low cost and environmentally compatible removal of toxic heavy metals from aqueous solutions can be better achieved by bio-sorption. The present bio-adsorbent, pericarp of *Pongamia pinnata* fits into this, since it is not only an agricultural waste but also an effective adsorbent. Hence, the use of pericarp of *Pongamia pinnata*, as a bio-adsorbent for Cr (VI) removal, is of both economically viable and environmentally safe.

6 Conclusion

The experimental data of the present investigation indicate that pericarp of *Pongamia pinnata* waste has the potential to remove Cr (VI) from wastewaters of electroplating industry. The adsorption data have shown good agreement with both Langmuir and Freundlich adsorption isotherms indicating its utility as an effective adsorbent for the removal of Cr (VI) from wastewater. The adsorption data of Cr (VI) on pericarp of *Pongamia pinnata* also suggest that adsorption of Cr (VI) on pericarp of *Pongamia pinnata* follows first-order rate expression. The present study provides a viable economic option for the removal of Cr (VI) using that pericarp of *Pongamia pinnata* as a bio-adsorbent. Further studies are needed to establish its commercial utility for the removal of Cr (VI) from wastewaters of small and medium scale industries.

Acknowledgements The infrastructural facilities provided by the Dept. of Environmental Sciences, Andhra University, to carry out this work are gratefully acknowledged.

References

Ali, M., & Deo, N. (1992). Use of low-cost material as an adsorbent in the removal of Cr (VI) from dilute aqueous solution. *Indian Journal Environmental Protection, 12,* 439–444.

APHA. (2005). Standard methods for the examination of water and wastewater, 21st ed. Water Environment Federation/American Water Works Association/American Public Health Association, Washington DC, USA.

AquaFed. (2013). *Working paper on post 2015 global goals towards waste water2013-01-17.docx.*

Chakravarti, A. K., Chowdhury, S. B., Chakrabarty, S., Chakrabarty, T., & Mukherjee, D. C. (1995). Liquid membrane multiple emulsion process of Chromium (VI) separation from Wastewaters. *Colloids and Surfaces A: Physicochemical Engineering Aspects, 103,* 59–71.

Costa, M., & Klein, C. B. (2006). Toxicity and carcinogenicity of chromium compounds in humans. *Critical Reviews in Toxicology, 36,* 155–163.

Dahbi, S., Azzi, M., & De la Guardia, M. (1999). Removal of hexavalent chromium from wastewaters by bone charcoal. *Fresenius Journal of Analytical Chemistry, 363,* 404–407.

Freundlich, H. (1906). Űber die adsorption in lősungen. *Zeitschrift für Physikalische Chemie, 57,* 384–470.

Gupta, S., & Babu, B. V. (2006). Adsorption of Cr (VI) by a low-cost adsorbent prepared from Neem leaves. In *Proceeding of National Conference on Environmental Conservation* (pp. 175–180), BITS, Pilani.

Gupta, G. S., Prasad, G., & Singh, V. (1988). Removal of chrome dye from water by fly ash. IAWPC, Tech. Annual, 15, 98.

Henefield. (1982). *Referred in Indian Journal Environmental Protection, 12*(12), 916–922.

Ingole, N. W., & Dharpal, S. V. (2012). State of art of bio sorption technique for treatment of heavy metals bearing wastes. *Journal of Advanced Engineering Technology, 3*(2), 143–153.

Jacobs, J. A., & Testa, S. M. (2004). Overview of Chromium (VI) in the environment: Background and history. In *Chromium (VI) Handbook*, CRC press.

Joshi, N., Ahluwalia, S., & SandGoyal, D. (2003). Removal of heavy metals by different bio-waste materials. *Research Journal of Chemistry Environment, 7*(4), 26–30.

Junyapoon, S., & Weerapong, S. (2006). Removal of Chromium (VI) from aqueous solution by scrap iron fillings. *KMITL Science and Technology Journal*, 6(1), Jan–Jun.

Kadirvelu, K., & Namasivayam, C. (2003). Activated carbon from coconut coir pith as metal adsorbent: Adsorption of Cd (II) from aqueous solution. *Advances in Environmental Research, 7,* 471–478.

Kannan, N., & Balamurugan, J. (2004). Removal of Copper ions by adsorption onto coconut shell and dates nut carbons—A comparative study. *Indian Journal Environmental Protection, 24*(5), 371–378.

Kiefer, D. (2013). Chromium for the support of blood sugar and blood lipids levels. In *Optimal health, weight loss.*

Kongsricharoern, N., & Polprasert, C. (1996). Chromium removal by bipolar electro-chemical precipitation process. *Water Science and Technology, 34,* 109–116.

Kumar, A., Mohanta, K., Kumar, D., & Parkash, Om. (2012). Properties and industrial applications of rice husk: A review. *International Journal of Emerging Technology and Advanced Engineering, 2*(10), 86–90.

Langmuir, I. J. (1916). The constitution and fundamental properties of solids and liquids, Part I. Solids'. *American Chemical Society, 38,* 2221–2295.

Lin, C. F., Rou, W., & Lo, K. S. (1992). Treatment strategy for Cr (VI)-bearing wastes. *Water Science Technology, 26,* 2301.
Manonmani, S. (2002). Ph. D. Thesis, Bharathiar University, Coimbatore, India.
McKay, G., Blair, H. S., & Gardner, J. R. (1982). Adsorption of dyes on chitin equilibrium studies. *Journal of Applied Polymer Science, 27,* 3043–3057.
Muthusamy, K., Gopalakrishnan, S., Ravi, T. K., & Siva Chidambaram, P. (2008). Biosurfactants: Properties, commercial production and application. *Current Science, 94,* 736–747.
Nigam, A., & Rama, O. P. (2002). Corncob-A promising adsorbent for the removal of Chromium (VI) from wastewater. *Indian Journal Environmental Protection, 22*(5), 550–553.
Oliveira, H. (2012). Chromium as an environmental pollutant: Insights on induced plant toxicity. *Journal of Botany, 2012,* ID 375843.
Pagilla, K. R., & Canter, L. W. (1999). Laboratory studies on remediation of chromium—contaminated soils. *Journal of Environmental Engineering, 125,* 243–248.
Parlayıcı, S., & Pehlivan, E. (2012). Elimination of Cr (VI) from aqueous solution by a new agro-waste material. In *International Symposium on Sustainable Development (ISSD)*, Sarajevo.
Rao, M., Parwate, A. V., & Deshpande, V. P. (2003). Removal of heavy metals by adsorption process using Bagasse. *Indian J. Environmental Protection, 23*(4), 428–433.
Sawada, A., Mori, K., Tanaka, S., Fukushima, M., & Tatsumi, K. (2004). Removal of Cr (VI) from contaminated soil by electro kinetic remediation. *Waste Management, 24*(5), 483–490.
Seaman, J. C., Bertsch, P. M., & Schwallie, L. (1999). In-situ Cr (VI) reduction within coarse-textured, oxide-coated soil and aquifer systems using Fe(II) solutions. *ES & T, 33*(6), 938–944.
Selvaraj, K., Chandramohan, V., & Pattabhi, S. (1998). Removal of Cr (VI) from aqueous solutions and chromium plating industry wastewater using photo film waste sludge. *Indian Journal Environmental Protection, 18*(9), 641–646.
Sen, M., Dastidar, M. G., & Roychoudhury, P. K. (2005). Biosorption of Cr (VI) using fungal strain of Fusarium sp. *American Society of Civil Engineers (ASCE), Practice Periodical of Hazardous, Toxic, and Radioactive Waste Management*, *9*(3), 147–151.
Singh, K. K., Rupainwar, D. C., & Hassan, S. H. (2003) *Chemistry Environmental Research, 12,* 243–254.
Singh, K. K., Singh, A. K., & Hasan, S. H. (2006). Low cost bio-sorbent wheat bran for the removal of cadmium from wastewater: Kinetic and equilibrium studies. *Bioresource Technology, 97,* 994–1001.
Singh, R. P., Vaishya, R. C., & Prasad, S. C. (1992). A study on removal of Hexavalent Chromium by adsorption on saw dust. *Indian Journal Environmental Protection, 12*(12), 916–922.
Srinivasan, K., & Geetamani, G. (2004). Studies on Chromium (VI) removal by activated Tamarind Nut Carbon. *Oriental Journal of Chemistry, 20*(2), 335–340.
Tiravanti, G., Petruzelli, D., & Passion, R. (1997). Pretreatment of tannery wastewaters by an ion exchange process for Cr (III) removal and recovery. *Water Science and Technology, 36*(2–3), 197–207.
Toxics Use Research Institute. (2006). Hexavalent chromium. In *Five chemicals alternatives assessment Study*. Final report.
United States Environmental Protection Agency. (2003). *National primary drinking water regulations 40 CFR14162*. Available at http://edocket.access.gpo.gov/cfr_2003/julqtr/40cfr141.62.htm.
Vasanty, M., Sangeeta, M., & Kavita, C. (2003). Removal of Chromium from aqueous solutions using a mixture of flyash and activated carbon. *Indian Journal Environmental Protection, 23*(12), 1321–1325.
Venkateswarulu, P., Durga, G. V., Chittibabu, N., & Venkateswara, Rao M. (2008). Biosorption of Zn (II) from an aqueous solution by Erythrina Variegata Orientlis leaf powder. *International Journal of Physical Sciences, 3*(9), 197–204.
Vincent, J. B. (2010). Chromium: Celebrating 50 years as an essential element? *Dalton Transactions, 39*(16), 3787–3794.

Weber, W. J., & Morris, J. C. (1963). Kinetics of Adsorption on Carbon from Solution. *Journal of the Sanitary Engineering Division ASCE, 89,* 31.

Zhou, X., Korenaga, T., Takahashi, T., & Moriwake, Shinoda S. (1993). A Process monitoring/controlling system for treatment of waste water containing Chromium (VI). *Water Research, 27,* 1049–1054.

A Comparative Study of the Fuel Characteristics Between Algal Biodiesel and Petro-Diesel

R. Karmakar, A. Rajor, K. Kundu and N. Kumar

Abstract The concept of using biodiesel and biodiesel-petro-diesel blends in existing engines is not very successful yet because of two reasons—(i) high market price of the crude oil and biodiesel, (ii) incompatibility of these biodiesels with the existing engines. To cope up with this situation, experiments with biodiesel produced from indigenous mixed algal biomass, assuming that the biodiesel would be cheaper in price as the raw material has no price value, were done. Several properties like viscosity, FFA content, etc., were measured along with the emission properties of the algal biodiesel. It was found that the biodiesel has very high calorific value (almost 41,000 kJ/kg) and very low viscosity (3.10 mm^2/s). Along with these two, all other properties of this biodiesel were found to be within the prescribed limits of ASTM/BIS standards. Apart from the emission of NO_x, other emissions for the combustion of B100 algal biodiesel in a CI engine were found be much less than that of petro-diesel.

Keywords Biodiesel · Emission · CI engine

1 Introduction

As most of the people are unaware of exact availability of the non-renewable fossil fuels and these fuels are available at an affordable cost, the biofuels are not getting used widely. High market price (Amano-Boadu et al. 2014; Roychowdhury et al.

R. Karmakar (✉) · A. Rajor
Thapar Institute of Engineering and Technology, Patiala, India
e-mail: rachan.in.air@gmail.com

A. Rajor
e-mail: arajor@thapar.edu

K. Kundu
CSIR CMERI CoEFM, Ludhiana, India
e-mail: k_kundu@cmeri.res.in

N. Kumar
Roorkee Institute of Technology, Roorkee, Uttarakhand, India
e-mail: nitin.saini90@gmail.com

S. K. Ghosh et al. (eds.), *Bioresource Utilization and Bioprocess*,
https://doi.org/10.1007/978-981-15-1607-8_5

2011) of the oils is also becoming an obstacle for the biofuel to be popular. But in a situation when fossil fuels are going to be depleted completely in next hundred years (Shafiee and Topal 2009) and use of liquid fuels increases by 50% after 2030 (B. P. Energy Outlook 2030 2011), usage of fuels derived from the most renewable bioresources is very much recommendable. Though biodiesels are the most compatible source of energy for existing diesel engines and for their high thermal efficiency and low price, these engines are of high importance in the energy sector (An et al. 2012), and use of 100% of these (B100) in engines has not been successful till now for some their properties like viscosity, density, etc.

It is also well known that the emission of greenhouse gases from petro-diesel causes serious environmental pollution (Smith et al. 1993; Riahi et al. 2011) and global warming (Hoel and Kverndokk 1996; Steinberg 1999). Therefore, as fuels from algae fix 0.6% of carbon dioxide (Ponnusamy et al. 2014) and carbon dioxide actually helps to increase the rate of growth of algae (Holbrook et al. 2014; Dassey et al. 2014; Widjaja et al. 2009), algal biodiesel was planned to be used in this experiment as a reference biofuel. Other reason for using algal biodiesel was fast growth rate and at the same time requirement of less land area for the growth of the same (Chisti 2007; Dauta et al. 1990)

2 Materials and Method

Algae (*Scenedesmus* sp., *Chlorella* sp., *Closterium* sp., *Gomphonema* sp., *Spirulina* sp., *Ocillatoria* sp., *Navicula* sp., *Pinullaria* sp., *Zygnema* sp., *Spyrogyra* sp.), for this experiment, were collected from the pisciculture ponds of Guru Angad Dev Veterinary and Animal Sciences University at Punjab Agricultural University campus. All the chemicals and glassware were collected from different shops at Patiala and Ludhiana, Punjab, India.

2.1 Production of Biodiesel from Algae

Algae, collected from the pisciculture, were brought to CSIR MERI CoEFM, Ludhiana, India, and stored in the laboratory ponds. Algae, grown under optimized conditions, were kept in the sun for several days (depending on the atmospheric temperature and humidity) for drying (Karmakar et al. 2018).

Algal oil was extracted from the algae which were dried by the procedure of solvent extraction (Matthäus 2011). Soxhlet apparatus was used for this job, and n-Hexane was used as the solvent. Algae oil extracted from the sun-dried algae was used for biodiesel production. Transesterification was used for this purpose. As the free fatty acid content (FFA) of the algal oil was very high, a two-step biodiesel production procedure was used for this purpose (Pandey et al. 2013). FFA of the biodiesel, at first, reduced down by using an acid-catalysed reaction. Sulphuric acid (H_2SO_4) was

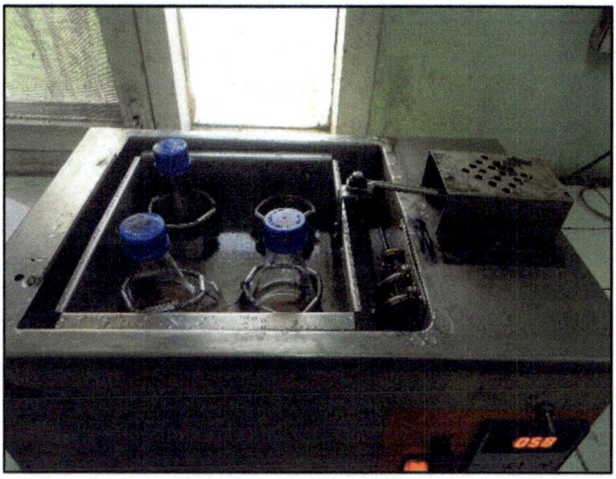

Fig. 1 Biodiesel production in water bath shaker

used to catalyse the reaction. Alkali-catalysed reaction was done by using potassium hydroxide (KOH) as the catalyst. The reactions, under optimized conditions, i.e. 6:1 molar ratio, 3% catalyst concentration, 60 °C of reaction temperature and 60 min of reaction time, were carried out in a water bath shaker (Fig. 1) (Kumar et al. 2017), and the oil was reacted with methanol (Ragit et al. 2013) to produce algal fatty acid methyl ester (FAME) or algal biodiesel in other words.

2.2 *Properties of the Algal Biodiesel*

Properties of this biodiesel like gross heat of combustion (calorific value), density, viscosity, cloud point, pour point, flash point, fire point, carbon residue content, ash content, acid number were measured by standard methods (Karmakar et al. 2018; Lin and Li 2009) and apparatus (Table 1). All the properties of the biodiesel were compared with those of petro-diesel and limits of ASTM/BIS standards.

2.3 *Emission Characteristics of Algal Biodiesel*

Biodiesel produced from the above said experiment was used to run a compression ignition engine (CI). The probe of a Testo made AVL 444 digas analyser (Fig. 2) was placed at the exhaust of the engine, the presence of different gases like carbon monoxide (CO), carbon dioxide (CO_2), oxides of nitrogen (NO_x), unburned hydrocarbon (HC) was detected, and their quantity was measured accordingly. In the same engine, petro-diesel was used to compare the emissions of it with those of algal FAME.

Table 1 List of apparatus used for finding out the properties of algal biodiesel

Fuel property	Testing apparatus	Standard
Gross heat of combustion	Digital bomb calorimeter	IS: 1448 [P:6]: 1984
Density	Pycnometer	IS: 1448 [P:32]: 1992
Viscosity	Redwood viscometer	IS: 1448 [P:25] 1976
Cloud and pour point	Cloud and pour point apparatus	IS: 1448 [P:10]: 1970
Flash and fire point	Flash and fire point apparatus	IS: 1448 [P:32]: 1992
Carbon residue	Carbon residue content apparatus	ASTM D189–IP 13 of institute of petroleum
Ash content	Muffle furnace	ASTM D482-IP 4 of Institute
Acid number	Burette and glassware (Titration)	–

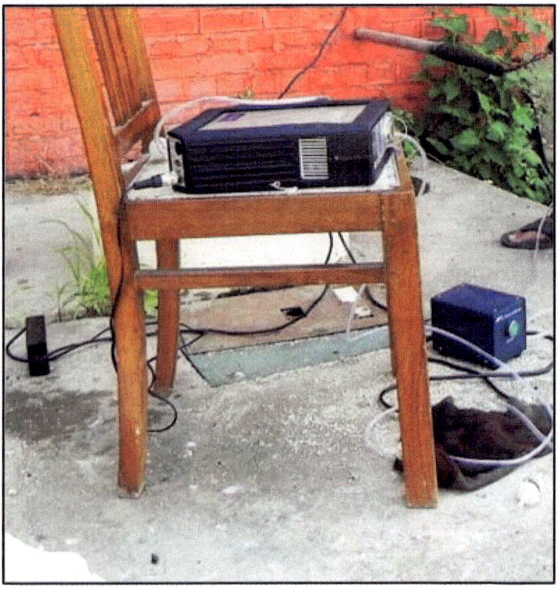

Fig. 2 Detection and measurement of emission using AVL 444 digas analyser

3 Results and Discussion

3.1 Properties of Algal Biodiesel

All the properties of this biodiesel were found to be within the limits of ASTM/BIS standards (McCurdy et al. 2014) (Table 2). Therefore, there is no problem to use this algal fatty acid methyl ester in existing internal combustion engines.

Biodiesel produced from unused indigenous mixed algal biomass was found to have very high calorific value (40,963 kJ/kg). High calorific value of any fuel is a

Table 2 Properties of algal biodiesel and related limits of ASTM/BIS standard

Fuel property	Value	ASTM/BIS standard
Gross heat of combustion	40,963 kJ/kg	–
Density (kg/m^3)	863	860–900
Viscosity (mm^2/s)	3.10	1.9–6
Cloud point	−1 °C	3 °C(Maximum)
Pour point	−7°C	–
Flash point	155 °C	93 °C (minimum)
Fire point	161 °C	–
Carbon residue	0.027%	0.050% (maximum)
Ash content	0.01%	0.01% (maximum)
Acid number	0.29 g of KOH/g	0.50 mg of KOH/g (maximum)

good sign as it gives a clear idea of the generation of power in engine when it gets used in it. This fuel was found to have higher calorific value than many popular biodiesels (Fig. 3). This value of this fuel was found to be almost similar to petro-diesel (44,800 kJ/kg). Most importantly, the value of gross heat of combustion of this fuel was found to be higher than that of coal (35,000 kJ/kg).

Density of this biodiesel was as low as 863 kg/m^3 at 15 °C, whereas the viscosity of this fuel was 3.10 mm^2/s at 40 °C. So, the chance of choking of engine, gum generation, etc., will be very less in this case as the density and viscosity of petro-diesel vary from 820 to 845 kg/m^3. Cloud point of this FAME was found to be very less (−1 and −7 °C, respectively). When the flash point of petro-diesel varies from 52 to 96 °C, both flash and fire points of algal FAME produced in this experiment were 155 and 161 °C, respectively. Carbon residue and ash content, which indicate the quantity of depositions on engines parts, were found to be low too (0.027 and 0.01%). Finally, the acid value of this biodiesel was 0.29 mg of KOH/g.

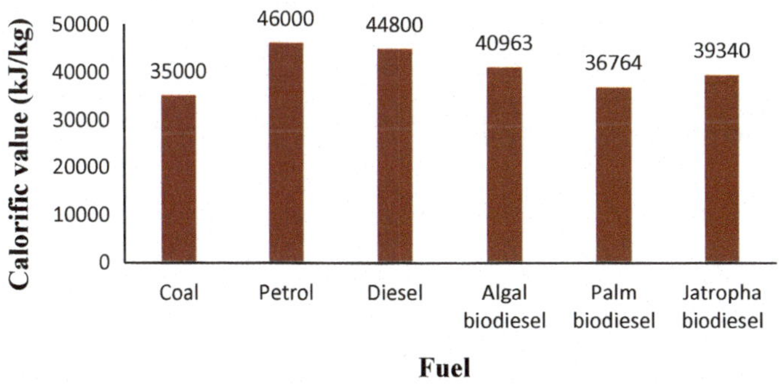

Fig. 3 Calorific values of different fuels and algal biodiesel

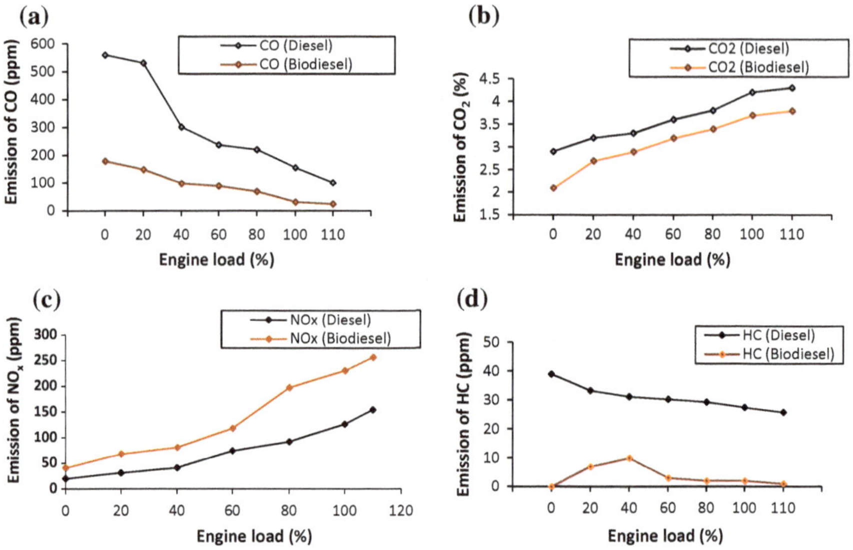

Fig. 4 Emission of **a** CO, **b** CO_2, **c** NO_x, **d** HC for the combustion of algal biodiesel and petro-diesel and their variation with engine load

3.2 *Emission Properties*

Apart from the emission of NO_x, emission of other gases like CO, CO_2, HC was found to be much less for the combustion of algal FAME than for the combustion of petro-diesel (Fig. 4). In all these cases, engine load was varied from 0 to 110%. When the emission of CO for the combustion of both types of fuel decreased simultaneously for the increase in load, CO_2 was found to be increasing with load on engine for those fuels. HC emission too was found to be decreased with increasing load on CI engines. As the biodiesel has in-built oxygen and higher cetane number, NO_x emission for the use of this biodiesel was higher because those properties help in fast ignition and generation of higher temperature.

4 Conclusion

Therefore, it can be concluded from this experiment that biodiesel produced from unused algae is suitable for the existing engine at least if applied as a blend with petro-diesel. This biodiesel is much more eco-friendly option than petro-diesel. The NO_x emission, which is higher for this algae-based FAME, can be reduced by using de-NO_x catalyser in engines.

References

Amano-Boadu, V., Pfromm, P. H., & Nelson, R. (2014). Economic feasibility of algal biodiesel under alternative public policies. *Renewable Energy, 67,* 136–142.

An, H., Wang, W. M., Chou, S. K., & Chua, K. J. (2012). Combustion of emissions characteristics of diesel engine fuelled by biodiesel at partial load conditions. *Applied Energy, 99,* 363–371.

B. P. Energy Outlook 2030. (2011). *Technology Report*, BP. https://www.bp.com.

Chisti, Y. (2007). Biodiesel from Microalgae. *Biotechnology Advances, 25,* 294–306.

Dassey, A. J., Hall, S. G., & Theegla, C. S. (2014). An analysis of energy consumption for algal biodiesel production: Comparing the literature with current estimates. *Algal Research, 4,* 89–95.

Dauta, A., Devaux, J., Piquemal, F., & Boumnich, L. (1990). Growth rate of four freshwater algae in relation to light and temperature. *Hydrobiologia, 207*(1), 221–226.

Hoel, M., & Kverndokk, S. (1996). Depletion of fossil fuels and the impacts of global warming. *Resource and Energy Economics, 18*(2), 115–136.

Holbrook, G. P., Davidson, Z., Tatara, R. A., Ziemer, N., Rosentrater, K. A., & Grayburn, W. S. (2014). Use of Microalga Monoraphidium sp. grown in waste water as a feedstock for biodiesel: Cultivation and fuel characterization. *Applied Energy, 131,* 386–393.

Karmakar, R., Kundu, K., & Rajor, A. (2018a). Fuel properties and emission characteristics of biodiesel produced from unused algae grown in India. *Petroleum Science., 15*(2), 385–395.

Karmakar, R., Rajor, A., Kundu, K., & Kumar, N. (2018b). Production of biodiesel from unused algal biomass in Punjab, India. *Petroleum Science., 15*(1), 164–175.

Kumar, N., Mohapatra, S. K., Ragit, S. S., Kundu, K., & Karmakar, R. (2017). Optimization of safflower oil transesterification method using Taguchi approach. *Petroleum Science, 14*(4), 798–805.

Lin, C. Y., & Li, R. J. (2009). Fuel properties of biodiesel produced from the crude fish oil from the soapstock of marine fish. *Fuel Processing Technology, 90*(1), 130–136.

Matthäus, B. (2011). Oil technology. *Technological Innovations in Major World Oil Crops, 2,* 23–92.

McCurdy, A. T., Higham, A. J., Morgan, M. R., Quinn, J. C., & Seefeldt, L. C. (2014). Two-step process for production of biodiesel blends from oleaginous yeast and microalgae. *Fuel, 137,* 269–276.

Pandey, R. K., Rai, A., Kundu, K., Karmakar, R., Roy, P., RoyChowdhury, A., et al. (2013). Effect of process parameters for standardization of esterification of cottonseed oil for production of biodiesel. *International Journal of Current Science, 8,* 74–78.

Ponnusamy, S., Reddy, H. K., Muppaneni, T., Downes, C. M., & Dong, S. (2014). Life Cycle Assessment of biodiesel production from algal bio- crude oils extracted under subcritical water conditions. *Bioresource Technology, 170,* 454–461.

Ragit, S. S., Mohapatra, S. K., Kundu, K., Karmakar, R. (2013). Methanolysis and ethanolysis of raw hemp oil: Biodiesel production and fuel characterization. *International Journal of Engineering Research & Technology, 2*(3).

Riahi, K., Rao, S., KreyV, Cho C., Chirkov, V., Fischer, G., Kindermann, G., et al. (2011). RCP 8.5—A scenario of comparatively high greenhouse gas emissions. *Climatic Change, 109*(1–2), 33–57.

Roychowdhury, A., Karmakar, R., Kundu, K., & Dahake, V. R. (2011). Algal biodiesel: Future prospects and problems. *Water and Energy International., 68*(11), 44–51.

Shafiee, S., & Topal, E. (2009). When will fossil fuel reserves be diminished? *Energy Policy., 37*(1), 181–189.

Smith, K. R., Khalil, M. A. K., Rasmussen, R. A., Thorneloe, S. A., Manegdeg, F., & Apte, M. (1993). Greenhouse gases from biomass and fossil fuel stoves in developing countries: A Manila pilot study. *Chemosphere, 26*(1–4), 479–505.

Steinberg, M. (1999). Fossil fuel decarbonization technology for mitigating global warming. *International Journal of Hydrogen Energy, 24*(8), 771–777.

Widjaja, A., Chien, C. C., & Ju, Y. H. (2009). Study of increasing lipid production from fresh water microalgae Chlorella vulgaris. *Journal of the Taiwan Institute of Chemical Engineers., 40,* 13–20.

Production of Bioethanol from Green Alga *Chlorella Vulgaris*: An Important Approach to Utilize Algal Feedstock or Waste

D. Varaprasad, N. Ragasudha, K. Paramesh, P. Chandramati Shankar, S. Nazaneen Parveen and T. Chandrasekhar

Abstract Regular depletion of fossil fuels urges human society to depend on renewable resources seriously and invest more on biofuels sector. Recently generation of bioethanol from algal feedstock or algal waste has been an interesting research. Unlike fossil fuels, production of bioethanol from algal feedstock or waste will take less time and expensive. In the present study, an important green alga *Chlorella vulgaris* (*C. vulgaris*) was selected for ethanol production. *Chlorella vulgaris* cultures were initiated under in vitro conditions using universal tris-acetate-phosphate (TAP) medium along with various concentrations and combinations of vitamins such as thiamin, biotin and cobalamin (B1, B7 and B12) to enhance the biomass in turn ethanol production. Optimal level of vitamins i.e. CV2 medium (TAP with 0.4 g/L of B1, 0.002 g/L of B7 and 0.002 g/L of B12) augmented the biomass production including lipid contents. Later all the algal feedstocks were used for production of ethanol in the company of *Saccharomyces cerevisiae* (*S. cerevisiae*) in both light and dark fermentations. Higher levels of ethanol production was achieved with the feedstock generated from CV2 medium at 48 h in dark fermentation and compared with other feedstocks as well with light fermentation yield at different time intervals. The results of the present investigation may grab the attention of investors in bioenergy sector for the production of bioethanol at commercial level from algal feedstock or algal waste.

Keywords *Chlorella* · Vitamins · Bioethanol · Yeast · Fermentation · Light and dark

D. Varaprasad · N. Ragasudha · K. Paramesh · S. Nazaneen Parveen · T. Chandrasekhar (✉)
Department of Environmental Science, Yogi Vemana University, Kadapa,
Andhra Pradesh 516005, India
e-mail: tcsbiotech@gmail.com

P. Chandramati Shankar
Department of Biotechnology, Yogi Vemana University, Kadapa,
Andhra Pradesh 516005, India

S. K. Ghosh et al. (eds.), *Bioresource Utilization and Bioprocess*,
https://doi.org/10.1007/978-981-15-1607-8_6

1 Introduction

The potential of ethanol generation from seaweed or algal feedstock or waste from algal industry is increasing regularly due to enriched sugars and lipids in algae, thereby cutting down the initial expenses (John et al. 2011). Moreover, several countries are not properly using algae and are remained as waste in coastal areas. In addition, research on renewable energy is one of the most concern issues recently due to exhausting fossil fuels. Hence, entire world is trying to invest more on alternate energy sources including production of ethanol from biological sources and their waste (Gray et al. 2006). Moreover, addition of bioethanol to conventional fuel reduces the cost and pollution led to sustainable development with respect to energy sector. Specifically, production and transportation use of the bioethanol will also reduce the greenhouse gas emissions (Hirayama et al. 1998; Balat 2009). In addition, production of biofuels from feedstock of advanced plants such as corn, sorghum and other agricultural wastes is common practice (Kim and Dale 2004; Prasad et al. 2007). Algae or fresh waste of algae is one of the best alternate sources for biofuel production in recent years apart from edible and medicinal uses (Daroch et al. 2013). Growth of algae varies depends on the species, their metabolic activities and environmental conditions. In addition, various algae possess significant carbohydrate and lipid content along with useful biomolecules (Sanchez and Cardona 2008).

Rate of biofuel production from algae or advanced plants through fermentation completely depends on quantity and quality of biomass (Wyman 1994; Hirano et al. 1997). Specifically, starch or sugar or cellulose level decides the bioethanol production. Various forms of carbohydrates such as starch, cellulose, other sugars and lipids are synthesized in algae, but these profiles are not same as in the case of land plants (Chen et al. 2013). Basic works have been done on bioethanol production using both fresh and marine algae for the improvement of output (Gfeller and Gibbs 1984; Kosaric and Velikonja 1995; Gupta et al. 2012; Hossain et al. 2015). It is a well-known fact that the addition of growth factors, vitamins and hormones improves the algal biomass in turn ethanol. In general, most of the vitamins have been found to act as coenzymes, and some also act as growth regulators (Croft et al. 2006). Among vitamins A, B, C, D, E and K, B-complex vitamins were focused more in the present study. Thiamin (vitamin B_1), biotin (vitamin B_7) and cobalamin (vitamin B_{12}) were available as prototroph or auxotroph for plants including primitive algae and are involved in biosynthesis of certain amino acids, acts as co-factors in metabolic and enzymatic activities (Grossman 2016; Hansoz et al. 2018; Ruangsomboon et al. 2018). The green alga *Chlorella vulgaris* from chlorellaceae family lives in both fresh and marine water conditions depends on the strain with unicellular in nature (Cha et al. 2010). It is a well-known fact that phototrophic organism such as algae and their fresh waste are useful for ethanol production through fermentation process in the presence of yeast accompanier. Few works on bioethanol production using feedstock of *C. vulgaris* through enzymatic hydrolysis and fermentation were carried out (Ho et al. 2013; Moncada et al. 2013; Kim et al. 2014). Similarly, Salman and Mohammed Ali (2014) also noticed the improved ethanol content in this alga.

But all the previous works were done without vitamin assistance. But till date, there was no information on biomass enhancement through exogenous supply of vitamin assistance in this species. For the first time, ethanol production has been carried out using various feedstocks of *C. vulgaris* obtained from different vitamin assisted tris-acetate-phosphate (TAP) media. Further, we also standardized the media to enhance the biomass leading to high-level production of ethanol from the feedstocks by yeast fermentation.

2 Materials and Methods

Chlorella vulgaris cultures were collected from University of Madras, Chennai, and preserved as per the standard protocols. The total glassware (Borosil, India) for media preparation was washed thoroughly using detergents/teepol (10%) solution and then cleaned with running tap water and rinsed with distilled water and kept in an oven for drying purpose (Kemi, K04.3, Ernakulam, India). Tris-acetate-phosphate (TAP) media with different doses of vitamins was prepared as per the procedure and used for in vitro culture. Various media such as CV0 (TAP without vitamins), CV1 (TAP with 0.2 g/L thiamin, 0.001 g/L biotin and 0.001 g/L cobalamin), CV2 (TAP with 0.4 g/L thiamin, 0.002 g/L biotin and 0.002 g/L cobalamin), CV3 (TAP with 0.8 g/L thiamin, 0.004 g/L biotin and 0.004 g/L cobalamin) and CV4 (TAP with 1.6 g/L thiamin, 0.008 g/L biotin and 0.008 g/L cobalamin) were designated based on the vitamin composition (Table 1). The media was adjusted to pH 7.0 using pH meter (Elico limited, India), and all the culture vessels containing media were autoclaved for 15 min at 15 lbs/in^2 in an autoclave (Inlab Equipment, Madras, India). Sterilized cultures vessels were removed from the autoclave and cooled down to room temperature and finally were kept in laminar airflow chamber [(LAF) (Hitech products, Chennai, India)] for inoculation. Before going to inoculation, the LAF chamber was sterilized by switching on the ultraviolet lamp for 20 min and later smeared with 70% ethanol.

Inoculation was carried out using sterilized loops and wood sticks. At the time of inoculation, hands were cleaned with 70% frequently. All the inoculated samples were kept in the orbital shaker (Remi Elektrotechnik Limited, Vasai, India) at 120 rpm in continuous light and cultures were grown at 25 ± 1 °C. Haemocytometer was used

Table 1 Different media used for present study under in vitro conditions

Medium	TAP	Thiamin (B_1) g/L	Biotin (B_7) g/L	Cobalamin (B_{12}) g/L
CV0	Yes	–	–	–
CV1	Yes	0.2	0.001	0.001
CV2	Yes	0.4	0.002	0.002
CV3	Yes	0.8	0.004	0.004
CV4	Yes	1.6	0.008	0.008

to know the growth condition of *C. vulgaris* using cell count experiments. Estimation of chlorophyll was carried out by Arnon's (1949) spectrophotometric method using UV-Visible spectrophotometer (Shimadzu UV- 1800, India) at 663 and 645 nm. Based on total chlorophyll content, all the samples were set to equal biomass using respective medium (CV0, CV1, CV2, CV3 and CV4). Further, all the algal samples were subcultured by adding 1 ml of culture into 99 ml of respective medium. Lipid contents of *C. vulgaris* grown in various media were estimated by method of Bligh and Dyer (1959) with minor modification. 10 ml of each algal culture were collected in 15 ml vials and centrifuge at 5000 rpm up to 10 min. Later, add 10 ml mixture of chloroform and methanol (1:1 v/v) to supernatant and kept in ultrasonication for 10 min and finally centrifuge it. Filtrate collected into the pre-weighed beaker and evaporated until the solvent evaporates completely. The weight of the remained lipid content was measured using the standard formula and expressed in g/100 g of algae (percentage).

During the early stationary growth phase, all the *C. vulgaris* cultures were collected and heated at 100 °C for 2 h on the hot plate and later allowed to cool at room temperature. Enzyme hydrolysis was done by adding alpha-amylase enzyme (0.6 g/L) and incubated for 90 min as mentioned by Sulfahri et al. (2011). Later cultures were centrifuged at 5000 rpm and 5 °C for 15 min. The supernatant was collected into fresh tube and filter sterilized and used for further fermentation process. Yeast (*S. cerevisiae*) was inoculated in yeast peptone medium (10 g/L yeast extract, 20 g/L peptone and 200 g/L glucose) and kept in the orbital shaker at 120 rpm and 27 °C for 24 h. Later, 30 ml serum vials contain 5.0 ml of algal solution along with 10% yeast cultures were used for fermentation. Vials were sealed and maintained in anaerobic conditions with the help of purging process using nitrogen gas. Fermentation was conducted at different time intervals (0, 12, 24, 48, 72 and 96 h) by rotating the vials at 27 °C and 120 rpm in an orbital shaker in both light and dark conditions. Both light and dark cultures were mixed with potassium dichromate and concentrated sulfuric acid, which were later used for test samples. Standard curve was prepared with commercially available ethanol, and final ethanol concentration was calculated by using the standard formula, and units were expressed in g/L.

3 Results and Discussion

By seeing the algal waste in coastal areas, the present work has been initiated and the results obtained were documented as follows. In the first stage, *C. vulgaris* cultures were initiated under in vitro conditions using various TAP media contain different concentrations and combinations of B_1, B_7 and B_{12} vitamins (Table 1). All the media used were tested to know the capacity of improvement of biomass from these *C. vulgaris* cultures. All the cultures in CV0, CV1, CV2, CV3 and CV4 media were grown well without any microbial contamination, and visually, there was minor differences were observed between the cultures (Fig. 1).

Fig. 1 *C. vulgaris* cultures generated in different media under in vitro conditions

Variations in terms of chlorophyll (biomass) were also noticed in all the cultures grown in different media used. Among all the media tested, *C. vulgaris* cultures grown in CV2 (TAP with 0.4 g/L of B_1, 0.002 g/L of B_7 and 0.002 g/L of B_{12}) exhibited more total chlorophyll content when compared to cultures grown in other media (Fig. 2a). In agreement with these results, Croft et al. (2006) emphasized these B vitamins roles in growth and development. Cultures grown in high concentration of vitamins (CV4) displayed less total chlorophyll content when compared to cultures grown without vitamin assistance. Similarly, lipid contents were also varied in *C. vulgaris* cultures grown in different media. With this experiment, also CV2 medium promotes the lipid content levels (Fig. 2b). The lowest content of lipid level was observed in *C. vulgaris* grown in CV4 medium. Similarly, Tandon et al. (2017) improved the microalgae productivity with exogenous help of vitamins. It is well-known fact that the inclusion of growth additives/enhancers such as vitamins improves the growth and biomass of both primitive and advanced plants which in turn lead to enhancement in biofuel production in most of the species (Croft et al. 2006; Daroch et al. 2013). To the best of our knowledge, this is the first report on addition of vitamins along with universal TAP medium exhibiting enhanced biomass production in *C. vulgaris*. Estimation of biomass is the crucial step in the process of ethanol production and comparative studies.

Further, these algal feedstocks obtained from various media were used to produce ethanol in the presence of *S. cerevisiae* under dark and light fermentation conditions.

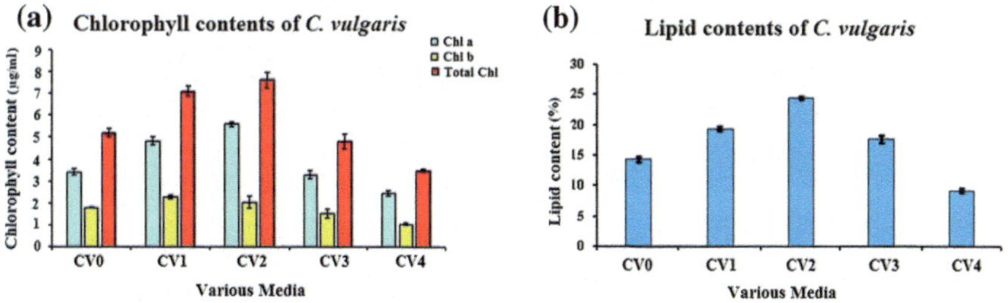

Fig. 2 Contents of chlorophylls (**a**) and lipids (**b**) in *C. vulgaris* grown in various media. Data represents three replicates, and bars indicate standard error

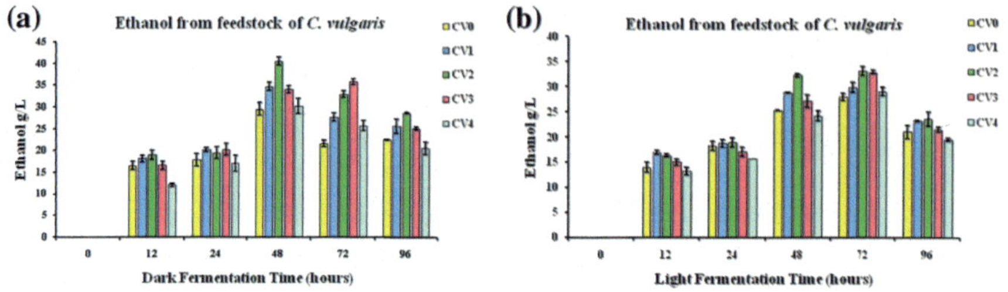

Fig. 3 Levels of ethanol production from feedstocks of *C. vulgaris* grown in different media in dark (**a**) and light (**b**) fermentation conditions. Data represents three replicates, and bars indicate standard error

In light and dark fermentations, variation in production of ethanol was observed at different time intervals (Fig. 3). In dark fermentation, content of ethanol production was more at 48 h in feedstock of *C. vulgaris* grown in CV2 medium when compared to feedstocks grown in other media (Fig. 3a). Surprisingly, at 72 h, feedstock grown in CV3 medium resulted more ethanol production. Probably, this may be due to variations in cellulose, reducing sugars, lipids, etc., in this feedstock.

For example, if one feedstock contains xylose, which is difficult to convert into ethanol using *S. cerevisiae*. So, the type of yeast is also an important factor to convert sugars into ethanol by fermentation process (Bettiga et al. 2009). In light fermentation, feedstock generated from CV2 medium yielded more ethanol at 72 h, and here also CV3 feedstock produced high levels of ethanol (Fig. 3b). Previous works on ethanol production from this alga using various methods were not with vitamin assistance (Moncada et al. 2013; Kim et al. 2014; Salman and Mohammed Ali 2014). Probably, this is the one of the reasons for augmentation in ethanol production in the present investigation. Overall in dark and light fermentations, feedstock of CV2 is best for ethanol production at 48 and 72 h. High vitamin content (CV4 feedstock) and increased fermentation time (96 h) lead to reduced production of ethanol in both light and dark fermentations. Many studies were conducted on ethanol production from bacteria and yeast with the assistance of vitamins but not using algal cultures (Sato et al. 1992; Alfenore et al. 2002; Liu et al. 2018). Recently, Ruangsomboon et al. (2018) used the same combination of vitamins for the production of biodiesel from green alga *Botrycoccus braaunii* and succeeded. Moreover, in the present work, the standardized media, i.e., CV2 for *C. vulgaris* can be useful for biomass production at industrial level. Figure 4 illustrates the method developed for ethanol production from algal feedstock, and the same may also benefit for fresh algal waste in coastal areas.

Fig. 4 Process involved in bioethanol production from algae

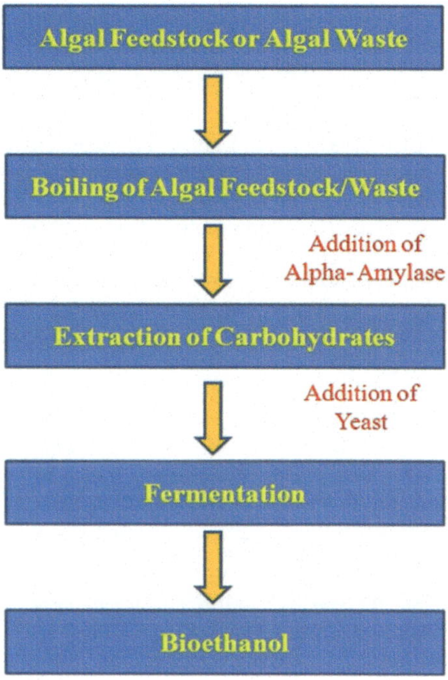

4 Conclusions

Present work demonstrated that addition of optimal level of vitamins such as thiamin, biotin and cobalamin to TAP medium assisted to produce high amount of biomass in *C. vulgaris*. The overall content of ethanol produced from feedstock of *C. vulgaris* grown in CV2 medium through dark fermentation was high at 48 h when compared to feedstock of other media even in light fermentations at different time intervals. The present protocol is also useful for bioethanol production from algal waste.

Acknowledgements The authors are thankful to Prof. Mathivanan, University of Madras, Chennai, India, for algal culture and also Dr. L. V. Reddy and Dr. Veda, Department of Microbiology, Yogi Vemana University, Kadapa, India, for providing yeast and valuable suggestions in this work.

References

Alfenore, S., Molina-Jouve, C., Guillouet, S., Uribelarrea, J. L., Goma, G., & Benbadis, L. (2002). Improving ethanol production and viability of *Saccharomyces cerevisiae* by a vitamin feeding strategy during fed-batch process. *Applied Microbiology and Biotechnology, 60,* 67–72.

Arnon, D. (1949). Estimation of total chlorophyll. *Plant Physiology, 24,* 1–15.

Balat, H. (2009). Prospects of biofuels for a sustainable energy future: A critical assessment. *Energy Education Science and Technology, 24,* 85–111.

Bettiga, M., Bengtsson, O., Hahn-Hägerdal, B., & Gorwa-Grauslund, M. F. (2009). Arabinose and xylose fermentation by recombinant *Saccharomyces cerevisiae* expressing a fungal pentose utilization pathway. *Microbial Cell Factories, 24,* 40.

Bligh, E. G., & Dyer, W. J. (1959). A rapid method of total lipid extraction and purification. *Canadian Journal of Biochemistry and Physiology, 37,* 911–917.

Cha, K. H., Lee, H. J., Koo, S. Y., Song, D. G., Lee, D. U., & Pan, C. H. (2010). Optimization of pressurized liquid extraction of carotenoids and chlorophylls from *Chlorella vulgaris*. *Journal of Agricultural and Food Chemistry, 58,* 793–797.

Chen, C. Y., Zhao, X. Q., Yen, H. W., Ho, S. H., Cheng, C. L., Lee, D. J., et al. (2013). Microalgae-based carbohydrates for biofuel production. *Biochemical Engineering Journal, 78,* 1–10.

Croft, M. T., Warren, M. J., & Smith, A. G. (2006). Algae need their vitamins. *Eukaryotic Cell, 5,* 1175–1183.

Daroch, M., Geng, S., & Wang, G. (2013). Recent advances in liquid biofuel production from algal feedstocks. *Applied Energy, 102,* 1371–1381.

Gfeller, R. P., & Gibbs, M. (1984). Fermentative metabolism of *Chlamydomonas reinhardtii*: I. Analysis of fermentative products from starch in dark and light. *Plant Physiology, 75,* 212–218.

Gray, K. A., Zhao, L., & Emptage, M. (2006). Bioethanol. *Current Opinion in Chemical Biology, 10,* 141–146.

Grossman, A. (2016). Nutrient acquisition: The generation of bioactive vitamin B12 by microalgae. *Current Biology, 26,* 319–321.

Gupta, R., Biswas, K., Mishra, I., & Suthidiran, K. (2012). Ethanol production from marine algae using yeast fermentation. *Research Desk, 1,* 17–22.

Hansoz, A. D., Amthor, J. S., JiayiSun, Niehaus T. D., GregoryIII, J. F., Bruner, S. D., & Ding, Y. (2018). Redesigning thiamin synthesis: Prospects and potential payoff. *Plant Science, 273,* 92–99.

Hirano, A., Ueda, R., Hirayama, S., & Ogushi, Y. (1997). CO_2 fixation and ethanol production with microalgal photosynthesis and intracellular anaerobic fermentation. *Energy, 22,* 137–142.

Hirayama, S., Ueda, R., Ogushi, Y., Hirano, A., Samejima, Y., Hon-Nami, K., et al. (1998). Ethanol production from carbon dioxide by fermentative microalgae. *Studies in Surface Science and Catalysis, 114,* 657–660.

Ho, S. H., Huang, S. W., Chen, C. Y., Hasunuma, T., Kondo, A., & Chang, J. S. (2013). Bioethanol production using carbohydrate-rich microalgae biomass as feedstock. *Bioresource Technology, 135,* 191–198.

Hossain, M. N. B., Basu, J. K., & Mamun, M. (2015). The production of ethanol from micro-algae spirulina. *Procedia Engineering, 105,* 733–738.

John, R. P., Anisha, G. S., Nampoothiri, K. M., & Pandey, A. (2011). Micro and macroalgal biomass: A renewable source for bioethanol. *Bioresource Technology, 102,* 186–193.

Kim, K. H., Choi, I. S., Kim, H. M., & Bae, H. J. (2014). Bioethanol production from the nutrient stress-induced microalga *Chlorella vulgaris* by enzymatic hydrolysis and immobilized yeast fermentation. *Bioresource Technology, 153,* 47–54.

Kim, S., & Dale, B. E. (2004). Global potential bioethanol production from wasted crops and crop residues. *Biomass and Bioenergy, 26,* 361–375.

Kosaric, N., & Velikonja, J. (1995). Liquid and gaseous fuels from biotechnology: Challenge and opportunities. *FEMS Microbiology Reviews, 16,* 111–142.

Liu, H., Sun, J., Chang, J. S., & Shukla, P. (2018). Engineering microbes for direct fermentation of cellulose to bioethanol. *Critical Reviews in Biotechnology, 10,* 1–17.

Moncada, J., Jaramillo, J., Higuita, J. C., Younes, C., & Cardona, C. A. (2013). Production of bioethanol using *Chlorella vulgaris* Cake: A techno economic and environmental assessment in the Columbian context. *Industrial and Engineering Chemistry Research, 52,* 16786–17794.

Prasad, S., Singh, A., & Joshi, H. C. (2007). Ethanol as an alternative fuel from agricultural, industrial and urban residues. *Resources, Conservation and Recycling, 50,* 1–39.

Ruangsomboon, S., Sornchai, P., & Prachom, N. (2018). Enhanced hydrocarbon production and improved biodiesel qualities of *Botryococcus braunii* KMITL 5 by vitamins thiamine, biotin and cobalamin supplementation. *Algal Research, 29,* 159–169.

Salman, J. M., & Mohammed Ali, M. A. (2014). Bioethanol production from green alga *Chlorella Vulgaris* under different concentrations of nitrogen. *Asian Journal of Natural and Applied Sciences, 3,* 27–36.

Sanchez, O. J., & Cardona, C. A. (2008). Trends in biotechnological production of fuel ethanol from different feedstocks. *Bioresource Technology, 99,* 5270–5295.

Sato, K., Goto, S., Yonemura, S., Sekine, K., Okuma, E., Takagi, Y., et al. (1992). Effect of yeast extract and vitamin B12 on ethanol production from cellulose by *Clostridium thermocellum* I-1-B. *Applied and Environmental Microbiology, 58,* 734–736.

Sulfahri, M. S., Sunarto, E., Irvansyah, M. Y., Utami, R. S., & Mangkoedihardjo, S. (2011). Ethanol production from algae Spirogyra with fermentation by *Zymomonas mobilis* and *Saccharomyces cerevisiae*. *Journal of Basic and Applied Scientific Research, 1,* 589–593.

Tandon, P., Jin, Q., & Huang, L. (2017). A promising approach to enhance microalgae productivity by exogenous supply of vitamins. *Microbial Cell Factories, 16,* 219.

Wyman, C. E. (1994). Alternative fuels from biomass and their impact on carbon dioxide accumulation. *Applied Biochemistry and Biotechnology, 45,* 897–915.

Biogas Production from Fat, Oil and Grease and Effect of Pre-treatment

Sreesha Malayil and Hoysall N. Chanakya

Abstract The feasibility of using fat, oil and grease (FOG), a typical waste discarded from wastewater treatment plants, as substrate for anaerobic digestion with various enzyme treatments was investigated using series of biochemical methane potential (BMP) tests. The ranges of an ideal substrate to inoculum (S/I) ratio were determined for FOG (0.25, 0.5 and 1%) with various pre-treatments. The results indicate an increased biogas production with both detergent enzyme formulation and enzyme alone ('Gelzyme' and lipase). The ideal inoculum to substrate concentration for both treatments was found to be 0.25% beyond which LCFA toxicity was encountered. In the case of 'Gelzyme' treatment, the maximum biogas production was observed at 0.25% substrate concentration and 0.25% 'Gelzyme' treatment for 6 h (389 ml/gTS). With lipase treatment, the optimum inoculum to substrate ratio was also found to be 0.25% with a lipase concentration of 0.5% and treatment time of 24 h (892 ml biogas/gTS) which accounts for 85.5% of the theoretical biogas yield from lipids. Of the two enzyme pre-treatments studied, lipase was found to give higher gas yields.

Keywords Fat oil and grease · Biogas from FOG · FOG pretreatment

1 Introduction

Fat, oil and grease (referred as FOG) arising from fatty food remains and oils used for cooking is a lipid-rich waste generated at the initial stage in wastewater treatment processes. FOG being insoluble in water poses several problems for treatment and aggregates to form large 'fatbergs' in sewerage systems. The presence of FOG in wastewater results in clogging of sewerage and drain pipes causing disturbances in sewage flow through the drains (sometimes reaching tens of tons), scum formation and sludge flotation caused by the adsorption of the lipid layer around the microbial

S. Malayil (✉) · H. N. Chanakya
Centre for Sustainable Technologies (Formerly ASTRA), Indian Institute of Science, Bengaluru 560012, India
e-mail: srisham@iisc.ac.in

H. N. Chanakya
e-mail: chanakya@iisc.ac.in

S. K. Ghosh et al. (eds.), *Bioresource Utilization and Bioprocess*,
https://doi.org/10.1007/978-981-15-1607-8_7

surfaces (Hwu et al. 1998). Fatbergs and clogging of sewage system lead to a reduction/blockage in conveyance capacity and ultimately to sanitary sewer overflow that cost municipalities millions of dollars each year in cleaning, repairing and maintenance fees (US EPA 2011). In order to prevent this situation, municipalities often implement pre-treatment steps to aid in the removal of grease from kitchen waste streams. The commonly practised FOG removal mechanism is to employ 'grease traps' or 'grease interceptors'. Further, direct disposal of FOG is considered illegal and harmful to land and water since this forms a layer on surface and the time taken for FOG to degrade is reasonably long. The common modes of FOG disposal include landfilling, composting, rendering for manufacturing of lubricants/industrial soaps, incineration or biodiesel production, etc.

FOG collected from the foodservice industry and wastewater treatment plants is a good substrate for biogas production and has the potential to increase biogas production by >30% when added directly to the anaerobic digester. Biogas production from FOG recovered through grease traps at wastewater treatment plants can meet over 50% of their electricity demand through on-site power generation (Suto et al. 2006; Bailey 2007; Kabouris et al. 2008, 2009; Davidsson et al. 2008; York et al. 2008; Parry et al. 2008; Luostarinen et al. 2009; Muller et al. 2010). FOG as a biogas substrate is rich in lipids and is found to yield highest methane (theoretical biogas production potential of 1014 l/kg volatile solids, Buswell and Neave 1930) which is high when compared to carbohydrates (glucose gives 380 l/kg VS, and proteins give 530 l/kg VS).

The lipids in FOG when subjected to anaerobic digestion lead to the significant increase of long-chain fatty acids (LCFAs) which is known to inhibit the methanogens when populations of syntrophic organisms are low, and this in turn leads to the failure of reactors. LCFAs the primary component of FOG's early breakdown stages are further degraded anaerobically via the β-oxidation pathway to acetate and H_2. β-oxidation begins when the fatty acid is activated with coenzyme A and the resulting oxidation leads to the release of acetyl-CoA and the formation of a fatty acid chain, which is shortened by two carbons. Acetyl-CoA is oxidized by way of the citric acid cycle, and the process of β-oxidation is repeated (Madigan et al. 2006) under aerobic systems. The degradation of LCFAs via the β-oxidation pathway follows the reaction

$$CH_3(CH_2)_nCOOH + 2H_2O \rightarrow CH_3(CH_2)_{n-2}COOH + CH_3COOH + 2H_2$$

Several mechanisms have been proposed for the inhibitory effect of LCFAs on methanogens, and the exact mechanism responsible for this action is not known completely. The most commonly cited reason for this inhibition is the surfactant effect of the LCFAs on the cell membranes accompanied by cell lysis. The LCFAs are found to be adsorbed on the cell membranes or encapsulate the cell membranes which in turn lead to the halting of mass transfer over the membranes (Long et al. 2011). It is often cited that the impact of the LCFAs is directly proportional to the degree of unsaturated fats in the substrate (Kim et al. 2004) and is often accompanied by a lag phase in methane production (Angelidaki and Ahring 1992; Pereira et al.

2004; Long et al. 2011). To overcome this issue of inhibition, the most commonly preferred techniques are

(1) Co-digestion with the abundance of buffering components such as bicarbonates or micronutrients (e.g. S, P and N) necessary for cellular growth and activity and
(2) pre-treatment.

Various reports on co-digestion of sludge and FOG at lab, pilot and full scale are available such as Kabouris et al. (2008), Kabouris et al. (2009a, b) and Wan et al. (2011) who report overall biogas production to increase up to 200%. Pre-treatment of FOG using direct enzymes leads to the release of short-chain fatty acids (acetate, etc.) which when fed into the anaerobic digesters has the capacity to reduce the lag phase, reduce the dependence on slow-growing syntrophic bacterial species and finally increase biogas production.

The objective of this paper is to compare the biogas production potential of FOG using different enzyme pre-treatments such as a primary lipase and a detergent-lipase complex (Gelzyme-a highly concentrated enzyme-based liquid laundry detergent) which have the potential to decrease the overall cost of FOG treatment and generate higher methane yields.

2 Materials and Methods

FOG samples used in this experiment were obtained from the grease traps of wastewater treatment facility at Reethi Beach Resort, Maldives. The inoculum for biochemical methane potential (BMP) was collected from the outlet of a functioning plug flow type biogas plant fed with leafy biomass operating at CST, IISc (Jagadish et al. 1998). 'Gelzyme' is a highly concentrated, enzyme-based liquid laundry detergent and is formulated to remove body oils and oily cosmetic stains from the clothes which were purchased from Amway Corporation. Lipase with activity 40–70 units per mg protein was purchased from SRL laboratory. The FOG substrate was treated with lipase and Gelzyme separately at different concentrations of 0.5, 1 and 2% for various time intervals such as 6, 12, 18 and 24 h and examined for biogas production potential. This pre-treated FOG was used for BMP assay, and the experiments were carried out in batches for 90 days in 135-ml serum bottles which were sealed with rubber stoppers and aluminium crimps. Different substrate to inoculum ratios of 0.25, 0.5 and 1% were tried and tested for above-mentioned treatments. The biogas produced inside the bottles was measured by the downward displacement of water. The total solids and volatile solids were estimated using the standard methods (APHA-AWWA 1985).

3 Results and Discussion

3.1 0.25% Substrate Concentrations

The TS content of the FOG sample collected was found to be 75% with a VS content of 94% of TS, and this was found to be close to the value reported in the literature for FOG collected from various restaurants and wastewater treatment plants (Kabouris et al. 2008, 2009; Long et al. 2011). FOG sample at 0.25% concentration was treated with 0.25, 0.5 and 1% lipase and 'Gelzyme', respectively. These samples were incubated at various time intervals of 6, 12, 18 and 24 h. BMP and cumulative gas production levels were plotted against various concentrations of 'Gelzyme' and lipase obtained for different time intervals of pre-treatment. The maximum biogas yield for FOG not subjected to any treatment was found to be 384 ml/gTS at 0.25% substrate to inoculum ratio (Fig. 1). However, for the biogas production to initiate, there was a lag period of 17d when no treatment was given. Similar lag phase for biogas production to occur has been reported in the literature (Kabouris et al. 2008, 2009; Long et al. 2011). With 'Gelzyme' treatment, this lag phase was found to be reduced to 6d and the maximum biogas yield achieved was 390 mg/gTS with 6 h of treatment efficiency with 0.5% 'Gelzyme' addition. In the case of 'Gelzyme', an increase in treatment time was found to have a negative effect on biogas production at 0.5% Gelzyme concentration (Fig. 1a). With 12 h of treatment, the biogas production fell from 390 ml/gTS to 343 ml/gTS and further to 236 ml/gTS with 18 h at 0.5%

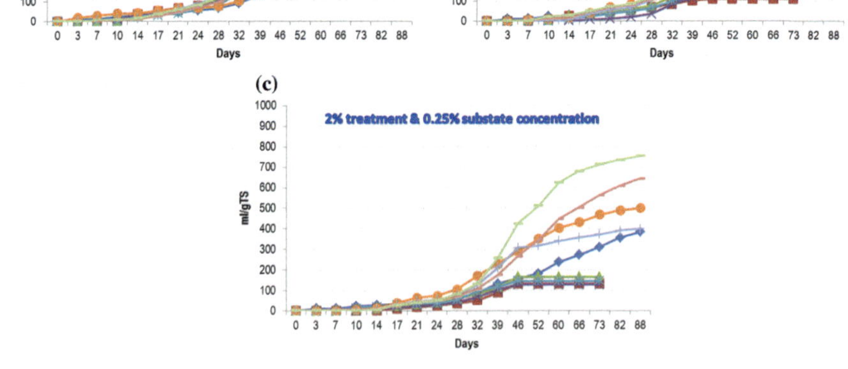

Fig. 1 BMP of 0.25% FOG at various concentrations of lipase and 'Gelzyme' **a** 0.5%, **b** 1% and **c** 2%

'Gelzyme' concentration (Fig. 1a). This suggests that increasing the treatment time could not increase the biogas production and 6 h treatment time was found to be the most effective. The reasons for these observations were not apparent and could not be discerned from the data collected.

Lipase addition at 0.5% was found to increase the gas production substantially from 384 ml/gTS without pre-treatment to 768 ml/gTS which was almost double when compared to the untreated sample. However, just as in the case of 'Gelzyme', an increase in treatment time was found to have a comparatively lower negative effect on biogas production. The biogas production decreased from 768 ml/gTS (6 h treatment) to 459 ml/gTS (18 h treatment). However, prolonging the pre-treatment to 24 h was found to increase biogas production from 768 to 892 ml/gTS which needs further investigation.

Increasing Gelzyme and lipase was found to reduce biogas production up to a certain level. At 1% 'Gelzyme' concentration, there was a reduction in biogas production from 390 to 108 ml/gTS with 6 h treatment and a similar reduction was observed at 12, 18 and 24 h. This clearly indicated that a higher concentration of 'Gelzyme' and treatment time had a negative effect on biogas production. The reasons were not easily discernible. One possible reason could be the surfactant concentration in Gelzyme, and its interference with methanogenesis is not known and needs further investigation. It is thus indicated that a slower pre-treatment with lower concentrations of these pre-treatment compounds needs to be tried out.

3.2 0.5% Substrate Concentrations

Increasing substrate concentration from 0.25 to 0.5% resulted in a decrease of biogas production in untreated and 'Gelzyme' and lipase treatment. In the case of untreated FOG samples, the biogas potential was reduced from 383 ml/gTS at 0.25% substrate concentration to 96 ml/gTS which indicates reduced methanogenic activity due to possible LCFA toxicity. A similar lag phase of 17d was observed at 0.25 and 0.5%. In the case of 'Gelzyme', 0.5% treatment at 6 h (409 ml/gTS) was found to be the best with 0.5% substrate concentration. In the case of lipase treatment, the maximum biogas production was observed at 0.5% treatment at 18 h (653 ml/gTS). In case of lipase, an increase in biogas production was observed with treatment time (24 h at 0.25% substrate concentration and 18 h at 0.5% substrate concentration) which needs further understanding. An important observation made was that in case of lipase treatment, there was no lag phase observed for various treatments tried and in case of 'Gelzyme', a lag phase of 3d was observed for most of the treatments (Fig. 2).

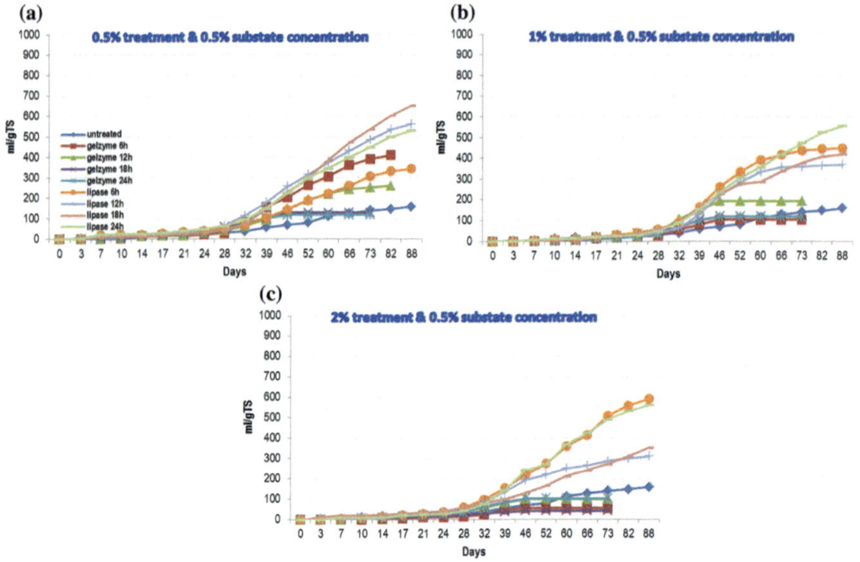

Fig. 2 BMP of 0.5% FOG at various concentrations of lipase and 'Gelzyme' **a** 0.5%, **b** 1% and **c** 2%

3.3 1% Substrate Concentration

The biogas production fell drastically with 1% substrate concentration for all treatments. In the case of untreated FOG, the biogas production was found to be 95 ml/gTS at 1% substrate concentration with a lag phase of 20d. Pre-treatment with both 'Gelzyme' and lipase was found to increase the biogas production to 140 mlg/gTS in case of 'Gelzyme' (12 h) and 351 ml/gTS (18 h), respectively. Increasing the 'Gelzyme' concentration decreased biogas production, and this pattern was similar to the other substrate concentrations of 0.25 and 0.5%; whereas in the case of lipase, increasing lipase concentration slightly enhanced biogas production levels (Fig. 3).

4 Conclusion

From the above data, it may be concluded that the gas yield obtained from untreated FOG samples is lower than the treated FOG samples and is suggestive of an accumulation of a toxic intermediate or an interference with methanogens. Different gas yields were observed for varying concentrations, and the best substrate to inoculum ratio for anaerobic digestion of FOG was found to be 0.25%. Pre-treatment using both 'Gelzyme' and lipase increased biogas production potential of FOG and also reduced the lag phase. Comparing 'Gelzyme' and lipase treatments, lipase gave the highest gas yield for 0.25% substrate concentration, 0.5% lipase addition with a

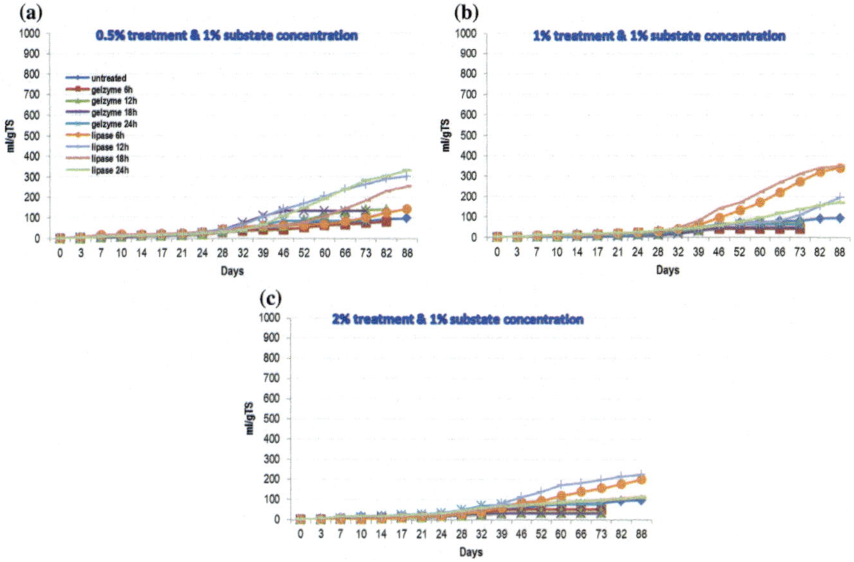

Fig. 3 BMP of 1% FOG at various concentrations of lipase and 'Gelzyme' **a** 0.5%, **b** 1% and **c** 2%

treatment time of 24 h (892 ml/gTS) which accounts for 85.5% of theoretical lipid gas yields. In the case of 'Gelzyme', the presence of surfactants in it is believed to decrease in gas production and needs to be researched further. The maximum biogas yields were observed for 0.25% substrate to inoculum, 0.25% 'Gelzyme' addition with 6 h treatment time (389 ml/gTS). In this approach, FOG need not be taken out of WWTPs and can be treated on the premises providing energy and power for in-plant operations.

References

Angelidaki, I., & Ahring, B. (1992). Effects of free long-chain fatty acids on thermophilic anaerobic-digestion. *Applied Microbiology and Biotechnology, 37*(6), 808–812.

APHA-AWWA. (1998). Standard methods for the examinaton of the water and wastewater, 20th ed. American public health association, Washington DC, USA.

Bailey, R. S. (2007). Anaerobic digestion of restaurant grease wastewater to improve methane gas production and electrical power generation potential. In *Proceedings of the 80th Annual Technical Exhibition and Conference of the Water Environment Federation* (pp. 6793–6805), San Diego, CA, 13–17 October 2007.

Buswell, A. M., & Neave, S. L. (1930). *Laboratory studies of sludge digestion*. Illinois Division of State Water Survey, Bulletin No. 30.

Davidsson, A., Lovstedt, C., Jansen, J. C., Gruvberger, C., & Aspegren, H. (2008). Co-digestion of grease trap sludge and sewage sludge. *Waste Management, 28*(6), 986–992.

EPA. (2011, January 15). http://www.epa.gov/npdes/sso/control/index.htm.

Hwu, C. S., Tseng, S. K., Yuan, C. Y., Kulik, Z., & Lettinga, G. (1998). Biosorption of long-chain fatty acids in UASB treatment process. *Water Research, 32*(5), 1571–1579.
Jagadish, K. S., Chanakya, H. N., Rajabapaiah, P., & Anand, V. (1998). Plug flow digesters for biogas generation from leaf biomass. *Biomass and Bioenergy, 14,* 415–423.
Kabouris, J. C., Tezel, U., Pavlostathis, S. G., Engelmann, M., Todd, A. C., & Gillette, R. A. (2008). The anaerobic biodegradability of municipal sludge and fat, oil, and grease at mesophilic conditions. *Water Environment Research, 80*(3), 212–221.
Kabouris, J. C., Tezel, U., Pavlostathis, S. G., Englemann, M., Dulaney, J. A., Todd, A. C., et al. (2009a). Mesophilic and thermophilic anaerobic digestion of municipal sludge and fat, oil, and grease. *Water Environment Research, 81*(5), 476–485.
Kabouris, J. C., Tezel, U., Pavlostathis, S. G., Englemann, M., Dulaney, J., Gillette, R. A., et al. (2009b). Methane recovery from the anaerobic codigestion of municipal sludge and FOG. *Bioresource technology, 100*(15), 3701–3705.
Kim, S. H., Han, S. K., & Shin, H. S. (2004). Kinetics of LCFA inhibition on acetoclastic methanogenesis, propionate degradation and beta-oxidation. *Journal of Environmental Science and Health A: Toxic/Hazardous Substances and Environmental Engineering, 39*(4), 1025–1036.
Long, J. H., Aziz, T. N., de los Reyes, F. L., III, & Ducoste, J. J. (2011). Anaerobic co-digestion of fat, oil and grease (fog): A review of gas production and process limitations. *Process Safety and Environmental Protection, 90,* 231–245.
Luostarinen, S., Luste, S., & Sillanpaa, M. (2009). Increased biogas production at wastewater treatment plants through co-digestion of sewage sludge with grease trap sludge from a meat processing plant. *Bioresource Technology, 100*(1), 79–85.
Madigan, M. T., Martinko, J. M., Brock, T. D. (2006). Biology of Microorganisms Pearson Prentice Hall, Upper Saddle River, NJ
Muller, C., Lam, P., Lin, E., Chapman, T., Devin-Clark, D., Belknap-Williamson, J., & Krugel, S. (2010). Co-digestion at Annacis Island WWTP: Metro Vancouver's path to renewable energy and greenhouse gas emission reductions. In *Proceedings of the Water Environment Federation.*
Parry, D. L., Vandenburgh S., Salerno, M. (2008). Making methane: co-digestion of organic waste with wastewater solids. In *Proceedings of the Water Environment Federation*(pp. 1045–1062).
Pereira, M. A., Sousa, D. Z., Mota, M., & Alves, M. M. (2004). Mineralization of LCFA associated with anaerobic sludge: Kinetics, enhancement of methanogenic activity, and effect of VFA. *Biotechnology and Bioengineering, 88*(4), 502–511.
Suto, P., Gray, D. M. D., Larsen, E., & Hake, J. (2006). Innovative anaerobic digestion investigation of fats, oils and grease. In *Proceedings of the Water Environment Federation* (pp. 858–879).
Wan, C., Zhou, Q., Fu, G., Li, Y. (2008). Semi-continous anaerobic co-digestion of thickened waste activated sludge and fat, oil and grease. *Waste Management, 31* 1752–1758.
York, R. V., Magner, J. A., & Chung, G. (2008). Increasing CHP productivity while reducing biosolids volume and climate changing gasses. In *Proceedings of the Water Environment Federation* (pp. 6744–6755).

Predicting Biomethanation Pattern from Feedstock Composition for Biomass Residues

D. Ravikumar, Chanakya N. Hoysall and S. Dasappa

Abstract Biomass residues form a significant part of MSW today, and predicting their biogas production pattern and, consequently, optimizing biomethanation plants are of immense importance. Five ligno-cellulosic feedstocks of significantly varying composition were chosen for the study and their biological methane potential (rate of biogas production) was monitored and correlated with the composition (extractives, hemicellulose, cellulose and lignin). Most of the feedstocks studied gave a BMP in the range of 200–300L/kg TS. Results also show that modified Gompertz fit gave best results achieving R2 > 0.99 in all cases. This could provide lag, specific gas production rates and cumulative gas production rates with high correlation.

Keywords Agro-feedstock · Biochemical methane potential · Gompertz kinetics · Composition

1 Introduction

Biomass resources are uniformly available across India (Hiloidhari et al. 2014) although the types may differ significantly. India has expertise in biomethanation for the use of agro-residue (Balachandra 2011). Many types of biomass species have been tested (Hiloidhari et al. 2014), although there is a lot of diversity due to spatio-temporal condition involved. It is estimated that a total of 626 MT gross residues of agro-residues are available from 39 crop residues (Ravindranath et al. 2005) while others have estimated primary productivity to be around. Majority of the rice straw is left on the field or burnt leading to widespread air pollution and particulate mass travelling thousands of miles across North India and cause air pollution far away. And also some fraction of the unburnt straw of cereals finds alternate usage such as fodder, briquetting, packaging for urban commodities and composting (Hiloidhari et al. 2014; Jagadish et al. 1998). Biomass such as straws can be a good source of biogas and as noted above holds an immense potential (Chanakya et al. 2009; Hiloidhari et al. 2014). Decentralized use of straw for biogas generation could be

D. Ravikumar · C. N. Hoysall (✉) · S. Dasappa
Centre for Sustainable Technologies, Indian Institute of Science, Bengaluru, India
e-mail: chanakya@iisc.ac.in

S. K. Ghosh et al. (eds.), *Bioresource Utilization and Bioprocess*,
https://doi.org/10.1007/978-981-15-1607-8_8

a potential to recover energy for electrification, industry and value-added products, etc. (Chanakya and Malayil 2012).

Many countries actively encourage biomethanation of various forms of urban and rural residues (MSW, agro-residues, agro-industry wastes, biomass-based packaging wastes, etc.), and thus, it is important to predict and understand the biomethanation patterns of various biomass residue inputs within MSW and know their influence on the biogas production pattern as well as the degradation kinetics in order to optimize the reactors. Further, as we shift from plastics to biodegradable packaging, the proportion of biomass residues in MSW is bound to increase as we go more towards biodegradable packaging systems. Agro-straws and straw-based paper and cardboard become predominant in packaging in the new tele and Internet marketing era. In this particular study, the potential for biogas for selected residues has been tested through the use of a modified biochemical methane potential (BMP) assay. Also Gompertz parameters were arrived at. This was important in designing the reactor.

2 Materials and Methods

Choice of feedstock: the feedstocks were collected from various parts of India depending on their availability.

2.1 Biochemical Methane Potential (BMP)

A serum bottle of 135 ml which was used with a working volume of 70 ml was used for BMP assay. A substrate to inoculum ratio of 0.25 was chosen to avoid the most common problems of acid build-up. Substrate to inoculum (S/I) was deemed to be important (Raposo et al. 2011; Hashimoto 1982). The volume of gas produced was measured at frequent intervals through the use of downward displacement of water. Methane and carbon dioxide were determined using gas chromatography fitted with a thermal conductivity detector and a Hayesep A column. Hydrogen was used as carrier gas.

2.2 Gompertz kinetics

Gompertz kinetics: Modified Gompertz model was used to fit the gas production data. Specific gas production and lag phase were also determined as follows:

$$Y = A * e^\wedge\{-e^\wedge(\mu * e/A(\lambda - t) + 1)\}$$

A = maximum gas production, μ = specific gas production rate and λ = lag time/phase.

3 Results and Discussion

In this section, the results are discussed and elaborated. The composition of the feedstock is discussed in Table 1. Gas production and Gompertz parameters are also discussed for the gas production.

From Table 1, it may be observed that the sesame and jowar stalks have highest amount extractives as compared to other substrates. Hemicellulose content in the jowar feedstock is higher (25.8%) as compared to mustard stalk which has the least hemicellulose content at 15%, and cellulose content in all the feedstocks is around an average of 35%. Further, since the samples were freshly harvested, the lignin content was comparatively lower for a stalk or a husk. Bajra has the highest lignin content at 15%. Mustard stalks have comparatively low lignin content at 11%. The average lignin content is 13%.

Jowar stalk has higher gas production at approximately 300 ml/gVS, followed by corncobs at about 290–300 ml/gVS. Mustard stalk, husk, sesame stalk and bajra stalk have the same gas production extent and rate (gas production of 200 ml/gVS). When compared with the composition, it is interesting to note that the stalks which have the least amount of lignin have the highest amount of gas production. Jowar and corncob with high extractives seem to have high gas production rate and extent. Also interesting to note is the initial rates of gas production, which are higher in jowar and corncob. Apart from lignin concentration, ash seems to be playing an important role in the gas production (Fig. 1; Table 2).

Gompertz parameters inference and statistics: the cumulative gas production in case of jowar stalk is highest at 309 ml/gVS. Lowest gas production is recorded in the case of mustard husk at 210 ml/gVS. It is interesting to note that the lag time of the feedstocks is based on the extractive hemicellulose and lignin concentration. The specific gas production rate is on an average 7 (/day) meaning once the system overcomes the initial lag phase, the gas production is linear and at the same rate. The

Table 1 Composition of the biomass feedstock

Substrate	Extractives (%VS)	Hemi-cellulose (%VS)	Cellulose (%VS)	Lignin (%VS)	Ash (%VS)
Mustard husk	19.8	17.6	38.8	12.7	11.1
Mustard stalks	22.7	15.1	37.1	10.9	14.2
Sesame husk	27.9	16.2	33.8	12.3	9.8
Jowar stalks	26.7	25.8	31.9	9.9	5.4
Bajra talks	23.1	23.6	33.4	14.8	5.1

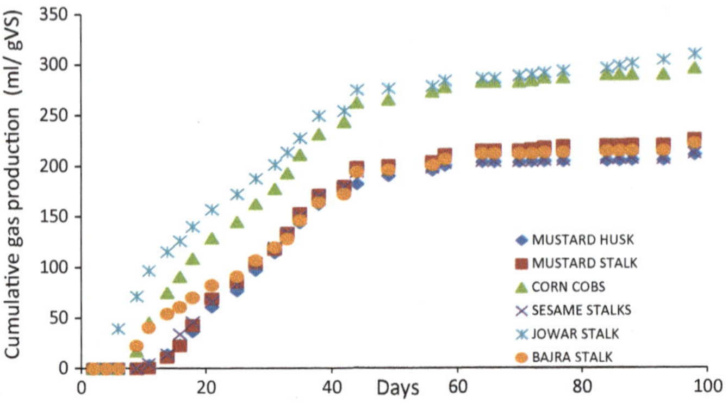

Fig. 1 Gas production of various biomass feedstock over a period of 100 days

Table 2 Gompertz parameters and their values

Substrate	Cumulative gas production (ml/gVS)	Specific gas production rate (μ/day)	Lag time (λ, day)
Mustard husk	210	7.14	13.7
Mustard stalk	226	7.4	13.5
Corn cobs	296	8.19	6.4
Sesame husk	212	7.4	12.9
Jowar stalk	309	7.8	1
Bajra stalk	222	5.4	6.3

Gompertz fit or correlation is significant at an R^2 of 0.99 in all the cases, as seen in Table 3.

From this difference in the lag time and specific gas production rate, it can be seen that the structure plays an important role (Chanakya and Malayil 2012). The rates can be used for the development of reactor design (sizing and operation).

Table 3 Gompertz statistics

	Mustard husk	Mustard stalk	Corn cobs	Sesame husk	Jowar stalk	Bajra stalk
Number of points	31	31	31	31	31	31
Degrees of freedom	28	28	28	28	28	28
Reduced Chi-Sqr	10.43524	24.32813	69.22792	23.8853	126.24367	56.71307
Adj. *R*-square	0.99848	0.9969	0.99369	0.99653	0.98635	0.99077

4 Conclusion

From this preliminary study, it becomes clear that the modified Gompertz fit predicts reasonably well the lag, decomposition rates and cumulative gas production potential for various agro-residues and consequently various packaging products that arise from them. This approach can be extended to other feedstocks common in MSW to predict upfront the likely biogas production pattern and plan marketing and sales of biogas in the immediate days following knowing what feedstocks were fed to biomethanation plants. Similarly, it is also possible to predict the quantum of undegraded residues that is likely to emerge so as to plan the sales potential of the anaerobic compost generated from them.

References

Balachandra, P. (2011). Modern energy access to all in rural India: An integrated implementation strategy. *Energy Policy, 39,* 7803–7814. https://doi.org/10.1016/j.enpol.2011.09.026.

Chanakya, H. N., & Malayil, S. (2012). Anaerobic digestion for bioenergy from agro-residues and other solid wastes—An overview of science, technology and sustainability. *Journal of the Indian Institute of Science, 92,* 111–144.

Chanakya, H. N., Sharma, I., & Ramachandra, T. V. (2009). Micro-scale anaerobic digestion of point source components of organic fraction of municipal solid waste. *Waste Management, 29,* 1306–1312. https://doi.org/10.1016/j.wasman.2008.09.014.

Hashimoto, A. G. (1982). Effect of inoculum/substrate ratio on methane yield and production rate from straw. *Biotechnology and Bioengineering, 24,* 9–23. https://doi.org/10.1002/bit.260240103.

Hiloidhari, M., Das, D., & Baruah, D. C. (2014). Bioenergy potential from crop residue biomass in India. *Renewable and Sustainable Energy Reviews, 32,* 504–512. https://doi.org/10.1016/j.rser.2014.01.025.

Jagadish, K. S., Chanakya, H. N., Rajabapaiah, P., & Anand, V. (1998). Plug flow digestors for biogas generation from leaf biomass. *Biomass and Bioenergy, 14,* 415–423. https://doi.org/10.1016/S0961-9534(98)00003-8.

Raposo, F., Fernández-Cegrí, V., De la Rubia, M. A., Borja, R., Béline, F., Cavinato, C., et al. (2011). Biochemical methane potential (BMP) of solid organic substrates: evaluation of anaerobic biodegradability using data from an international interlaboratory study. *Journal of Chemical Technology & Biotechnology, 86,* 1088–1098. https://doi.org/10.1002/jctb.2622.

Ravindranath, N. H., Somashekar, H. I., Nagaraja, M. S., Sudha, P., Sangeetha, G., Bhattacharya, S. C., et al. (2005). Assessment of sustainable non-plantation biomass resources potential for energy in India. *Biomass and Bioenergy, 29,* 178–190. https://doi.org/10.1016/j.biombioe.2005.03.005.

Synthesis of Cellulose from Peanut Shell Waste and Its Use in Bioethanol Production

Preetha Ganguly, Shubhalakshmi Sengupta, Papita Das and Avijit Bhowal

Abstract The present worldwide situation on energy is emphasizing the researchers to investigate new recyclable resources and techniques able to produce biofuels from renewable resources, such as Arachis hypogaea (groundnut) shell processing. The main advantage of lignocellulosic material is lying on its wide distribution; availability in large amount and its low cost makes it extremely popular for study and development. Peanut trinomial name is Arachis hypogeae and it belongs to the family of Leguminosae. The countries which are the chief producers of groundnut are China and then India (1.5–2 million tons). From the literature survey, it can be concluded that peanut shell contains lignin (26.4%), hemicelluloses (14.7%), and cellulose (40.5%) and are rich in cellulosic substance, because of these features, the peanut can be used ideally for bioethanol production. Isolation of microcrystalline cellulose is carried in a single procedure due to its potential significance using alkaline treatment followed by bleaching. Concentration of cellulose was estimated by anthrone test and it is reported to be 11.41 mg/ml. The cellulose extracted was then hydrolyzed to its monomeric units by *Aspergillus* sp. The total reducing sugar of the sample was estimated by dinitrosalicylic acid test and it is found to be 2.34 mg/ml. After hydrolysis, the sample was filtered and then inoculated with pre-selected yeast strain at 30 °C for 72 h. Sample was taken at regular intervals and supernatant were taken for estimation of ethanol. Residue after the ethanol recovery was filtered, washed, and used for the isolation of other products which are valuable.

Keywords Biofuel · Microcrystalline cellulose · Lignocelluloses · Peanut shell

1 Introduction

The diminution of present non-renewable sources of energy has become the center of concern in terms of both fuel and material's viewpoint. As the reservoirs of petroleum are depleting day by day, increase in the prices of petroleum and its derived fuel's cost has been evaluated by the end of the twenty-first decade. Addition to this, various

P. Ganguly · S. Sengupta · P. Das (✉) · A. Bhowal
Department of Chemical Engineering, Jadavpur University, Kolkata, India
e-mail: papitasaha@gmail.com

S. K. Ghosh et al. (eds.), *Bioresource Utilization and Bioprocess*,
https://doi.org/10.1007/978-981-15-1607-8_10

materials which are presently in use such as plastics are also the derivative products from the petroleum sources (Mosier et al. 2005).

Biomass economy on the other side may be defined as, an economy which is based on transforming the biomass into biorefineries and then into their other valuable products such as biofuel, new chemicals and materials, which have acquired immense significance due to their reliability on the recyclable resources and also being a petroleum-based alternative. The biomass economy concept lies on biorefinery, defined as unified process in which diverse new products are produced like fuel, energy, materials, and chemicals from biomass conversion by utilizing various combinations of techniques (FitzPatrick et al. 2010).

One of the best examples of concept of biorefinery is the bioethanol generation. In this theory, ethanol generation is dependent mainly upon the derived sugar from different sources such as sugar beet, sugarcane, and starch-based grains like corn starch. However, the raw material used for the procedure are usually materials of food, and utilizing these food materials for biofuel production creates competition of food versus biofuel (Vancov et al. 2012; Agbor et al. 2011).

One of the most important industrial as well as laboratory-based chemical, ethanol, can be produced extensively from many lignocellulosic biomasses such as bagasse, maize cob, rice husk, and groundnut shell. The main component of this lignocellulosic feedstock is cellulose and hemicelluloses, the lignocelluloses that can be excellent energy sources. According to a report submitted in the year 2011, it was evaluated that worldwide ethanol production was about 52.0 billion liters by Brazil and USA (which was about 87.1% of world production of ethanol by both countries). The ethanol production in India is lagging even though availability of feedstock is abundant (only 4% ethanol is produced per year which is approximately around 2 billion liters using bagasse as a substrate).

Ethanol production from lignocellulosic waste has gained significance because the feedstock is renewable, low-priced, furthermore for biofuel generation, about one billion tons of biomass waste is available (Li et al. 2010). Another advantage of using lignocellulosic waste as a feedstock is that the other parts of the cell wall such as lignin and hemicelluloses can be exploited. The main disadvantage in ethanol production is the expensive pretreatment and hydrolysis steps on lignocellulosic feedstock.

Hence, to decrease the food versus fuel competition, another biorefinery concept was developed. In this concept, hydrolyzed sugar are used for the process of fermentation which in turn produce ethanol from the lignocellulosic feedstock, such as sugarcane molasses, jute, cotton waste, paper waste, switch grass, wood chips, etc. The most abundant biopolymer on the earth is cellulose. One of the main components of the cell wall in plants is the cellulose. The cellulose is a carbohydrate containing number of sugar molecules bonded to each other, made up of polymers consisting D-glucose subunits bonded with glycosidic bond of β (1–4) linkage. It is crystalline in structure and water insoluble.

Peanut botanical name (Arachis hypogaea) is an agrarian plant, whose pod can be used in the food industry as important raw material. The peanut production worldwide annually in the years 2009 or 2010 was estimated to be 33.36 metric million tons (this report was submitted by the Agricultural Department of USA). The leading producer countries of Arachis hypogaea are china and then India about 60% of the total world production. The groundnut seeds/fruit are used as a raw material in food industries. The products produced from groundnut are oil from groundnut, groundnut flour, snacks, and groundnut butter. The shells in which the fruits develop contain most amount of cellulose. These pods or shells are cellulose containing agricultural, domestic, and industrial wastes.

About one-third mass of the groundnut pod is its shell, which is estimated to be about 25,000 tons of peanut shells. According to the reports represented by Department of Agriculture of Turkey in their statistics showed that this waste groundnut shell can be processed annually for different purposes. The shell of groundnut is an important farm waste because of its greater content of cellulose, and also because of its high lignin content. A report given by Zhang et al. had shown that the peanut shell contains approximately about 36% cellulose, 30% lignin, and 19% hemicellulose (Zhang et al. 2006). Therefore, from the above report, we can conclude that by using peanut shell, there is a probability for operating about cellulose of 9000 tons, 4800 tons of hemicelluloses and 7500 tons of lignin annually.

In this study presented, the groundnut shell is utilized for generation of bioethanol as a fuel. In this study, both physical and chemical pretreatment has been applied on the peanut shell for the production of bioethanol. Saccharification study was done by *Aspergillus* sp. and for the production of ethanol isolated fungal strain *Sacchromyces cerevisiae* has been used.

2 Literature Survey

In the paper given by Rani Krishnan Punnadiyil et al. suggested the method of extracting microcrystalline cellulose (MCC) from peanut shell. The MCC which was extracted from the lignocellulosic feedstock have a potential to be used pharmaceutical industry. The cellulose is extracted using alkaline treatment followed by bleaching.

Firstly, it is important to consider the amount of cellulose present in the lignocellulosic feedstock for the bioethanol production. Priyamwada Bharthare et al. found that peanut contains 65.5% cellulose (dry weight). In the study, it was revealed that chemical pretreatment method was much better than the physical pretreatment method, and the best results came from steam explosion on the peanut shell (Cheng et al. 2011). The chemical pretreatment with acid (0.25 N HCl) showed better results than that of with the alkaline (0.25 N NaOH) treatment.

In various studies, it is showed that ground nut shell has the potential to be used as a source for ethanol production (Zeeshan Quader et al.). Tejas Suryawanshi performed a study in which it was shown that the maximum amount of carbohydrate and glucose that was found after the saccharification process was about 3.9 mg/ml and 3.2 mg/ml, respectively. *S. cerevisiae* was used for fermentation of the media. The alcohol estimation done on the fourth day of the experiment carried out using dichromate method and maximum alcohol content from the sample was found to be

0.5 mg/ml. Cellulose was also extracted in the study and it was found that 0.41 g of cellulose/gram of groundnut shell powder was obtained. The groundnut shell ash was also tested to see the effect on increasing crop yield. And from the experiment, it was found that efficiency of GNS ash for increase in carbohydrate and protein content was about 12.5% and 71.11%, respectively.

A reasonable amount of sugar (glucose) is present in peanut shell and maize cob. (U. G. Akpan et al.) found that when the analyses of both the sugar contents are done, then the ratio was found to be 3:1. If the product (glucose) is fermented under experimental condition with (Baker's Yeast), then a significant amount of ethanol can be produced from the chemically pretreated feedstock.

Sheelandra Mangal Bhatt in his study concentrated on the biofuel generation by utilizing groundnut shell via enzymatic hydrolysis and then fermentation of the sugar was done by using cellulolytic thermophilic bacteria strain and yeast sp. The yeast strain was *S. cerevisiae* while the cellulolytic thermophilic bacterial strain used for the experiment was *Bacillus stearothermophilus*. Biochemical analysis of groundnut shell reveled that it contains 35% cellulose, 22.10% lignin, and 4.3% other substances. It also contains organic carbon about 27.7%, and nitrogen content was evaluated about 23.4%. Pretreatment of the peanut shell was done using HCl acid (0.25 N) for the ejection of the lignin and then simultaneously hydrolysis and fermentation are done in the experiment. After the process of fermentation, distillation process is followed for the production of ethanol from the fermented sample. In the experiment above, it is shown that combination of *Bacillus stearothermophilus* and *S. cerevisiae* can produce better amount of ethanol.

3 Material and Method

Peanut shell were obtained from a local merchant in Behala was used as the raw material. NaOH, NaOCl, HNO_3, and ethanol were the chemicals used in the experiment. All the chemicals used for the isolation and identification are of reagent grade and obtained from Jadavpur University laboratory.

4 Method

4.1 Isolation of Microcrystalline Cellulose

The groundnut shell was dried under sun and then was grinded into fine powder. The fine powder was sieved in 100 um mesh. Treatment with NaOH (1000 ml, 0.5 M) at 80–90 °C with continuous stirring at (720 rpm) of 30 g sample was done for 2 h. After 2 h, a dark slurry is obtained which is filtered out. The cake obtained after filtration is washed several times with the distill water. The cake is then refluxed with

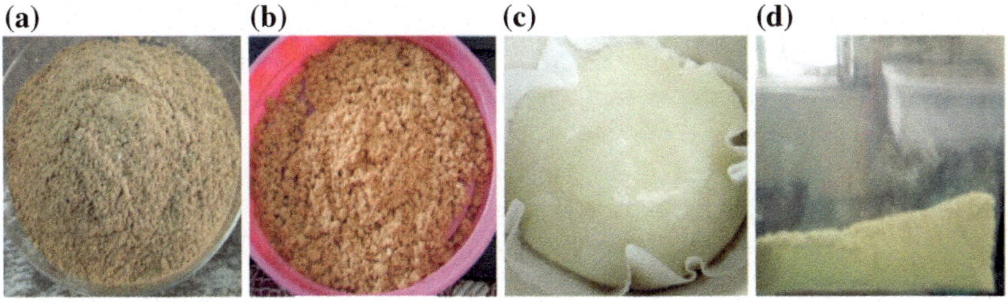

Fig. 1 **a** Groundnut powder, **b** alkali treated, **c** sodium hypochlorite treated, **d** MCC

a mixture containing nitric acid in ethanol (20% v/v). The treatment is carried out for twice to thrice till the color changes successively from brown to yellow.

The residue obtained after the treatment was continuously washed with distills water until 6.5–7 pH was reached. The yellow colored residue obtained from the above step was then further bleached using sodium hypochlorite to get off the yellow color. The white colored cellulose is obtained after the procedure. The cellulose in then died in the oven for 2–3 h at 50 °C. Lyophilization is done to get powder cellulose. The powdered cellulose obtained was kept in polythene bag for further use in bioethanol production (Fig. 1).

4.2 Anthrone Test for Cellulose Estimation

Cellulose is one of the major constituents of lignocellulosic feedstock. Cellulose is a polysaccharide consisting of long linear chain of glucose unit. To analyze the amount of cellulose present in the above extracted sample, anthrone method is followed. The method is as follows.

4.2.1 Standard Solution

A standard stock solution is prepared by dissolving of cellulose in the distill water. Standard curve was prepared by taking 0–1 ml of range of working solution.

4.2.2 Anthrone Reagent Preparation

Anthrone was dissolved in sulphuric acid and it was kept in ice bath for 2 hr.

4.2.3 Experiment

For anthrone test, each sample was taken in the beaker. (67%) H_2SO_4 was added to the beaker and kept at rest. 1 ml of sample were taken in different beakers and the volume was made up. Again the diluted sample was taken and then anthrone reagent was added to them. The sample in the test tube was heated to the boiling point for about 10 min for proper mixing of the reagents in the sample. The solution was cooled and the absorbance was measured at 630 nm using UV/VIS spectrophotometer (Perkin Elmer, USA)

4.3 Biological Treatments

4.3.1 Basal Media Preparation

Basal media was prepared by adding the chemicals—Sodium nitrate, potassium chloride, magnesium sulfate, iron sulfate, potassium hydrogen phosphate, and peptone (nitrogen source) in a definite ratio in distill water.

4.3.2 Inoculum Preparation

The isolated fungal culture (*Aspergillus* sp.) was subcultured on Czapek-modified medium with adding 2% Agar in the Czapek media. The subculture was incubated at 30 °C. Fully sporulated culture plate was obtained after 5 days of incubation. The sporulated spores were collected from the culture plate by flooding the culture with (20 ml) 0.1% Tween 80. Spores were displaced by pipetting them out from the culture media. The suspension was centrifuged and washed three times to remove the Tween 80. The resulting suspension was used as inoculum.

4.3.3 Enzyme Production

3 g cellulose, raw groundnut shell, and roasted groundnut shell were weighted and moistened with basal media. The substrate was autoclaved. After cooling, the substrate was inoculated with spore suspension. The inoculated substrate was incubated at 30 °C for five days until the media became turbid. The media containing the inoculums and substrate were then autoclaved to kill the fungal strain. For the estimation of the hydrolyzed enzymatic products produced by the fungal strain (*Aspergillus* sp.), aliquots of the solution were taken.

4.3.4 Estimation and Analysis of Reducing Sugar

Lignocellulosic feedstock mainly consists of lignin, hemicelluloses, and cellulose. Hemicelluloses and cellulose are long chains of polysaccharides that are needed to be broken into smaller components before the fermentation. *Aspergillus* sp. used in the experiment helps to break the long chains into their monomeric units—glucose, fructose maltose, xylose, galactose, etc. The amount of reducing sugar produced by the fungal strain was analyzed by the DNS test.

4.4 DNS Test

4.4.1 Standard Solution

A standard stock solution is prepared by dissolving of glucose in distill water. Standard curve was prepared by taking 0.2–1 ml of working solution.

4.4.2 DNS Preparation

1% NaOH solution was prepared, and then di-nitro salicylic acid (DNS), sodium sulfite, and crystalline phenol were dissolved one after the other in the NaOH solution by stirring.

4.4.3 Experiment

1 ml of the media containing the substrate was taken from the extracted cellulose and groundnut shell (raw, roasted) samples which has already been inoculated with (*Aspergillus* sp.), and then the remaining media were autoclaved to kill the fungal strain. The volume of the media in test tube was made up to 3 ml by adding distill water to the sample. DNS was added to the test tubes and was heated at boiling. After the evolution of color, 40% (1 ml) Rochella salt solution was added in the tubes and was meticulously mixed.

4.5 Bioethanol Production

From Sheelendra M. Bhatt study, the yield of ethanol was calculated about (16.11%) by utilizing the strain of *Bacillus stearothermophilus*, **and** *S. cerevisiae* **utilizing previously pretreated groundnut shell about (2% w/v) after 14 days** (Bhatt and Shilpa 2014). **From this study, the ethanol yield was obtained about (17%) utilizing** *Saccharomyces cerevisiae* **within 6 days**.

4.5.1 Fermentation

Microorganism: *S. cerevisiae* (yeast) is the universal microorganism used generally for the production of bioethanol. This organism is capable of fermenting mainly the glucose sugar.

4.5.2 Experiment

3 g of cellulose and groundnut shell powders which was formerly pretreated with (*Aspergillus* sp.) and autoclaved at 121 °C and 15 psi pressure to kill the fungal culture is filtrated. The same filtrate was again treated with S. cervisiae. Fermentation of filtrate by S. cerevisiae was shown after 3 days at 35 °C.

4.5.3 Retrieval of Bioethanol

In the rotary evaporator, extraction of ethanol which was produced by *S. cerevisiae* by fermenting the samples of cellulose and peanut shell was taken place.

4.5.4 Ethanol Estimation

The ethanol extracted from the filtrate was further utilized for ethanol estimation.

4.5.5 Standard Solution

Ethanol standard curve was made by using ethanol solution in the concentration of 0, 5, 10, 15, and 20% ethanol (v/v). Different concentrations of samples were prepared by adding water in range from [100–80% (v/v)] in volumetric flask. 1 ml aliquots of the above samples were taken in test tubes containing potassium dichromate solution.

4.5.6 Preparation of Potassium Dichromate Solution

Solution was prepared by adding distill water in volumetric flask and then 325 mlconc. H_2SO_4 is added to the flask carefully. After mixing and cooling up to (80–90 °C), potassium dichromate was added and the volume was made up to 1 L and then was kept for cooling.

4.5.7 Preparation of Standard Curve

Standard curve was prepared by taking a small amount of aliquot from each concentration of the standard solution range from [0–20% (v/v)] in a volumetric flask containing of potassium dichromate solution. The samples were heated at 60 °C for 25 min in a water bath and then cooled and then finally diluted with distilled water. Absorbance was recorded at a wavelength of 600 nm using spectrophotometer (Srivastava et al. 2017).

4.5.8 Estimation of Alcohol Present in Samples

1 ml of each fermented sample was taken in conical flasks and the volume was made up. 25 ml $K_2Cr_2O_7$ was also added to each of the flask and was heated at 60 °C in waterbath for 20 min. The volume was adjusted up to 50 ml by adding water after cooling of the samples.

In UV-Vis spectrophotometer absorbance was noted at 600 nm.

5 Result and Discussion

5.1 Anthrone Test Results

The percentage of cellulose before and after the pretreatment which showed the best result by UV-Vis spectrophotometer study are represented in Fig. 2.

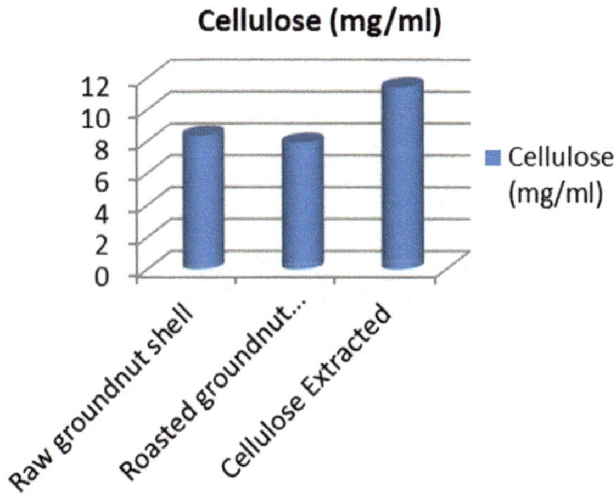

Fig. 2 Cellulose concentration in treated and untreated samples

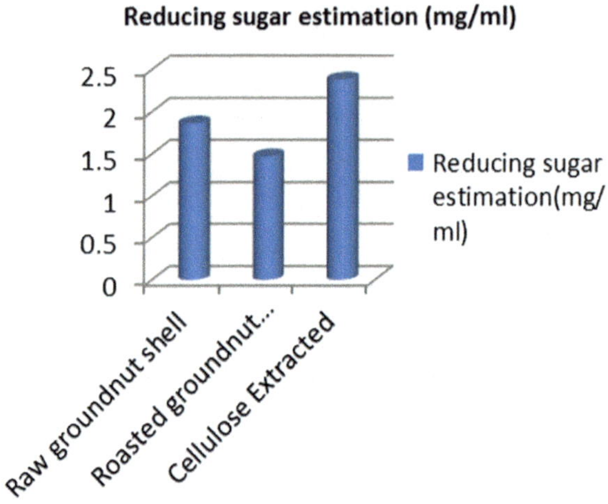

Fig. 3 Amount of reducing sugar in treated and untreated samples

5.2 *DNS Result*

The total amount of the reducing sugar produced from different sources is analyzed by di-nitro salicylic acid test. The best optimized result is represented by Fig. 3. It was observed that reducing sugar content was higher in case of extracted cellulose than untreated samples (2.38 mg/ml).

5.3 *Bioethanol Production*

Bioethanol production was carried out with the fermentable sugar from cellulose and raw-roasted groundnut shell. The maximum bioethanol was obtained as 16.97% from cellulose (Fig. 4).

6 Conclusion

This experiment leads to the conclusion that bioethanol can be produced from waste lignocellulosic materials like peanut shell using pretreatment method, in which hemicelluloses and lignin fraction can be removed. It can also be concluded that pretreatment of the feedstock results in better production of the ethanol. The highest ethanol percentage in ethanol–water mix obtained from the study of this experiment is about 17%.

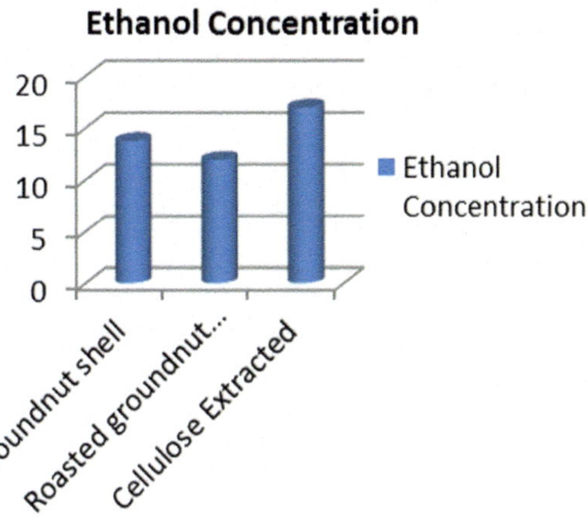

Fig. 4 Ethanol extraction from treated and untreated samples

References

Agbor, V. B., Cicek, N., Sparling, R., Berlin, A., & Levin, D. B. (2011, November–December). Biomass pretreatment: Fundamentals toward application. *Biotechnology Advances, 29*, 675–685.

Bharthare, P., Shrivastava, P., Singh, P., & Ttiwari, A. (2014, April). Peanut shell as renewable energy source and their utility in production of ethanol. 2320-9178.

Bhatt, S. M., & Shilpa. (2014, December). Bioethanol production from economical agro waste (groundnut shell) in SSF mode. 0975-8585.

Cheng, G., Varansi, P., Li, C., Liu, H., Melnichenko, Y. B., Simmons, B. A., et al. (2011). Transition of cellulose crystalline structure and surface morphology of biomass as a function of ionic liquid pretreatment and its relation to enzymatic hydrolysis. *Biomacromolecules, 12,* 933–941.

FitzPatrick, M., Champagne, P., Cunningham, M. F., & Whitney, R. A. (2010, December). A biorefinery processing perspective: Treatment of lignocelluosic materials for the production of value-added products. *Bioresource Technology, 101,* 8915–8922.

Li, C., Knierim, B., Manisseri, C., Arora, R., Scheller, H., Auer, M., et al. (2010, July). Comparison of dilute acid and ionic liquid pretreatment of switchgrass: Biomass recalcitrance, delignification and enzymatic saccharification. *Bioresource Technology, 101,* 4900–4906.

Mosier, N., Wyman, C., Dale, B., Elander, R., Lee, Y.Y., Hotzapple, M., & Ladisch, M. (2005, April). Features of promising technologies for pretreatment of lignocellulosic biomass. *Bioresource Technology, 96,* 673–686.

Punnadiyil, R. K., Sreejith, M. P., & Purushothaman, E. (2016, November). Isolation of microcrystalline and nano cellulose from peanut shells. 0974-2115.

Srivastava, M., Sengupta, S., Das, P., & Datta, S. (2017, December). Novel pre treatment techniques for extraction of fermentable sugars from natural waste materials for bioethanol production. 2572–1119.

Suryawanshi, T., Nair, V., Patel, P., & Gupta, A. D. (2017, May). Extraction of cellulose and biofuel production from groundnut shells and its application to increase crop yield. 2278–4357.

Tsukamoto, J., Durán, N., & Tasic, L. (2013, September). Nanocellulose and bioethanol production from orange waste using isolated microorganisms. 0103–5053.

Vancov, T., Alston, A., Brown, T., & McIntosh, S. (2012). Use of ionic liquids in converting lignocellulosic material to biofuels. *Renewable Energy, 45,* 1–6.

Jatropha Biodiesel Blends as Renewable Diesel Fuel Additives

S. K. Dash, S. B. Chavan, A. Kumar, M. S. Ahamed and P. Lingfa

Abstract Worldwide, biodiesel has proved promising to a larger extent amid growing concerns about energy demand and environmental pollution. However, from the key perspectives as a supplementary or substitute fuel, it has generated many more debates and the industries yet to accept the derived fuel positively. In this paper, the popular Jatropha biodiesel blends from the key perspectives as a renewable diesel fuel additive is explored. Three different lower biodiesel concentration blends (B2, B5, and B10) have been prepared using eggshell-derived CaO as base catalyst and tested in a single cylinder four strokes variable compression ratio diesel engine having power output of 3.5 kW. Lower proportion of Jatropha biodiesel blends not only reduced pollutant emissions but also improved performance of the diesel engine.

Keywords Jatropha biodiesel · Transesterification · Fuel additive · Performance · Emission · Combustion

1 Introduction

Energy is crucial for the growth and progress of mankind. However, for the developing nations, its importance greatly outweighs than the already developed ones. India is a fastest rising economy in the world. However, dependency on other countries for major shares of the energy consumption in transportation sector raises questions on the energy security of the country. The paucity of conventional energy sources, environmental degradation, and ecological imbalance is the main hurdle developing economies are facing. In this respect, biofuel (especially, biodiesel) appears to be suitable. Biodiesel is prepared by chemically reacting plant oil or animal fats with an alcohol in the presence of a suitable catalyst. Biodiesel offers better physico-chemical properties, such as the higher flash point, higher cetane numbers, and inherent lubricity, compared to the conventional diesel fuel. Moreover, when used in a

S. K. Dash (✉) · A. Kumar · M. S. Ahamed · P. Lingfa
Department of Mechanical Engineering, NERIST, Nirjuli, India
e-mail: 90nash@gmail.com

S. B. Chavan
Indian Biodiesel Corporation, Baramati, Maharashtra, India

S. K. Ghosh et al. (eds.), *Bioresource Utilization and Bioprocess*,
https://doi.org/10.1007/978-981-15-1607-8_11

diesel engine, it does not emit pollutants like diesel fuel. A lot of articles support the reduction of particulate matter, HC, CO, and smoke (Agarwal 2007; No 2011; Xue et al. 2011; Dash and Lingfa 2018a). However, researchers opined contradictory views on NOx emission (Lapuerta et al. 2008; Varatharajan and Cheralathan 2012). Different physico-chemical properties of diverse feedstocks, engine types and engine operating conditions are made responsible for the different emission and performance results.

In India, non-edible oilseeds are already established as a viable solution for biodiesel production (Azam et al. 2005; Dash and Lingfa 2017, 2018b). However, cost and feedstock scarcity have mounted pressure on its large-scale use. To solve those impediments, researchers have tried using underutilized feedstocks for biodiesel production with the help of heterogeneous catalyst. Heterogeneous catalyst and underutilized feedstock is the optimum mix for India (Lee et al. 2015). Among several non-edible feedstocks, Jatropha has well received as a supplementary fuel for diesel engine in India. There are many articles published pertaining to the performance and emission aspect of a diesel engine powered by Jatropha biodiesel and its different diesel blends (Chukwuezie et al. 2014). However, no article has been reported on its key use as a diesel fuel additive. Moreover, lower concentration biodiesel blends as a fuel additive will encourage its widespread use and can be easily accepted by diesel vehicle manufacturers. Additionally, feedstock scarcity and low-temperature operability issues of pure biodiesel or higher-concentration blends will be solved to a greater degree when the diesel engine operated with lower-concentration blends as fuel additives. This study endeavors to experimentally investigate various performance, emission, and combustion characteristics of a single cylinder four strokes direct injection VCR diesel engine powered by prepared lower-concentration biodiesel blends (B2, B5, and B10). The results are compared with diesel operation on the same engine.

2 Materials and Methods

Jatropha oil is characterized in NERIST, Arunachal Pradesh, and India. Methanol, anhydrous sodium sulfate, isopropyl alcohol, and sulfuric acid are procured from Merck Pvt. Ltd, Maharashtra, India. Low-cost calcium oxide (CaO) powder derived from waste eggshell has been purchased from IBDC, Baramati, Pune, India. Characterization of the eggshell catalyst is reported in an earlier study (Chavan et al. 2015).

2.1 Production of Jatropha Biodiesel

Hot plate and magnetic stirrer were used for the biodiesel synthesis. As the free fatty acid of Jatropha oil observed to be higher than the limit for single-stage transesterification, a pretreatment is done prior to the transesterification. Calcium oxide (CaO) is

Table 1 Properties of Jatropha oil and Jatropha biodiesel

Property	Jatropha oil	Jatropha biodiesel	Diesel	Test methods
Acid value (mg KOH/g)	17.8	0.36	–	ASTM D664
Kinematic viscosity (cSt@40 °C)	39.46	4.86	2.7	ASTM D445
Density (g/cc)	0.920	0.880	0.830	ASTM D1298
Calorific value (kJ/kg)	37,794	38,143	42,000	ASTM D4809
Cetane number	54	52	48	ASTM D613
Flash point (°C)	210	186	64	ASTM D93
Pour point (°C)	5.2	3.1	−8.6	ASTM D97

used for the transesterification as a base catalyst. The optimum reaction conditions of different parameters for transesterification are given as follows: reaction time—2 h, reaction temperature—65 °C, molar ratio of oil to alcohol—1:6, catalyst amount—2 wt.%, and reaction speed—600 rpm. The yield and conversion was obtained as 91% and 97.97%, respectively.

$$\text{Biodiesel Yield (\%)} = (\text{Weight of biodiesel obtained/Weight of raw oil used}) \times 100$$

$$\text{Biodiesel conversion (\%)} = \left(\frac{p - q}{p}\right) \times 100,$$

where p = initial acid value and q = final acid value.

Three different biodiesel diesel blends (B2, B5, and B10) have been prepared by mixing biodiesel with high-speed diesel in a stirrer with a mixing speed of 1000 rpm. Here, B2 means 2% Jatropha biodiesel mixed with 98% high-speed diesel. The miscibility and stability of prepared blends have been checked by storing the blends for a period of 20 days. The properties of Jatropha oil, Jatropha biodiesel, and diesel are shown in Table 1.

3 Experimentation

A single cylinder four strokes VCR diesel engine of power output 3.5 kW at 1500 rpm used for the engine trial. The diesel engine experimental setup is shown in Fig. 1. The specification of the test engine is given in Table 2. All the tests were carried out at a stable ambient air temperature 28 °C. The experiments were repeated six times for a particular load and average of six values is presented in this article. The pressure transducer was calibrated before taking any readings.

720 data points were recorded for a particular cycle pressure and volume. A total 60 cycle pressure data was recorded for a load and averaged to get the cylinder pressure. AVL ditest 1000 five gas analyzer and smoke meter were used to measure

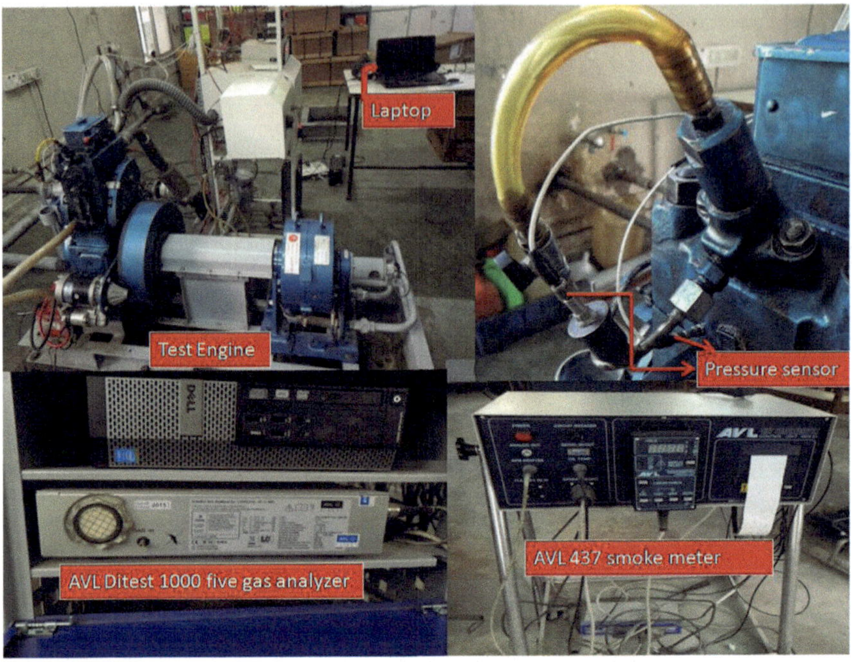

Fig. 1 Experimental setup

Table 2 Specification of the test engine setup

Engine parameter	Specification
Make	Kirloskar
No. of cylinder	1
No. of stroke	4
Injection	Direct injection
Bore and stroke	87.5 mm × 110 mm
Capacity	661 cc
Rated power output	3.5 kW at 1500 rpm
Dynamometer	Eddy current type, radius 185 mm
Data acquisition device	NI USB-6210, 16 bit

the pollutant emissions. The uncertainties of various parameters were calculated by Kline, and McClintock method (Holman 2011) is presented in Table 3.

Table 3 Uncertainty of instruments

Instruments	Measured parameter	Uncertainty (%)
Loading unit	Load (kg)	±0.21
Speed sensor	Engine speed	±2.1
Pressure transmitter	Cylinder pressure, bar	±0.1
Crank angle encoder	Crank angle position	±0.01
Burette	Fuel consumption	±0.5
AVL gas analyzer	HC	±0.7
	CO	±0.03
	NO_x	1.29
AVL smoke meter	Smoke density	±0.1

4 Results and Discussions

4.1 Mechanical Efficiency

The variation of mechanical efficiency with load for all test fuels is shown in Fig. 2. The mechanical efficiency (ME) is nothing but the ratio of engines brake power developed to the indicated power obtained from the fuel combustion. Mechanical efficiency is a complete engine phenomena and it is the ability of an engine to produce net output from the energy obtained from the fuel combustion after dissipating some amount as friction, radiation, and other unaccounted losses.

$$\eta_{\text{mech}} = \frac{\text{BP}}{\text{IP}}$$

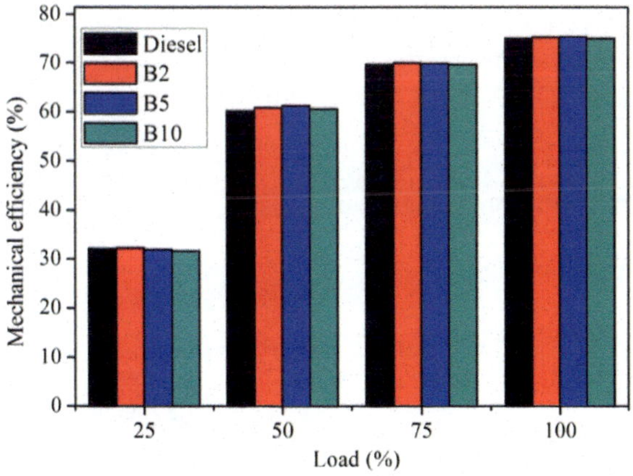

Fig. 2 Effect of load level on mechanical efficiency

where BP = brake power and IP = indicated power.

$$BP = IP - FP \quad (FP = \text{friction power})$$

From the definition, it is clear that the mechanical efficiency does not depend on the fuel consumption. However, earlier literature attributed the lower calorific value affects the ME (Islam et al. 2014). It is seen from the figure that the ME for the blends B2 and B5 is slightly higher, and for B10, it is almost equal compared to diesel fuel operation. This may be attributed to the lower frictional losses due to the enhancement of lubrication as a result of biodiesel addition. However, more percentage of biodiesel addition to the blend resulted in more frictional losses as a result of intermolecular collision of denser mixture. Hence, from the fuel additive perspective, B5 is better and can be used as a natural diesel fuel additive to improve performance of the engine.

4.2 Brake Thermal Efficiency

Figure 3 shows the variation of brake thermal efficiency with loads for all test fuels. It is seen that the BTE increases with increase in load. However, the increase is very significant for change in load from 25 to 50% and comparatively slower pace for the subsequent load level. Surprisingly, it is seen that the BTE for the blends is improved compared to commercial diesel fuel. This may be due to the improvement in mechanical efficiency as a result of lower exergy destruction when the engine run by lower biodiesel concentration blends. In other words, the test engine becomes more efficient when run by lower proportion of biodiesel added to diesel fuel. At full load, the BTE of diesel, B2, B5, and B10 is observed to be 32.48%, 32.51%, 32.55%, and 32.58%, respectively.

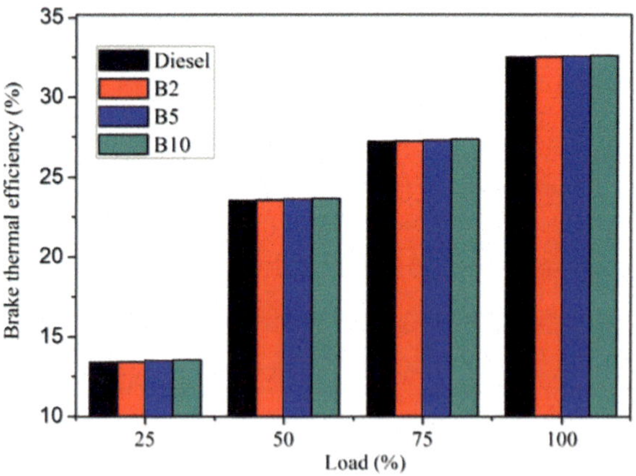

Fig. 3 Effect of load level on the brake thermal efficiency

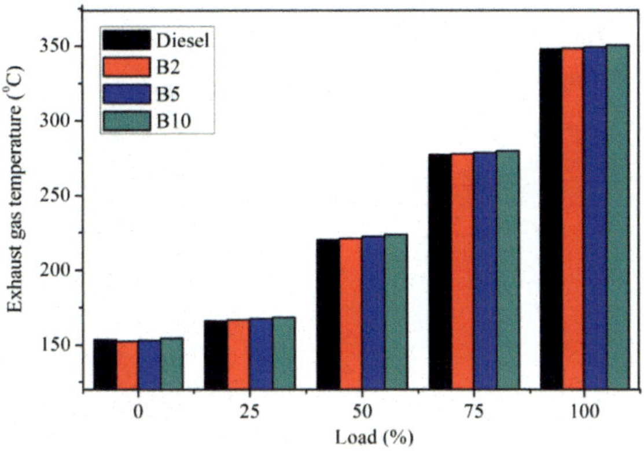

Fig. 4 Effect of load level on exhaust gas temperature

4.3 *Exhaust Gas Temperature*

The variation of exhaust gas temperature (EGT) with load is shown in Fig. 4. With increase in load, the EGT increases for all test fuels. With an increase in load, the combustion activity increases as a result of more fuel supplied to the cylinder chamber, which resulted in higher EGT. At zero load, the EGT for diesel, B2, B5, and B10 is observed to be 153.46 °C, 152.37 °C, 153.18 °C, 154.58 °C, respectively. It is seen that for all load level, B2 showed lowest EGT and higher-order blend B10 showed highest EGT. This may be attributed to the different physico-chemical properties of different blends. Additionally, higher density and viscosity put resistance on the combustion, for which the EGT of the higher blends increased (Dash et al. 2017).

4.4 *CO Emission*

The variation of brake specific carbon monoxide (CO) emission with load for all test fuels is presented in Fig. 5. It is seen that with an increase in load, the specific CO emission significantly reduced for all test fuels. This may be due to the fact that with an increase in load, brake power increases and the cylinder temperature increases for which the oxidation rate of fuel particles enhanced, which ultimately resulted in lower CO level at higher load. Interestingly, at all load level, the CO emission decreased with blend concentration and minimum specific CO emission is obtained for B10. This is attributed to the higher oxygen concentration of biodiesel fuel (Mofijur et al. 2013). At full load, specific CO emission for diesel, B2, B5, and B10 is obtained as 2.46 g/kWh, 2.29 g/kWh, 2.14 g/kWh, and 2.03 g/kWh, respectively.

Fig. 5 Effect of load level on CO emission

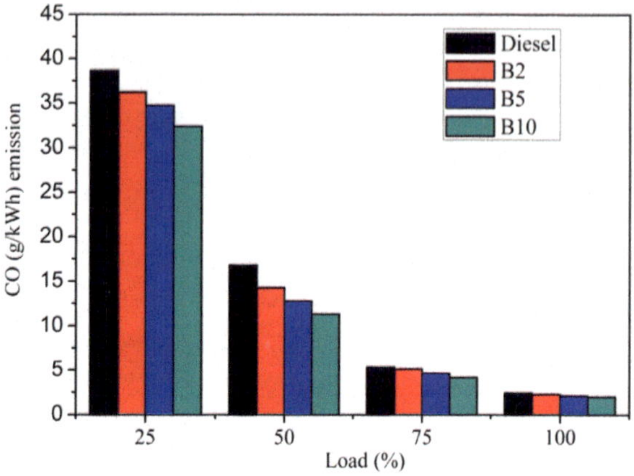

4.5 HC Emission

The variation of unburned hydrocarbon (HC) emission with load is given in Fig. 6. It is observed that with an increase in load, the trend of HC emission follows similar to that of CO emission. This is because both CO and HC emission strongly dependent on a single factor, which is incomplete combustion (Dash and Lingfa 2018c). At any load, HC emission found to be lower for higher biodiesel proportion blends. At full load, HC emission for B2, B5, and B10 is observed to be lowered by 12.81, 26.08, and 33.33% compared to diesel fuel.

Fig. 6 Effect of load level on HC emission

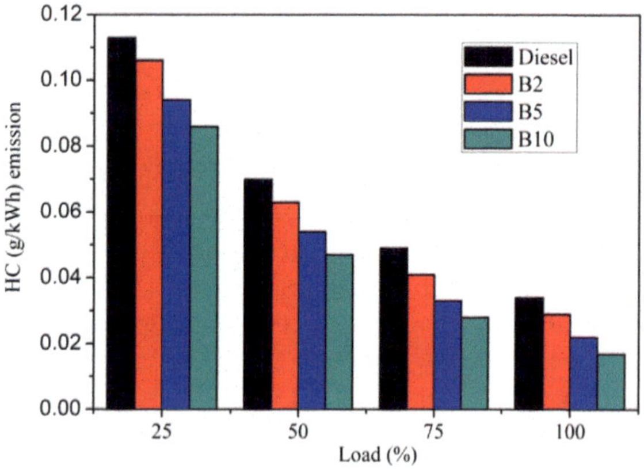

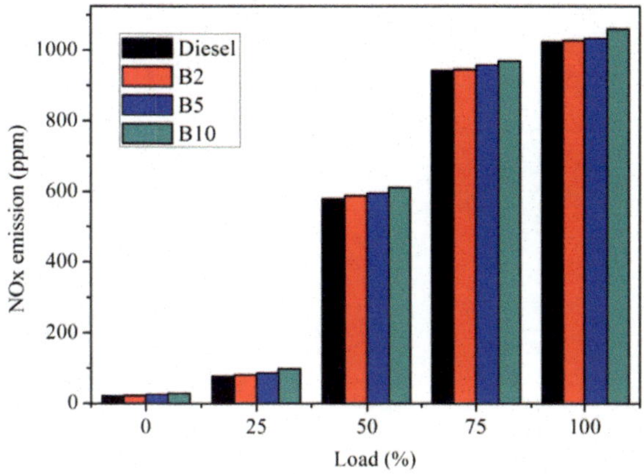

Fig. 7 Effect of load level on NO*x* emission

4.6 NOx Emission

Nitric oxide emission is a major air pollutant, which causes smog and is responsible for environmental contamination. Figure 7 shows the variation of NO*x* emission with respect to load. It is seen that the NO*x* emission greatly increases with increase in load. As the load increases, the cylinder temperature increases, which helps in the reaction of the nitrogen present in the intake air with the oxygen present in the mixture. It is also established that the high temperature of the combustion chamber accelerates the formation of NO*x*. However, the NO*x* emissions of the blends are observed to be slightly higher compared to the diesel fuel. Several earlier literature accused oxygen concentration of biodiesel fuel is responsible for more NO*x* emission (Agarwal 2007; Lapuerta et al. 2008; Varatharajan and Cheralathan 2012). However, oxygen concentration up to B10 blend is not that significant compared to diesel fuel in this investigation. Higher adiabatic flame temperature and more bulk modulus of the blends are suspected for more NO*x* emission behavior. For all load condition, B2 blend witnessed minimum amount of NO*x* emission and beyond B2, with increase in biodiesel concentration, the NO*x* emission increases. At full load, the NO*x* emission for diesel, B2, B5, and B10 are observed to be 1023 ppm, 1027 ppm, 1033 ppm, and 1061 ppm, respectively.

4.7 Smoke Emission

The effect of load level on the smoke emission of all test fuels is presented in Fig. 8. It is observed that with an increase in load, the smoke emission increases for both diesel fuel and biodiesel diesel blends. This is because more fuel enters into the combustion chamber with an increase in load (Dash et al. 2018). However, at lower load level, no substantial difference is observed for the smoke emission behavior

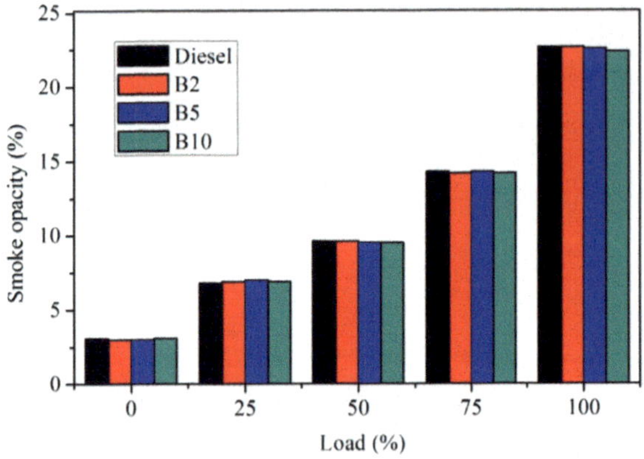

Fig. 8 Effect of load level on smoke opacity

of diesel and blends. At full load, the smoke emission for blends decreases slightly compared to diesel fuel. This may be due to the inbuilt oxygenated characteristics of blended fuels.

4.8 Peak Cylinder Pressure

Cylinder pressure is an important parameter in the study of an IC engine, which should not be overlooked. It is observed from Fig. 9 that at all load level, the peak cylinder pressure (PCP) is not deviated much. This is attributed to the smooth and healthy combustion of fuel mixture. From 25% load to 50% load, the cylinder pressure increased at a faster rate compared to the increase rate in subsequent higher load level. At full load, it is seen that the PCP increased with blend concentration and is maximum for B10. The higher PCP may be ascribed to the earlier start of combustion,

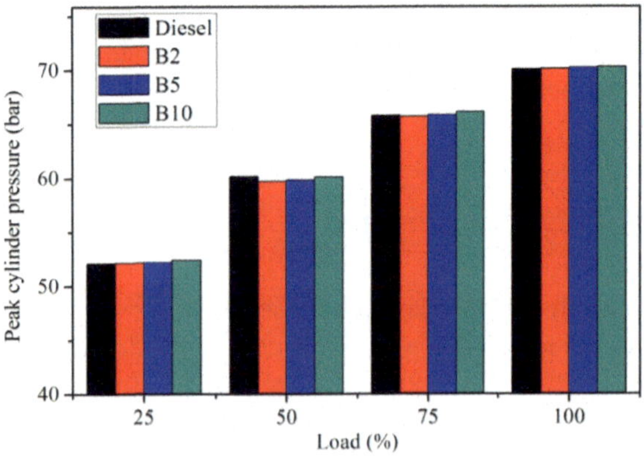

Fig. 9 Effect of load level on peak cylinder pressure

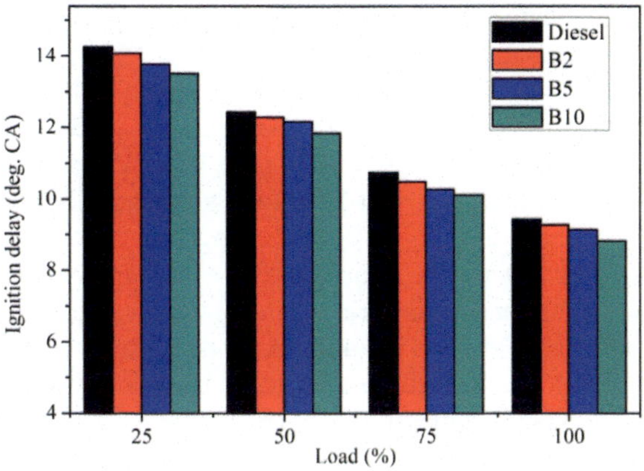

Fig. 10 Effect of load level on ignition delay

due to higher cetane number of blends. It is also seen that with an increase in load, the peak got closer toward the TDC. At full load, the peak observed 5–6° CA after TDC for all test fuels.

4.9 Ignition Delay

Ignition delay is the time gap between the start of fuel combustion and start of fuel injection into the cylinder. The effect of load on the ignition delay period of the engine for all test fuels is shown in Fig. 10. It is observed that the ignition delay decreased with increase in load level for all test fuels. This is mainly due to the higher cylinder temperature as a result of an increase in load level. The ID decreased with increase in biodiesel concentration for all load. This is ascribed to the higher cetane number of blends. At full load, the B10 blend witnessed lowest ignition delay period (8.82° CA), which is 0.61° CA lower compared to diesel fuel.

5 Conclusions

Jatropha biodiesel has been prepared from Jatropha raw oil using CaO powder as heterogeneous catalyst. The properties of the prepared biodiesel blends (B2, B5, and B10) are observed to be equivalent to that of diesel fuel. In this study, lower concentration of biodiesel blends as diesel engine fuel additives are explored. A single cylinder four strokes variable compression ratio diesel engine having a rated power output of 3.5 kW at 1500 rpm is used for the engine trial of all test fuels. Based on the results obtained, the following conclusions are made:

- The mechanical efficiency improved with blended fuel operation compared to diesel fuel operation.

- The brake thermal efficiency increased slightly with increase in biodiesel proportion up to B10 blend.
- The exhaust gas temperature observed to be lowest for B2 and highest for B10.
- At full load, the NOx emission for diesel, B2, B5, and B10 are observed to be 1023 ppm, 1027 ppm, 1033 ppm, and 1061 ppm, respectively.
- At full load, the smoke emission for blends decreased slightly compared to diesel fuel.
- Both CO and HC emission decreased significantly for blends.
- At full load, peak cylinder pressure increased with blend concentration.
- For all load-level ignitions, delay decreased with increase in biodiesel addition. At full load, the B10 blend witnessed a delay of 8.82° CA, which is 0.61° CA lower compared to diesel fuel.

Hence, it is concluded that lower biodiesel concentration blends such as B2, B5, and B10 can be used as renewable diesel fuel additives, which would serve various desired purposes.

References

Agarwal, A. K. (2007). Biofuels (alcohols and biodiesel) applications as fuels for internal combustion engines. *Progress in Energy and Combustion Science, 33,* 233–271.

Azam, M. M., Waris, A., & Nahar, N. M. (2005). Prospects and potential of fatty acid methyl esters of some non-traditional seed oils for use as biodiesel in India. *Biomass and Bioenergy, 29,* 293–302.

Chavan, S. B., Kumbhar, R. R., Madhu, D., Singh, B., & Sharma, Y. C. (2015). Synthesis of biodiesel from *Jatropha curcas* oil using waste eggshell and study of its fuel properties. *RSC Advances, 5,* 63596–63604.

Chukwuezie, O. C., Nwaigwe, K. N., Asoegwu, S. N., & Anyanwu, E. E. (2014). Diesel engine performance of Jatropha biodiesel: A review. *Biofuels, 5,* 415–430.

Dash, S. K., Dash, S., & Lingfa, P. (2017). Comparative assessment of performance and emission analysis of a diesel engine fueled with biodiesel prepared from different sources. *Journal of Industrial Pollution Control, 33*(2), 1114–1119.

Dash, S. K., & Lingfa, P. (2017). A review on production of biodiesel using catalyzed transesterification. *AIP Conference Proceedings, 1859,* 020100. https://doi.org/10.1063/1.4990253.

Dash, S. K., & Lingfa, P. (2018a). An overview of biodiesel production and its utilization in diesel engines. *IOP Conference series: Material Science and Engineering, 377,* 012006. https://doi.org/10.1088/1757-899x/377/1/012006.

Dash, S. K., & Lingfa, P. (2018b). Production of biodiesel from high FFA non-edible Nahar oil and optimization of yield. In *Advanced manufacturing and materials science* (pp. 431–439). https://doi.org/10.1007/978-3-319-76276-0_44.

Dash, S. K., & Lingfa, P. (2018c). Performance evaluation of Nahar oil-diesel blends in a single cylinder direct injection diesel engine. *International Journal of Green Energy, 15,* 400–405.

Dash, S. K., Lingfa, P., & Chavan, S. B. (2018). An experimental investigation on the application potential of heterogeneous catalyzed Nahar biodiesel and its diesel blends as diesel engine fuels. *Energy Sources, Part A: Recovery, Utilization, and Environmental Effects, 40,* 2923–2932. https://doi.org/10.1080/15567036.2018.1514433.

Holman, J. P. (2011). *Experimental methods for engineers* (8th ed.). New Delhi: Tata McGraw Hill.

Islam, M. S., Ahmed, A. S., Islam, A., Aziz, S. A., Xian, L. C., & Mridha, M. (2014). Study on emission and performance of diesel engine using castor biodiesel. *Journal of Chemistry*, Article ID 451526. https://doi.org/10.1155/2014/451526.

Lapuerta, M., Armas, O., & Rodriguez-Fernandez, J. (2008). Effect of biodiesel fuels on diesel engine emissions. *Progress in Energy and Combustion Science, 34,* 198–223.

Lee, H. V., Juan, J. C., Taufiq-yap, Y. H., Kong, P. S., & Rahman, N. A. (2015). Advancement in heterogeneous base catalyzed technology: An efficient production of biodiesel fuels. *Journal of Renewable and Sustainable Energy, 7*(032701), 1–46. https://doi.org/10.1063/1.4919082.

Mofijur, M., Atabani, A. E., Masjuki, H. H., Kalam, M. A., & Masum, B. M. (2013). A study on the effects of promising edible and non-edible biodiesel feedstocks on engine performance and emissions production: A comparative evaluation. *Renewable and Sustainable Energy Reviews, 23,* 391–404.

No, S. Y. (2011). Inedible vegetable oils and their derivatives for alternative diesel fuels in CI engines: A review. *Renewable and Sustainable Reviews, 15,* 131–149.

Varatharajan, K., & Cheralathan, M. (2012). Influence of fuel properties and composition of NO*x* emissions from biodiesel powered diesel engines: A review. *Renewable and Sustainable Energy Reviews, 16,* 3702–3710.

Xue, J., Grift, T. E., & Hansen, A. C. (2011). Effect of biodiesel on engine performances and emissions. *Renewable and Sustainable Energy Reviews, 15,* 1098–1116.

A Critical Review on the Production of Biodiesel from Jatropha, Karanja and Castor Feedstocks

M. S. Ahamed, S. K. Dash, A. Kumar and P. Lingfa

Abstract With rapid industrialization and modernization of the production sector, the demand for energy has been growing at a rapid rate. Large part of this energy has been derived from coal, petroleum and natural gas. The depletion of fossil fuels and environmental concerns like air pollution and global warming caused by combustion of fossil fuels is the main concern for researcher to find an alternative fuels which can be produced from renewable feedstock. Unlike the other countries, India import a huge amount of diesel fuel from Saudi Arabia, Iraq, Iran, Nigeria, UAE, Venezuela, Kuwait and Qatar hence looking for alternative renewable fuel to replace diesel fuel is a natural choice. Biodiesel can be the best alternative fuel to replace diesel as it does not require engine modification and it is produced from renewable raw materials. This paper focuses on review of biodiesel production from three non-edible oils such as Jatropha oil, Karanja oil and Castor oil, which are abundantly available in India.

Keywords Jatropha · Transesterification · Karanja · Castor · Non-edible oil · Biodiesel

1 Introduction

Energy is the basic need on every aspect of our life. It is one of the most important sectors for all the countries, especially for the developing countries from economical point of view. It plays an important role in global economy. The rising price of non-renewable fuel, depletion of fossil fuel and environmental problems make a great impact on the economies of the developing nations and has been playing crucial roles in shaping them. So there is an urgent need to find an alternative source of energy. Many researchers found that biodiesel is very promising fuel, the properties of biodiesel are comparable to diesel fuel, and it is environmentally friendly and renewable energy source (Dash and Lingfa 2017, 2018a, b; Dash et al. 2017). It caught the global attention during the fuel crisis between the periods 1973 and 1979 which hit many countries. Ministry of New and Renewable Energy made a National

M. S. Ahamed (✉) · S. K. Dash · A. Kumar · P. Lingfa
Department of Mechanical Engineering, NERIST, Nirjuli, India
e-mail: sajitahamed56@gmail.com

S. K. Ghosh et al. (eds.), *Bioresource Utilization and Bioprocess*,
https://doi.org/10.1007/978-981-15-1607-8_12

Biofuel Policy in 2009 to encourage the use of biofuels blends with diesel in the country. Government initiatives such as Make in India and Skill Development can be integrated to achieve various targets of air pollution control, employment generation, import reduction and increase of farmer's income. Under this policy, biofuels are categorised as basic biofuels and advanced biofuels. Appropriate financial incentives are given under different category. First-generation bioethanol and biodiesel are come under the category of basic biofuels and second-generation ethanol, third-generation biofuels, municipal solid waste, bio-CNG, etc., are under the category of advanced biofuels. Biodiesel is derived from lipids by transesterification, pyrolysis, and micro-emulsification (Dash and Lingfa 2018c; http://pib.nic.in/newsite/PrintRelease.aspx?relid=1793).

The transesterification reaction involves three steps; at first, the triglycerides are converted into diglycerides, and then, the diglycerides are converted into monoglycerides. Finally, the monoglycerides are converted into glycerol; each step gives one ester molecules as follows.

$$\text{Triglyceride (TG)} + \text{ROH} \leftrightarrow \text{Diglyceride (DG)} + \text{RCOOR}_1 \tag{1}$$

$$\text{Diglyceride (DG)} + \text{ROH} \leftrightarrow \text{Monoglyceride (MG)} + \text{RCOOR}_2 \tag{2}$$

$$\text{Monoglyceride (MG)} + \text{ROH} \leftrightarrow \text{Glycerol} + \text{RCOOR}_3 \tag{3}$$

Biodiesel is derived both from plant or animal. Plants can be edible as well as non-edible. Edible oils such as coconut oil, peanut oil, palm oil, sunflower oil, etc., can be used for biodiesel production. Non-edible oil includes castor oil, cottonseed oil, Jatropha oil and Karanja oil. Animal fat and oil from fishes are the animal feedstock which can be used for biodiesel production (Alptekin et al. 2014; Yuvaraj et al. 2016). Different countries use different feedstock for the production of biodiesel. European countries use sunflower oil and rapeseed, and USA uses soya bean as feedstock. Coconut oil and Palm oil are used in Philippines and Southeast Asia (Malaysia and Indonesia), respectively, as their main feedstock (Table 1).

There is a huge demand for biodiesel through the world. New technologies are being used for biodiesel production. The Government of India launched a 'National Mission on Biodiesel' to find a renewable, low-price alternative fuel based on renewable raw materials to replace biodiesel (Shukla 2005). Feedstock scarcity is the main factor which affects biodiesel production (Wani et al. 2006) (Table 2).

There are about 100–300 species which can produce oil, and of this species, 30 families can be used for the production of biodiesel. In many developed counties, many edible oils are used for the production of biodiesel. However, India cannot afford the use of edible oil for the production of biodiesel as India already import edible oil for consumption. So, to overcome this problem, Indian government identified the kernel of *Jatropha curcas* as a potential feedstock for biodiesel production with criteria not to compromise food security and to improve livelihoods in arid regions of the country (Reddy et al. 2008).

Table 1 Different feedstocks used in different countries for biodiesel production

Country	Sources
Spain	Linseed and olive oil
Europe	Sunflower oil, rapeseed oil
Italy	Sunflower oil
France	Sunflower oil
Ireland	Beef tallow, animal fat
India	Jatropha oil, sunflower oil
Malaysia	Palm oil
Indonesia	Palm oil
Australia	Animal fat, rapeseed oil and beef tallow
USA	Brazil soya bean
Canada	Vegetable oil, animal fat
Germany	Rapeseed oil

Table 2 Physical–chemical properties of Jatropha, Karanja and castor oils (Patil and Deng 2009)

Properties	Karanja	Jatropha	Canola
Viscosity (mm^2/s)	5.52–5.79	2.35–2.47	2.56–2.84
Specific gravity	0.88–0.89	0.86–0.88	0.88–0.90
Acid value (mg KOH/g)	0.72–0.76	0.61–0.66	–
Cetane number	59.68–60.90	60.74–63.27	55.90–56.10
Pour point (°C)	−3 to −1	−6 to 2	−4 to −1
Calorific value (MJ/kg)	37.8–39.69	39.65–41.63	44.65–44.93

The objectives of this review paper are given below:

(a) To find the most abundant feedstock out of Jatropha oil, Karanja oil and castor oil.
(b) To find the best biodiesel production method.
(c) To determine the best oil in terms of their characteristic properties and their oil contents.

2 *Jatropha curcas*

Jatropha curcas belongs to the family of Euphorbiaceae. It is abundantly available in India, Africa, South and Central America and in my Southeast Asia. Compared to diesel, Jatropha oil has higher cetane number. Various similar properties of Jatropha biodiesel with diesel fuel make Jatropha a good alternative fuel (Tapanes et al. 2008;

Divakara et al. 2010; Jain and Sharma 2010). Pandey et al. (2012) studied various benefits of *Jatropha curcas*. The seeds are used for soap production, and its leaves and other parts are used for the production of medicines. Recently, *Jatropha curcas* became popular for the production of biodiesel due to easy propagation, high oil content, rapid growth rate, ecological benefits and low capital investment. Deng et al. (2011) used solid basic nanosized particles as catalyst for biodiesel production from Jatropha oil after oil pre-treatment. They found that at molar ratio (methanol to oil) 4:1, 1 wt.% catalyst, 318 K temperature for 1.5 h, biodiesel yield of 95.2% was obtained. Akintayo (2004) found that there are about 72% of unsaturated fatty acids in Jatropha seed oil. Along with unsaturated fatty acid, there are oleic acid and linoleic acid. Viscosity of Jatropha is lower compared to soya bean (31 cSt), sunflower (43 cSt) and cottonseed (36 cSt) when tested at 30 °C, which shows biodiesel is more suitable to replace diesel fuel. Compare to oil content of palm, linseed and soya bean kernel which are 44.6%, 33.3% and 18.35%, respectively. Jatropha kernel has higher oil content which was determined to be 66.4 and 63.16% (Gunstone 2004; Akbar et al. 2009; Adebowale and Adedire 2006). Various techniques such as transesterification, pyrolysis, blending and micro-emulsification can be used for the production of biodiesel from Jatropha oil (Ma and Hanna 1999).

Bojan and Durairaj (2012) produced biodiesel from *Jatropha curcas* oil by using both one-step and two-step catalysed transesterification. In one-step transesterification, due to high FFA concentration (8.67%), biodiesel yield was reduced to 80.5%. In two-step transesterification, they could achieve 93% of the yield of biodiesel. They carried out the reaction at methanol-to-oil molar ratio of 5.41:1, 0.55% catalyst and 60 °C temperature. Haldar et al. (2009) investigate the production of alternative diesel fuel from Putranjiva, Karanja and Jatropha by degumming chemical process with the presence of concentrated phosphoric acid. They observed the best emissions and performance at high load and 45° TDC injection timing. Out of the three non-edible oils, Jatropha gives the best performance, emission and yield. Berchmans and Hirata (2008) could successfully reduce the FFA content to less than 1% by using methanol-to-oil ratio 60% w/w, 1% w/w of H_2SO_4 as catalyst for 1-h reaction time at 50 °C.

3 Castor Oil

The castor is a species of flowering plant, and it belongs to Euphorbiaceous family. Castor is abundantly available in India, East Africa and Mediterranean Basin. The growth period of castor is much shorter than that of Jatropha and Karanja. The oil content is in the range of 40–60%, which is rich in triglycerides, ricinoleic acid. The castor oil is toxic for consumption due to the presence of ricin, a toxic substance. Hot and humid tropical conditions are ideal for castor growth. It takes 4–5 months to grow to its maturity. Castor oil has a very low cloud and pour points which will make it suitable to use as alternative fuel in cold weather regions (Sattanathan 2013; Conceicüa and Fernandes 2007). da Silva et al. (2006) generated a statistical model (4) to predict ethanolysis of castor oil. For transesterification reaction, catalyst concentration is the most important influencing parameter. They observed higher ester was converted at 30 °C, when more catalyst up to 1.3% was and lower ethanol-to-castor oil molar ratio is used.

$$Y = 87.44 + 12.38 * X_2 - 7.03 * X_2^2 + 4.86 * X_3 - 6.35 * X_2 * X_3 \quad (4)$$

where X_1 represents temperature, X_2 represents catalyst concentration and X_3 represents ethanol-to-castor oil molar ratio.

4 Karanja Oil

Karanja (*Pongamia Pinnata*) can grow easily, and it completes its growth in 4-5 years. It belongs to the family of Leguminaceae. It can survive under harsh weather conditions. It is available abundantly along riverbanks, marginal lands and coastal areas. It has a height of about 18 m and a trunk diameter 50 cm approx. The colour of fresh oil is brown and darkens when keep on storage. The fresh extracted oil is yellowish orange to brown and rapidly darkens on storage (Baiju et al. 2009). As Karanja is non-edible, its seed is not used for any purpose. Oleic acid is the primary fatty acid of Karanja oil. Various researchers studied the possibilities of biodiesel production from Karanja oil by using different methods. Thiruvengadaravi et al. (2012) reduced the acid value of Karanja oil from 12.27 mg KOH/g to 1.3 mg KOH/g by using 1% sulphated zirconia (SZ) as a solid acid catalyst, molar ratio (methanol to oil) 9:1, at 60 °C temperature and 2-h reaction time. Verma and Sharma (2016) produced biodiesel from Karanja oil and also developed regression Eqs. (5) and (6) to determine the yield at various influencing parameters for the transesterification process with the help of Design Expert 9.0.6.2. They could achieve a yield of 91.05% for methanol, when used 10.44:1 molar ratio, 1.22% w/w catalyst (KOH) and reaction time 90.78 min at 66.8 °C temperature, whereas for ethanolysis, 77.4% could be achieved 8.42:1 molar ratio, 61.3 °C reaction temperature with 1.21% of catalyst and 120 min of reaction.

$$\begin{aligned} Y_{\text{KOME}} = {} & 90.96 + 2.65 * A - 0.91 * B - 1.51 * C + 0.033 * D + 1.75 * A * B \\ & + 4.33 * A * C + 3.22 * A * D - 0.25 * B * C - 3.52 * B * D \\ & + 3.1 * C * D - 10.45 * A^2 - 29.54 * B^2 - 12.31 * C^2 - 6.82 * D^2 \end{aligned} \quad (5)$$

$$\begin{aligned} Y_{\text{KOEE}} = {} & 74.98 - 0.62 * A - 5.78 * B - 1.2 * C + 1.08 * D + 2.7 * A * B \\ & + 0.38 * A * C - 0.05 * A * D - 0.25 * B * C + 1.25 + B + D \\ & + 0.98 * C * D - 3.92 * A^2 - 5.61 * B^2 - 1.86 * C^2 + 1.06 * D^2 \end{aligned} \quad (6)$$

where *A* is reaction temperature (°C), *B* is alcohol-to-oil molar ratio, *C* is catalyst amount (wt.%), *D* is reaction time (min).

Bajpai et al. (2009) studied performance and emission of the diesel engine different blends (K5, K10, K15 and K20) of Karanja oil with diesel engine. They have recommended using 10% blends of Karanja oil with diesel as a fuel for the diesel engine. The density of Karanja oil is around 10% higher than diesel. Therefore, separation of oil from the diesel is the main issue for blending technique. It is required to heat the mixture or stir it by a suitable mechanism for proper mixing. Hotti and Hebbal (2011) tested engine at different blend of diesel and Karanja oil (K10, K15, K20 and K100). They found K15 to be the optimum blend in terms of performance, thermal efficiency and combustion. The viscosity, density, carbon residue and flash point of K100 are higher than that of diesel. They recommend K15 to be the best blend to use with diesel fuel. Aniya et al. (2015) investigated transesterification reaction for Karanja oil in the presence of KOH catalyst. They found time and temperature to be the main influencing parameters for mass transfer and reaction rate. Volumetric mass transfer coefficients increase with time as 51×10^{-3}, 135×10^{-3} and 334×10^{-3} min^{-1} at temperatures of 35 °C, 45 °C and 55 °C, respectively. They used 1% w/w KOH as catalyst, 6:1 methanol-to-oil molar ratio and temperature between 35 and 55 °C. Kadu and Sarda (2010) have tested the performance of four-stroke, single-cylinder C.I engine at temperatures between 30 and 100 °C by using preheated Karanja oil. At high speed, BSFC for preheated oil is not much different from unheated oil. Brake thermal efficiency is almost similar for high-speed and low-speed engine. NO_x emission increases significantly at high speed. Prabhavathi Devi et al. (2014) employed esterification and transesterification method for the production of biodiesel from Karanja oil by using reusable solid acid catalyst. They could achieve more than 99% yield at reaction parameters such as 1:45 molar ratio, 20% catalyst, 160 °C temperature and 4-hour reaction time. Hossain and Davies (2012) tested engine using preheated Karanja oil. The Karanja oil was preheated at a temperature between 58 and 75 °C by hot jacket water. They observed higher peak cylinder pressure and fuel line injection pressure at full load. They found higher BSFC and emissions of CO, CO_2 and NO_x compared to diesel fuel. Verma and Sharma (2016) optimised process variables for transesterification process for methanolysis and ethanolysis for biodiesel production from Karanja oil. For methanolysis, they could achieve 91.05% when used molar ratio of 10.44:1, 1.22% w/w KOH as catalyst for 90.78-min reaction time at the temperature of 66.8 °C. For ethanolysis, 77.4% yield could be achieved at optimum reaction parameters such as 8.42:1 molar ratio, 61.3 °C reaction temperature with 1.21% of catalyst and 120 min of reaction time. Meher et al. (2006) optimised various reaction variables for biodiesel production from Karanja oil at 1% KOH as catalyst, 65 °C temperature and 6:1 molar ratio. They achieved a yield of 97–98% under optimised conditions. Naik et al. (2008) could obtain a yield of 96.6–97% biodiesel from Karanja oil containing 20% free fatty acid using esterification method. Sahoo and Das (2009) obtained maximum yield of 91% biodiesel from Karanja oil when used 11.5:1 molar ratio and 120-min reaction time.

5 Conclusions

This paper provides a comprehensive report on the important contributions of researcher worked on three non-edible feedstocks such as Jatropha, Karanja and castor oils. Transesterification reaction is reported most commonly used method for the reduction of kinematic viscosity and other free fatty acid value of the oil. Reaction with alkali catalyst is more faster and economical compared to acid catalyst. Catalyst concentration and molar ratio play an important role in transesterification reaction. It is concluded that the characteristic properties of the biodiesel produce from these three oils are comparable with diesel fuel. The biodiesel produced from Jatropha oil gives the best-quality biodiesel compared with Karanja and castor oils. The growth period of castor plant is much shorter than that of Jatropha and Karanja plants. However, compared to Jatropha and castor plants, Karanja plant can grow easily and can survive in the harsh weather conditions.

References

Adebowale, K. O., & Adedire, A. C. (2006). Chemical composition and insecticidal properties of the underutilized *Jatropha curcas* seed oil. *African Journal of Biotechnology, 5*(10), 901–906.

Akbar, E., Yaakob, Z., Kamarudin, S. K., Ismail, M., & Salimon, J. (2009). Characteristic and composition of *Jatropha curcas* oil seed from Malaysia and its potential as biodiesel feedstock. *European Journal of Scientific Research, 29*(3), 396–403.

Akintayo, E. T. (2004). Characteristics and composition of *Parkia biglobbossa* and *Jatropha curcas* oils and cakes. *Bioresource Technology, 92,* 307–310.

Alptekin, E., Canakci, M., & Sanli, H. (2014). Biodiesel production from vegetable oil and waste animal fats in a pilot plant. *Waste Management, 34,* 2146–2154.

Aniya, V. K., Muktham, R. K., Alka, K., & Satyavathi, B. (2015). Modeling and simulation of batch kinetics of non-edible Karanja oil for biodiesel production: A mass transfer study. *Fuel, 161,* 137–145.

Baiju, B., Naik, M. K., & Das, L. M. (2009). A comparative evaluation of compression ignition engine characteristics using methyl and ethyl esters of Karanja oil. *Renewable Energy, 34*(6), 1616–1621. https://doi.org/10.1016/j.renene.2008.11.020.

Bajpai, S., Sahoo, P. K., & Das, L. M. (2009). Feasibility of blending Karanja vegetable oil in petrodiesel and utilization in a direct injection diesel engine. *Fuel, 88,* 705–711.

Berchmans, H. J., & Hirata, S. (2008). Biodiesel production from crude *Jatropha curcas* L. seed oil with a high content of free fatty acids. *Bioresource Technology, 99,* 1716–1721.

Bojan, S. G., & Durairaj, S. K. (2012). Producing biodiesel from high free fatty acid *Jatropha curcas* oil by a two step method—An Indian case study. *Journal of Sustainable Energy & Environment, 3,* 63–66.

Conceicüa, M. M., & Fernandes, V. J. (2007). Thermal and oxidative degradation of castor oil biodiesel. *Energy & Fuels.*

da Silva, N. de L., Maciel, M. R. W., Batistella, C. B., & Filho, R. M. (2006). Optimization of biodiesel production from castor oil. *Applied Biochemistry and Biotechnology, 129–132.*

Dash, S. K., Dash, S., & Lingfa, P. (2017). Comparative assessment of performance and emission analysis of a diesel engine fuelled with biodiesel prepared from different sources. *Journal of Industrial Pollution Control, 33*(2), 1114–1119.

Dash, S. K., & Lingfa, P. (2017). A review on production of biodiesel using catalyzed transesterification. *AIP Conference Proceedings, 1859,* 020100. https://doi.org/10.1063/1.4990253.
Dash, S. K., & Lingfa, P. (2018a). An overview of biodiesel production and its utilization in diesel engines. *IOP Conference Series: Material Science and Engineering, 377,* 012006. https://doi.org/10.1088/1757-899X/377/1/012006.
Dash, S. K., & Lingfa, P. (2018b). Performance evaluation of Nahar oil-diesel blends in a single cylinder direct injection diesel engine. *International Journal of Green Energy, 15,* 400–405.
Dash, S. K., & Lingfa, P. (2018c). Production of biodiesel from high FFA non-edible Nahar oil and optimization of yield. In *Advanced manufacturing and materials science* (pp. 431–439). https://doi.org/10.1007/978-3-319-76276-0_44.
Deng, X., Fang, Z., Liu, Y-h, & Yu, C.-L. (2011). Production of biodiesel from Jatropha oil catalyzed by nanosized solid basic catalyst. *Energy, 36,* 777–784.
Divakara, B. N., Upadhyaya, H. D., Wani, S. P., Laxmipathi Gowda, C. L. (2010). Biology and genetic improvement of *Jatropha curcas* L.: A review. *Applied Energy, 87,* 732–742.
Gunstone, F. D. (2004). *The chemistry of oils and fats: Sources, composition, properties and uses.* London: Blackwell Publishing Ltd.
Haldar, S. K., Ghosh, B. B., & Nag, A. (2009). Studies on the comparison of performance and emission characteristics of a diesel engine using three degummed non-edible oils. *Biomass and Bioenergy, 33,* 1013–1018.
Hossain, A. K., & Davies, P. A. (2012). Performance emission and combustion characteristics of an indirect injection (IDI) multi-cylinder compression ignition (CI) engine operating on neat Jatropha and Karanja oils preheated by jacket water. *Biomass-Bioenergy, 46,* 332–342.
Hotti, S. R., & Hebbal, O. (2011). Performance and combustion characteristics of single cylinder diesel engine running on Karanja oil/diesel fuel blends. *Engineering, 3,* 371–375. https://doi.org/10.4236/eng.2011.34042.
Jain, S., & Sharma, M. P. (2010). Prospects of biodiesel from Jatropha in India: A review. *Renewable and Sustainable Energy Reviews, 14,* 763–771.
Kadu, S. P., & Sarda, R. H. (2010). Experimental investigations on the use of preheated neat Karanja oil as fuel in a compression ignition engine. *World Academy of Science, Engineering, and Technology, 48,* 540–544.
Ma, F., & Hanna, M. A. (1999). Biodiesel production: A review. *Bioresource Technology, 70,* 1–15.
Meher, L. C., Dharmagadda, S. S. V., & Naik, S. N. (2006). Optimization of alkali-catalyzed transesterification of *Pongamia pinnata* oil for production of biodiesel. *Bioresource Technology, 97,* 1392–1397.
Naik, M., Meher, L. C., Naik, S. N., & Das, L. M. (2008). Production of biodiesel from high free fatty acid Karanja (*Pongamia pinnata*) oil. *Biomass and Bioenergy, 32,* 354–357.
Pandey, V. C., Singh, K., Singh, J. S., Kumar, A., Singh, B., & Singh, R. P. (2012). *Jatropha curcas*: A potential biofuel plant for sustainable environmental development. *Renewable and Sustainable Energy Reviews, 16,* 2870–2883.
Patil, P. D., & Deng, S. (2009). Optimization of biodiesel production from edible and non-edible vegetable oils. *Fuel, 88,* 1302–1306.
Prabhavathi Devi, B. L. A., Reddy, T. V. K., Lakshmi, K. V., & Prasad, R. B. N. (2014). A green recyclable SO_3H-carbon catalyst derived from glycerol for the production of biodiesel from FFA-containing Karanja (*Pongamia glabra*) oil in a single step. *Bioresource Technology, 153,* 370–373.
Reddy, B. V. S., Ramesh, S., & Ashok, Kumar A. (2008). Bio-fuel crops research for energy security and rural development in developing countries. *BioEnergy Research, 1,* 248–258.
Sahoo, P. K., & Das, L. M. (2009). Process optimization for biodiesel production from Jatropha, Karanja and Polanga oils. *Fuel, 88,* 1588–1594.
Sattanathan, R. (2013). Production of biodiesel from castor oil with its performance and emission test. *IJSR.*
Shukla, S. K. (2005). Experiences of Chhattisgarh biofuel development authority. *Biofuels India, 3,* 12–13.

Tapanes, N. C. O., Gomes Aranda, D. A., De Mesquita Carneiro, J. W., & Ceva Antunes, O. A. (2008). Transesterification of *Jatropha curcas* oil glycerides: Theoretical and experimental studies of biodiesel reaction. *Fuel, 87,* 2286–2295.

Thiruvengadaravi, K. V., Nandagopala, J., Baskaralingamb, P., Sathya Selva Balab, V., & Sivanesanb, S. (2012). Acid-catalyzed esterification of Karanja (*Pongamia pinnata*) oil with high free fatty acids for biodiesel production. *Fuel, 98,* 1–4. https://doi.org/10.1016/j.fuel.2012.02.047.

Verma, P., & Sharma, M. P. (2016). Comparative analysis of effect of methanol and ethanol on Karanja biodiesel production and its optimisation. *Fuel, 180,* 164–174. https://doi.org/10.1016/j.fuel.2016.04.035.

Wani, S. P., Osman, M., & D'silva, E. (2006). Improved livelihoods and environmental protection through biodiesel plantations in Asia. *Asian Biotechnology and Development Review, 8,* 11–29.

Yuvaraj, D., Bharathiraja, B., Rithika, J., Dhanasree, S., Ezhilarasi, V., Lavanya, A., et al. (2016). Production of biofuels from fish wastes: An overview. *Biofuels*. https://doi.org/10.1080/17597269.2016.1231951.

Biodiesel—A Review on Recent Advancements in Production

N. Dhivya Priya and M. Thirumarimurugan

Abstract Biofuel, fuel produced from lipids extracted from animal or vegetable sources, has been accepted widely as an efficient and ecofriendly alternative for conventional fossil diesel. Technically, biodiesel may be defined as product of alcoholysis of long chain fatty acids in the presence of a catalyst. The use of vegetable oils in compression engines is a longstanding idea dating back to a century, when, people used it only in cases of emergency, due to the economic infeasibility. There are four primary methods of using vegetable oils in engines, namely, blend or direct usage, micro emulsification of vegetable oil, thermal cracking of vegetable oils or pyrolysis and transesterifictaion. Out of these methods, transesterifictaion has been significantly researched and commercially exploited. Transesterification of unsaturated fatty acids using acid or alkali as catalyst has been widely studied and reviewed in this paper. Various operating parameters such as temperature, molar ratio of oil to alcohol, reaction time are known to affect the quantity and quality of biodiesel yield. Lipids from animal sources primarily comprise of saturated fatty acids cannot be used for biodiesel production, as; on conventional transesterification wax would be produced instead of biodiesel. This paper also reviews production methods of biodiesel using saturated fatty acids catalyzed by acetyl chloride and metallic sodium in the presence of alcohol. The post production purification of biodiesel and associated byproduct recovery have been explained in this paper.

Keywords Biodiesel · Transesterification · Saturated fatty acids · Unsaturated fatty acids · Extraction and purification

1 Introduction

The existing energy scenario is quite a complex one, where significant differences exist in the production and consumption of energy between the first and the third worlds. Though the per capita consumption in the third world is relatively low (Pendse

N. Dhivya Priya (✉) · M. Thirumarimurugan
Department of Chemical Engineering, Coimbatore Institute of Technology, Coimbatore, Tamil Nadu, India
e-mail: dhivyapriya.n@cit.edu.in

S. K. Ghosh et al. (eds.), *Bioresource Utilization and Bioprocess*,
https://doi.org/10.1007/978-981-15-1607-8_13

2016), the production is also proportionately less, thereby striking an imbalance. Also, with the fast depleting fossil fuel resources and increased price of them, people are looking forward to a more sustainable and affordable form of energy. The sources of energy can be broadly categorized into two types, based on the intended use, as commercial and noncommercial (Pendse 2016). Commercial sources of energy include, but not limited to, coal, crude oil, and natural gas; noncommercial sources include firewood, animal waste, wind power, and human power, etc. The search for an alternative source of energy has turned significant attention towards biodiesel, as it can be used by both the commercial and noncommercial sectors.

Biodiesel, technically defined as a monoalkyl ester of long-chain fatty acids, produced by transesterification of long-chain fatty acid using a monohydric alcohol, in the presence of an acid, alkali or enzyme catalyst (Lin et al. 2009). This specification of biodiesel confirms with the definition of fuel (B100)—ASTM D6751-09 that has been dictated for use in compression ignition engines. In simple terms, biodiesel is an alternative fuel that is derived from biological sources. These sources include plant oils and animal fats. With the development of this technology, it has been found that biodiesel can also be derived from plant sugars such as cellulose and hemicellulose with the involvement of oleaginous microorganisms, which convert these sugars into cellular lipid and accumulate it as a part of their cellular metabolism. The evolution of biodiesel can be broadly distinguished into four generations, depending on the type of biological source involved.

1.1 First-Generation Biofuel

This is derived from plant sugars and plant oils. This includes bioethanol derived from food crops such as sugar beet, cereals, sugarcane, and biodiesel obtained from oil seeds such as sunflower, soya, and palm seeds (Chaturvedi and Verma 2013). This is mainly produced from food crops by extracting the oil through fermentation of sugars in feedstock to be used as biofuel. Crops such as wheat and rice, containing high cellulose and hemicelluloses content can also be employed for bioethanol production, while oil seeds such as *Jatropha curcas* and *Helianthus annuus* are used for biodiesel production. The major disadvantage of first-generation biofuel was the alarming competition it poses to the food industry. As the raw materials employed in food industry and biodiesel production are the same, it was perceived that the price hike in food industry might be due to the diversion of food crops from general consumer market to bioenergy production.

1.2 Second-Generation Biofuel

Biofuel derived from lignocellulose biomass such as wood and agricultural wastes is called as second-generation biofuel (Chaturvedi and Verma 2013). The feedstock employed here were predominantly non-food crops or inedible parts of food crops such as wood waste, organic waste due to various anthropological activities, such as rice husks and straws, eliminating the problem posed by the previous-generation biofuel. Assessments of production and usage patterns of second-generation biofuel have indicated that there is an increase in net energy gain, making them more desirable than first-generation biofuel (Sims et al. 2010).

1.3 Third-Generation Biofuel

Third-generation biofuel is based on improvements in the production of biomass. It possesses the advantage of employing specifically engineered algae as the source of lipids. The algae are cultured on suitable substrates to act as a low cost, high energy, and entirely renewable feedstock. It is predicted third-generation biofuel will have greater potential than its precursors in terms of its yield percentage (Agbor et al. 2011). Algae can also be grown on substrates such as land or water unsuitable for cultivation of food crops, or on other suitable organic wastes; thus, reducing the stress on quantity of resources (economic resources and feedstock) required for a commercial-scale production of biodiesel. An additional benefit of algae-based biofuel is that, the obtained fuel can be converted into a wide range of fuels such as diesel, petrol, and jet fuel.

1.4 Fourth-Generation Biofuel

Fourth-generation biofuel aims at not only being a sustainable source of energy, but also at capturing and storing carbon dioxide or carbon sequestration. Biomass which absorbs carbon dioxide during its growth is converted into fuel through process similar to second-generation biofuel. The major difference between the fourth-generation biodiesel and the earlier generation fuels is that, carbon dioxide is captured at all stages of production. This has been made possible by the process of oxy-fuel combustion. The carbon, thus acquired, can be into geological repositories such as gas fields or saline aquifers, the carbon capture and geo-sequestration leads to a carbon-negative fuel, better than a carbon-neutral fuel.

2 Production of Biodiesel

2.1 Early Techniques

The use of vegetable oil as a fuel dates back to the 1980s (Ma and Hanna 1999). When the use of pure biodiesel or vegetable oil (B100) was practically difficult, various blend mixtures were tried to attain maximum efficiency without altering the construction and working of the diesel engine. While, B20 mixture was used successfully on a practical level, short-duration use or research purpose included a B50 mixture also. The main advantage with the direct use of vegetable oil as fuel in the engine is its higher heat content which is almost 80% of that of the conventional diesel fuel. Easy portability, renewability of the oil, and its ready availability made it a desirable substitute for the conventional diesel. The main disadvantage of using biodiesel directly or as a blend was its higher viscosity and lower volatility. The reduced reactivity of unsaturated hydrocarbon chains and lower cetane number causes the incomplete combustion of the fuel, thereby, resulting in coking of the engine and trumpet formation (Pryde 1983; Peterson et al. 1983). This also led to gum formation, thickening, and gelling of the lubricating oil which on mixing with vegetable oil caused its contamination and deterioration (Kaya et al. 2009; Issariyakul et al. 2008; Kansedo et al. 2008; Tiwari et al. 2007; Rao et al. 2009).

As a solution to the high viscosity problems, microemulsions of the fuel with a solvent, generally alcohol, with an ionic or nonionic amphiphiles of the nanoscale size substituted the direct/blend fuels. Emulsions of vegetable oil on lower-chain alcohols, namely, methanol, ethanol, propanol, and butanol were used (Ziejewski et al. 1984). Emulsification of vegetable oil using alcohol brought down the viscosity of the fuel, and also displayed better spray patterns during combustion (Leung et al. 2010). 200 h tests were conducted on these emulsions and no significant deviation in the performance was observed. Though the problem of viscosity has been addressed, carbon deposition due to incomplete combustion of biodiesel still persisted. Carbon and lacquer deposition on injector tips, in-take valves, and top of cylinders was a significant problem (Goering and Fry 1984).

To achieve a complete combustion of the fuel and to produce a fuel similar in chemical nature to that of the conventional diesel fuel, a procedure similar to the production of conventional diesel, namely pyrolysis was employed. This involved the conversion of long-chain saturated fatty substances to biodiesel by thermal cracking. The obtained fuel was chemically similar to that of the petroleum-derived gasoline and diesel fuel. This fuel could be used in the compression—ignition engines without much change in the engine. The major drawback in this process is that it is energy intensive, thus, proving to be economically demanding (Alonso et al. 2008; Santos et al. 2009; Saraf and Thomas 2007; Singh and Singh 2010; Srivastava and Prasad 2000).

2.2 *Transesterification*

Most important and commercially exploited process for biodiesel production is transesterification. It is the alcoholysis of lipids to form fatty acid alkyl ester (FAAE) or biodiesel as product and glycerol as a by-product. Since, the reaction usually takes more than a day for complete conversion, it is generally aided by a catalyst. The reaction of transesterification shown in steps of sequential change is as follows:

$$1\text{TriG} + 1\text{EOH} \rightarrow 1\text{DiG} + 1\text{FAEE}$$
$$1\text{DiG} + 1\text{EOH} \rightarrow 1\text{MonoG} + 1\text{FAEE}$$
$$1\text{MonoG} + 1\text{EOH} \rightarrow 1\text{GLY} + 1\text{FAEE}$$

where, TriG, DiG, and MonoG indicate tri-, di-, and mono-glyceride, respectively. FAEE is biodiesel as fatty acid ethyl ester as ethanol (EOH) is the alcohol considered. Emphasis has been laid on ethanol to demonstrate the sequence of the reaction, as ethanol can also be derived from plant sources by fermentation, making it more suitable for production of a cleaner energy source.

The reaction system of transesterification is heterogeneous, as the alcoholic polar phase and the triglyceride non-polar phase are practically immiscible at normal conditions. The rate of the reaction depends primarily on rate of interaction between these two phases. The interaction between the phases was found to increase with the help of increase in temperature (Muniyappa et al. 1996), ultrasonic irradiation (Stavarache et al. 2003), addition of a co-solvent (Boocock et al. 1996), mechanical stirring (Ma and Hanna 1999) or supercritical condition (Saka and Kusdiana 2001). An inhibition period has been reported in several cases, which might be due to the limitation in mass transfer across the two phases initially (Freedman et al. 1984; Boocock et al. 1996; Darnoko and Cheryan 2000; Noureddini and Zhu 1997).

The influence of various operational parameters has been studied variedly in many cases, where, different feed stocks, catalysts, and reaction conditions were assessed (Goff et al. 2004; Freedman et al. 1984; Canacki and Gerpen 1999; Zheng et al. 2006; Freedman et al. 1986). It has been synthesized from literature, that the reaction temperature, catalyst, and reaction time were critical for the efficiency of the process (Freedman et al. 1984; Canacki and Gerpen 1999) than other factors such as the alcohol employed, reaction vessel used, etc. The conversion of triglyceride to ester was found to be directly proportional to the reaction temperature and the molar ratio of alcohol to oil (Canacki and Gerpen 1999). Though, catalyst does not take part directly in the reaction, the presence of catalyst was found to increase the rate of ester formation to a considerable extent (Canacki and Gerpen 1999).

2.2.1 Acid-Catalysed Transesterification

A larger amount of the money required for biodiesel production goes to procuring the feedstock. As a result of this, cheap and abundantly available resources are being explored to find suitable raw materials (Haas 2005). An inexpensive feedstock without pretreatment may not be transesterified using alkali catalyst, as moisture present even in small quantities leads to soap formation (Miao et al. 2009). The presence of free fatty acid can also cause water formation through esterification reaction. The water thus formed might also inhibit the catalytic action of alkalis. In such cases, liquid catalysts such as acids are found to be less sensitive to FFA than the homogenous solid alkali catalyst (Vicente et al. 2004; Lotero et al. 2005). Therefore, for feedstock having an FFA content of greater than 0.5%, an acid-catalyzed transesterification reaction is preferred.

Marchetti et al. (2007), Srivastava and Prasad (2000), Fukuda et al. (2001) this facilitates the use of low-cost feedstock whose FFA varies from 3 to 40% for the production of biodiesel (Marchetti and Errazu 2010). A liquid acid catalyst can effectively conduct esterification and transesterification (Goff et al. 2004). But the operating conditions involve higher temperatures and pressure. Also, the reaction is considerably slow, and might take almost 24 h to complete (Lotero et al. 2005; Marchetti et al. 2007). A wide variety of acids, including sulphuric acid, hydrochloric acid, trifluoroacetic acid have been employed for transesterification as homogeneous catalyst. Catalyst is generally introduced into the reaction mixture as alkoxides, where alcohol and catalyst (acid) are mixed completely in predetermined proportions. Though, an acid-catalyzed transesterification is effective, the extreme operating conditions prove to be a serious disadvantage while extrapolating the process to an industrial scale as it turns out to be energy intensive. This might also lead to equipment corrosion (Dizge et al. 2009; da Silva et al. 2008). Also the ethoxide (ethanol + catalyst) prepared, when added to the free fatty acids, a neutralization reaction is observed immediately, wasting the feedstock (Di Serio et al. 2007). In recent times, various heterogeneous catalysts such as niobic acid, sulphated zirconia ZnO/I2, ZrO2/SO42- and other carbon-based solid acid catalyst are also being explored. These heterogeneous catalysts are recyclable and ecofriendly. On using a heterogeneous catalyst, the disadvantages of harsh operating conditions is avoided. But, low acid site concentrations, low diffusivity, high cost and low micro porosity are to be addressed (Dizge and Keskinler 2008; Di Serio et al. 2007; Kawashima et al. 2009; Lou et al. 2008).

2.2.2 Alkali-Catalysed Transesterification

Basic catalysts for transesterification include alkali hydroxides such as sodium and potassium hydroxides, because of their low cost (Freedman et al. 1984) and are used in commercial-scale production. These are usually preferred as the reaction rate is faster and harsh operating conditions of acid-catalyzed reaction can be avoided (Antolin et al. 2002; Meher et al. 2006). But, the homogenous alkali catalyst mentioned above are hygroscopic and tend to absorb moisture during storage (Leung et al. 2010). They also form water while reaction with alcohol which leads to saponification (Dizge et al. 2009; da Silva et al. 2008; Di Serio et al. 2007; Kawashima et al. 2009;

Qian et al. 2008). Catalyst of weight 0.1–1.2% of weight of oil is generally used for transesterification (Demirbas 2003).

Various parameters such as reaction time, temperature, and agitation are known to affect the yield of the reaction. The transesterification system turns out to be a three-phase system (Leung et al. 2010), when a solid alkali catalyst is used. Limitations in mass transfer between the three phases might slower the rate of the reaction. The catalyst is prepared by adding a precalculated weight of sodium hydroxide to ethanol to form ethoxide, as in an acid-catalyzed reaction. As this forms a heterogeneous mixture, overnight mixing/dissolving is allowed.

Homogeneous alkali catalyst, inspite of having its advantages of low cost, requires the pretreatment of cheap feedstock to reduce the FFA content to facilitate transesterification again adds up to the cost of the process. Other heterogeneous catalysts such as CaO, $CaTiO_3$, KOH/Al_2O_3, and zeolite have been tried to replace the conventional catalysts (Di Serio et al. 2007; Kawashima et al. 2009; Liu et al. 2008; Shimada et al. 2002; Qian et al. 2008). Though the heterogeneous catalysts are recyclable, and prove to be more ecofriendly, the problem with the FFA content of the feedstock cannot be solved by these catalysts. Also, the separation and purification of biodiesel from the product mixture is difficult and requires more water wash than that from acid-catalyzed reaction (Liu et al. 2008; Shimada et al. 2002; Qian et al. 2008). The relation between reaction temperature and yield of biodiesel is directly proportional, i.e., the yield increases proportionately as the temperature increases till 65 °C, beyond which FAEE degrades (Chitra et al. 2005). The yield increases till 90 min of the reaction, beyond which almost all fatty acids would be converted to FAME (around 96%) (Chitra et al. 2005).

2.2.3 Two-Step Acid–Alkali-Catalysed Transesterification Process

A two-step process combining the acid- and alkali-catalyzed transesterification was then employed to overcome the difficulties faced on using them separately. This process involved a metallic salt of an acid such as ferrous sulphate (salt of sulphuric acid) in the first step, when the free fatty acid in the lipid/oil reacted with methanol. Following this, a basic salt catalyst such as sodium hydroxide or potassium hydroxide was added to the reaction mixture to catalyze the reaction of triglyceride with methanol. The insoluble ferrous salt added in the first step being insoluble with oil/lipid can be recovered from the mixture by centrifugation (Wang et al. 2007) and can be reused. The addition of catalyst (around 1% of weight of oil) significantly improved the rate of conversion of the free fatty acid to triglyceride, reducing the reaction time by 50% (Wang et al. 2007). The formation of soap in the second step while using alkali catalyst was negligible in amount and was separated easily due to the difference in the viscosity of the system (Wang et al. 2007).

2.2.4 Supercritical Methanol Treatment

The liability of free fatty acid (acid value greater than 1) (Freedman et al. 1984) and water content (greater than 0.6%) (Ma and Hanna 1999) in oil/lipid has been discussed earlier in alkali transesterification. The treatment is carried out in an excessive methanol environment to shift the equilibrium of the reaction towards formation of FAME. Typically, the molar oil: methanol ratio employed in supercritical methanol treatment varies from 1:6 to 1:40 (Demirbas 2005). The supercritical methanol treatment ensures the involvement of free fatty acid and the esterification of the same (Kusdiana and Saka 2004). Also, unlike in catalyzed methods, the presence of water in oil/lipid has positive effect on transesterification of FFA. The presence of larger quantities of water (about 50% of oil) has shown a reduction in the reaction time. The large quantity of water present hydrolyses the triglycerides to fatty acids, which are further transesterified. This is a striking advantage of SCM process over the other methods of transesterification. It is also important to note that the presence of water can also result in an unfavourable shift of equilibrium by hydrolyzing the methyl ester formed. Thus, in an SCM method, three reactions take place, namely esterification of triglycerides, esterification of free fatty acids, and hydrolysis of produced methyl esters. Kusdiana et al. (2004), have reported that a supercritical methanol treatment at 350 °C, under a pressure of 43 MPa, with a 1:42 oil: methanol ratio is optimum for the transesterification of rapeseed oil to biodiesel. Thus, SCM proves to be a simpler method, having no catalyst and producing maximum yield in a shorter reaction time (Saka and Kusdiana 2001). Though these methods are comparatively simple and efficient, the usage of larger amount of chemicals such as methanol is a possible disadvantage.

2.2.5 Enzyme-Catalysed Transesterification

With the increase in environmental concern in the industrial sector, replacement of chemical feedstock, catalysts, and solvents with more ecofriendly substitutions are being critically reviewed. On an alternative approach, the catalyst employed in transesterification can be alternated by using lipase as a biocatalyst. Enzymes are extremely thermally stable in any organic solvent, thus improving its catalytic activity (Goldberg et al. 1990; Nara et al. 2002). The enzymes are also substrate specific in action due to their confrontal rigidity. Similar to the two-step transesterification process, the presence of water has a positive effect by yielding more products within a short reaction time. The presence of water was reported as critical to activate the enzyme, as water forms hydrogen bonds with the functional groups of the polymer constituting it which were earlier linked to each other, thus activating it (Goldberg et al. 1990). In recent studies, composite catalysts incorporating an enzyme as the active site and an inert substrate for the enzyme has been used as catalysts, and have yielded positive results (Leung et al. 2010). Biodiesel obtained from enzyme catalyzed reaction is found to be pure when compared to the other yield and can be easily purified (Leung et al. 2010). The lipase catalyzed transesterification is an effective environment friendly alternative for biodiesel production, but the economy of the process is far too high than the yield. Synthesis of larger quantity of lipids by industrial-scale production or extraction of lipid from other unconventional cheaper sources can help improve the feasibility of the process.

2.2.6 Transesterification of Wax (or) Saturated Lipids

While the transesterification of oils has been widely explored, the esterification of wax is being looked into in the recent times (Hussein et al. 2014). Animal lipid sources such as insects in their pupal or larval stage are mostly comprised of wax than that of oil. Conventional transesterification process catalyzed by acid or alkali catalyst would lead to formation of soap. Transesterification using metallic sodium or acetyl chloride (Canoira et al. 2006) as catalyst has been reported to produce biodiesel of same quality as the conventionally transesterified product. The process involves the preparation of methoxide catalyst by adding metallic sodium to methanol. Similarly, catalyst can be prepared by adding acetyl chloride to methanol. The reaction takes place in two steps, where the catalyst prepared is added to the wax and initially treated at a temperature of around 75 °C. Further, extra methanol is added to shift the equilibrium desirably. This reaction ensured the esterification of oil as well as wax, ensuring a complete transformation of fatty acids to biodiesel. Acetyl chloride catalyzed esterification ensured a 96% yield (Canoira et al. 2006), but the reaction time was longer than the conventional methods. In Na-metal-catalyzed method, 90% yield was obtained (Canoira et al. 2006). But, the yield obtained was in its crude form and required large amounts of petroleum ether to purify the FAME obtained.

3 Purification of Products

Completion of transesterification yields two products, namely biodiesel and glycerol. Specific gravity of glycerol (approx. 1.3) (Bosart and Snoddy 1928) is greater than that of biodiesel (0.8–0.9) (EN 2003; ASTM). This facilitates gravity settling of glycerol, initiating a separation. This separation starts within 10 min of allowing the product mixture to settle and can take more than 24 h for complete separation. Instead of gravity settling, centrifugation may be employed to separate glycerol and biodiesel (Qian et al. 2008). Centrifugation to separate FAEE from glycerol has been used extensively in laboratory-scale researches.

There is a chance of contamination of glycerol and biodiesel with unreacted catalyst, alcohol, and water. Formation of water due to alcoholysis may also contaminate the products by saponification. It is reported that the percentage of contaminants present in the glycerol phase would be more than that of the biodiesel phase (Schumacher 2007). Refining of both the products is essential before commercialization of them.

Glycerol is a valuable by-product from the biodiesel industry. It has applications in various pharmaceutical, cosmetic, and medicine industries (Wang et al. 2001, Da Silva et al. 2009). Whittington et al. (2006), reported that the glycerol obtained on transesterification may be further fermented to produce ethanol itself (Whittington 2006). The layer of glycerol settled after reaction contains about 50% glycerol only (Whittington 2006). The remaining 50% consists of unreacted alcohol, catalyst and water. Any extra alkali can be removed by neutralization. Water and alcohol present can be removed by vacuum flash process or other types of solvent evaporation processes. The recovered alcohol can be reused in the process after passing through a distillation column to remove any traces of moisture present.

4 Conclusion

The general procedure of various methods of transesterification has been enlisted in this paper. The advantages and disadvantages of each method have also been discussed. Though biodiesel obtained on transesterification is relatively a clean fuel, the process of transesterification involves a wide variety of chemical feedstock and catalysts. The scope for future result lies in imbibing greener technologies in the production process as a whole, trying to economies the process ensuring continuous production.

References

Agbor, V. B., Cicek, N., Sparling, R., Berlin, A., & Levin, D. B. (2011). Biomass pretreatment: Fundamentals toward application. *Biotechnology Advances, 29*(6), 675–685.

Alonso, J. S. J., Sastre, J. L., Romero-Avila, C., & Lopez, E. (2008). A note on the combustion of blends of diesel and soya, sunflower and rapeseed vegetable oils in a light boiler. *Biomass and Bioenergy, 32*(9), 880–886.

Antolin, G., Tinaut, F. V., Briceno, Y., Castano, V., Perez, C., & Ramirez, A. I. (2002). Optimisation of biodiesel production by sunflower oil transesterification. *Bioresource Technology, 83,* 111–114.

ASTM. ASTM D6751-biodiesel blend stock specification (B100).

Boocock, D. G., Konar, S. K., Mao, V., & Sidi, H. (1996). Fast one-phase oil-rich processes for the preparation of vegetable oil methyl esters. *Biomass and Bioenergy, 11*(1), 43–50.

Bosart, L. W., & Snoddy, A. O. (1928). Specific gravity of glycerol. *Industrial and Engineering Chemistry, 20*(12), 1377–1379.

Canacki, M., & Gerpen, J. V. (1999). Biodiesel production via acid catalysis. *Transactions of the ASAE, 42*(5), 1203–1210.

Canoira, L., Alcantara, R., García-Martínez, M. J., & Carrasco, J. (2006). Biodiesel from Jojoba oil-wax: Transesterification with methanol and properties as a fuel. *Biomass and Bioenergy, 30*(1), 76–81.

Chaturvedi, V., & Verma, P. (2013). An overview of key pretreatment processes employed for bioconversion of lignocellulosic biomass into biofuels and value added products. *3 Biotech, 3*(5), 415–431.

Chitra, P., Venkatachalam, P., & Sampathrajan, A. (2005). Optimisation of experimental conditions for biodiesel production from alkali-catalysed transesterification of *Jatropha curcus* oil. *Energy for Sustainable Development, 9*(3), 13–18.

Da Silva, G. P., Mack, M., & Contiero, J. (2009). Glycerol: A promising and abundant carbon source for industrial microbiology. *Biotechnology Advances, 27*(1), 30–39.

da Silva, R. B., Neto, A. F. L., dos Santos, L. S. S., de Oliveira Lima, J. R., Chaves, M. H., dos Santos Jr, J. R., et al. (2008). Catalysts of Cu (II) and Co (II) ions adsorbed in chitosan used in transesterification of soy bean and babassu oils—A new route for biodiesel syntheses. *Bioresource Technology, 99*(15), 6793–6798.

Darnoko, D., & Cheryan, M. (2000). Kinetics of palm oil transesterification in a batch reactor. *Journal of the American Oil Chemists' Society, 77*(12), 1263–1267.

Demirbas, A. (2003). Biodiesel fuels from vegetable oils via catalytic and non-catalytic supercritical alcohol transesterifications and other methods: A survey. *Energy Conversion and Management, 44,* 2093–2109.

Demirbas, A. (2005). Biodiesel production from vegetable oils via catalytic and non-catalytic supercritical methanol transesterification methods. *Progress in Energy and Combustion Science, 31,* 466–487.

Di Serio, M., Tesser, R., Pengmei, L., & Santacesaria, E. (2007). Heterogeneous catalysts for biodiesel production. *Energy & Fuels, 22*(1), 207–217.

Dizge, N., Aydiner, C., Imer, D. Y., Bayramoglu, M., Tanriseven, A., & Keskinler, B. (2009). Biodiesel production from sunflower, soybean, and waste cooking oils by transesterification using lipase immobilized onto a novel microporous polymer. *Bioresource Technology, 100*(6), 1983–1991.

Dizge, N., & Keskinler, B. (2008). Enzymatic production of biodiesel from canola oil using immobilized lipase. *Biomass and Bioenergy, 32*(12), 1274–1278.

EN. (2003). The EN 14214 standard-specifications and test methods.

Freedman, B., Butterfield, R. O., & Pryde, E. H. (1986). Transesterification kinetics of soybean oil 1. *Journal of the American Oil Chemists' Society, 63*(10), 1375–1380.

Freedman, B., Pyrde, E. H., & Mounts, T. L. (1984). Variables affecting the yields of fatty esters from transesterified vegetable oils. *JAOCS, 61,* 1638–1643.

Fukuda, H., Kondo, A., & Noda, H. (2001). Biodiesel fuel production by transesterification of oils. *Journal of Bioscience and Bioengineering, 92*(5), 405–416.

Goering, C. E., & Fry, B. (1984). Engine durability screening test of a diesel oil/soy oil/alcohol microemulsion fuel. *JAOCS, 61,* 1627–1632.

Goff, M. J., Bauer, N. S., Lopes, S., Sutterlin, W. R., & Suppes, G. J. (2004). Acid-catalyzed alcoholysis of soybean oil. *Journal of the American Oil Chemists' Society, 81,* 415–420.

Goldberg, M., Thomas, D., & Lecoy, M.-D. (1990). The control of lipase-catalysed transesterification and esterification reaction rates: Effects of substrate polarity, water activity and water molecules on enzyme activity. *European Journal of Biochemistry, 190,* 603–609.

Haas, M. J. (2005). Improving the economics of biodiesel production through the use of low value lipids as feedstocks: Vegetable oil soapstock. *Fuel Processing Technology, 86*(10), 1087–1096.

Hussein, L., Elsabee, M. Z., Ismail, E. A., Naguib, H. F., Aziz, H. A., & Elsawy, M. A. (2014). Transesterification of jojoba oil-wax using microwave technique. *International Journal of Environmental & Earth Sciences, 8,* 248–252.

Issariyakul, T., Kulkarni, M. G., Meher, L. C., Dalai, A. K., & Bakhshi, N. N. (2008). Biodiesel production from mixtures of canola oil and used cooking oil. *Chemical Engineering Journal, 140*(1–3), 77–85.

Kansedo, J., Lee, K. T., & Bhatia, S. (2008). Feasibility of palm oil as the feedstock for biodiesel production via heterogeneous transesterification. *Chemical Engineering & Technology: Industrial Chemistry-Plant Equipment-Process Engineering-Biotechnology, 31*(7), 993–999.

Kawashima, A., Matsubara, K., & Honda, K. (2009). Acceleration of catalytic activity of calcium oxide for biodiesel production. *Bioresource Technology, 100*(2), 696–700.

Kaya, C., Hamamci, C., Baysal, A., Akba, O., Erdogan, S., & Saydut, A. (2009). Methyl ester of peanut (*Arachis hypogea* L.) seed oil as a potential feedstock for biodiesel production. *Renewable Energy, 34*(5), 1257–1260.

Kusdiana, D., & Saka, S. (2004). Effects of water on biodiesel fuel production by supercritical methanol treatment. *Bioresource Technology, 91,* 289–295.

Leung, D. Y., Wu, X., & Leung, M. K. H. (2010). A review on biodiesel production using catalyzed transesterification. *Applied Energy, 87*(4), 1083–1095.

Lin, L., Ying, D., Chaitep, S., & Vittayapadung, S. (2009). Biodiesel production from crude rice bran oil and properties as fuel. *Applied Energy, 86*(5), 681–688.

Liu, X., He, H., Wang, Y., Zhu, S., & Piao, X. (2008). Transesterification of soybean oil to biodiesel using CaO as a solid base catalyst. *Fuel, 87*(2), 216–221.

Lotero, E., Liu, Y., Lopez, D. E., Suwannakarn, K., Bruce, D. A., Goodwin, J. G., et al. (2005). Synthesis of biodiesel via acid catalysis. *Industrial & Engineering Chemistry Research, 44,* 5353–5363.

Lou, W. Y., Zong, M. H., & Duan, Z. Q. (2008). Efficient production of biodiesel from high free fatty acid-containing waste oils using various carbohydrate-derived solid acid catalysts. *Bioresource Technology, 99*(18), 8752–8758.

Ma, F., & Hanna, M. A. (1999). Biodiesel production: A review. *Bioresource Technology, 70*(1), 1–15.

Marchetti, J. M., & Errazu, A. F. (2010). Biodiesel production from acid oils and ethanol using a solid basic resin as catalyst. *Biomass and Bioenergy, 34*(3), 272–277.

Marchetti, J. M., Miguel, V. U., & Errazu, A. F. (2007). Possible methods for biodiesel production. *Renewable and Sustainable Energy Reviews, 11,* 1300–1311.

Meher, L. C., VidyaSagar, D., & Naik, S. N. (2006). Technical aspects of biodiesel production by transesterification—A review. *Renewable and Sustainable Energy Reviews, 10,* 248–268.

Miao, X., Li, R., & Yao, H. (2009). Effective acid-catalyzed transesterification for biodiesel production. *Energy Conversion and Management, 50,* 2680–2684.

Muniyappa, P. R., Brammer, S. C., & Noureddini, H. (1996). Improved conversion of plant oils and animal fats into biodiesel and co-product. *Bioresource Technology, 56*(1), 19–24.

Nara, S. J., Harjani, J. R., & Salunkhe, M. M. (2002). Lipase-catalysed transesterification in ionic liquids and organic solvents: A comparative study. *Tetrahedron Letters, 43,* 2979–2982.

Noureddini, H., & Zhu, D. (1997). Kinetics of transesterification of soybean oil. *Journal of the American Oil Chemists' Society, 74*(11), 1457–1463.

Pendse, D. R. (2016). The energy crisis and third world options. *Third World Quarterly, 1*(4), 69–88.

Peterson, C. L., Auld, D. L., & Korus, R. A. (1983). Winter rape oil fuel for diesel engines: Recovery and utilization. *JAOCS, 60,* 1579–1587.

Pryde, E. H. (1983). Vegetable oil as diesel fuel: Overview. *JAOCS, 60,* 1557–1558.

Qian, W. W., Yang, L., & Lu, X. P. (2008). Preparation of biodiesel catalyzed by KF/CaO with ultrasonic. In *International Conference on Biomass Energy Technologies*, Guangzhou, China (Vol. 92, pp. 2061–2067).

Rao, Y. H., Voleti, R. S., Raju, A. S., & Reddy, P. N. (2009). Experimental investigations on jatropha biodiesel and additive in diesel engine. *Indian Journal of Science and Technology, 2*(4), 25–31.

Saka, S., & Kusdiana, D. (2001). Biodiesel fuel from rapeseed oil as prepared in supercritical methanol. *Fuel, 80,* 225–231.

Santos, F. F., Rodrigues, S., & Fernandes, F. A. (2009). Optimization of the production of biodiesel from soybean oil by ultrasound assisted methanolysis. *Fuel Processing Technology, 90*(2), 312–316.

Saraf, S., & Thomas, B. (2007). Influence of feedstock and process chemistry on biodiesel quality. *Process Safety and Environmental Protection, 85*(5), 360–364.

Schumacher, J. (2007). Small scale biodiesel production: an overview. *Agricultural Marketing Policy Center, Montana State University, Montana, USA, Agricultural Marketing Policy Paper,* (22).

Shimada, Y., Watanabe, Y., Sugihara, A., & Tominaga, Y. (2002). Enzymatic alcoholysis for biodiesel fuel production and application of the reaction to oil processing. *Journal of Molecular Catalysis. B, Enzymatic, 17*(3–5), 133–142.

Sims, R. E. H., Mabee, W., Saddler, J. N., & Taylor, M. (2010). An overview of second generation biofuel technologies. *Bioresource Technology, 101,* 1570–1580.

Singh, S. P., & Singh, D. (2010). Biodiesel production through the use of different sources and characterization of oils and their esters as the substitute of diesel: A review. *Renewable and Sustainable Energy Reviews, 14*(1), 200–216.

Srivastava, A., & Prasad, R. (2000). Triglycerides-based diesel fuels. *Renewable and Sustainable Energy Reviews, 4*(2), 111–133.

Stavarache, C., Vinatoru, M., Nishimura, R., & Maeda, Y. (2003). Conversion of vegetable oil to biodiesel using ultrasonic irradiation. *Chemistry Letters, 32*(8), 716–717.

Tiwari, A. K., Kumar, A., & Raheman, H. (2007). Biodiesel production from jatropha oil (*Jatropha curcas*) with high free fatty acids: an optimized process. *Biomass and Bioenergy, 31*(8), 569–575.

Vicente, G., Martinez, M., & Aracil, J. (2004). Integrated biodiesel production: A comparison of different homogeneous catalysts systems. *Bioresource Technology, 92,* 297–305.

Wang, Y., Ou, S., Liu, P., & Zhang, Z. (2007). Preparation of biodiesel from waste cooking oil via two-step catalyzed process. *Energy Conversion and Management, 48,* 184–188. (2006).

Wang, Z., Zhuge, J., Fang, H., & Prior, B. A. (2001). Glycerol production by microbial fermentation: A review. *Biotechnology Advances, 19*(3), 201–223.

Whittington, T. (2006). *Biodiesel production and use by farmers: Is it worth considering*. Western Australia: Department of Agriculture and Food.

Zheng, S., Kates, M., Dubé, M. A., & McLean, D. D. (2006). Acid-catalyzed production of biodiesel from waste frying oil. *Biomass and Bioenergy, 30*(3), 267–272.

Ziejewski, M., Kaufman, K. R., Schwab, A. W., & Pryde, E. H. (1984). Diesel engine evaluation of a nonionic sunflower oil-aqueousethanol microemulsion. *JAOCS, 61,* 1620–1626.

Effect of Compression Ratio on Combustion, Performance and Emission Characteristics of DI Diesel Engine with Orange Oil Methyl Ester

V. Karthickeyan, B. Dhinesh and P. Balamurugan

Abstract Biodiesel was used as an effective alternative to the fossil fuel, as it was non-toxic, renewable and biodegradable resource. Engine parameters like compression ratio, injection timing and injection pressure play a key role in the combustion of biodiesel. This research work presents the effect of compression ratio on the combustion, performance and emission characteristics of 80% diesel blended with 20% orange oil methyl ester (20OME) in diesel engine. Experiments were conducted on a single-cylinder direct injection (DI) diesel engine with compression ratios of 17:1 and 18:1. Diesel exhibited higher Brake Thermal Efficiency (BTE) than 20OME at both compression ratios. At all loads, an increased Brake Specific Energy Consumption (BSEC) was observed for 20OME in both compression ratios. Due to the presence of more oxygen content in the prepared biodiesel sample, 20OME showed higher heat release rate than diesel at compression ratio 18. Consequently, 20OME showed higher rate of pressure than diesel at CR 18. 20OME showed higher Exhaust Gas Temperature (EGT) than diesel at all loads. At CR 18, both diesel and 20OME showed similar trend of HC and CO emissions whereas in CR 17, diesel exhibited lower CO and HC emissions than 20OME. At low loads, 20OME showed lower NO_x emission than diesel in compression ratio 17 whereas in CR 18, both fuels exhibited higher NO_x emission at all loads. Lower smoke emission was observed with 20OME than diesel at both compression ratios.

Keywords Diesel engine · Variable compression ratio · Biodiesel · Combustion · Performance and emission

---

V. Karthickeyan · P. Balamurugan
Department of Mechanical Engineering, Sri Krishna College of Engineering and Technology, Coimbatore, India

B. Dhinesh (✉)
Department of Mechanical Engineering, Mepco Schlenk Engineering College, Sivakasi, Tamil Nadu, India
e-mail: dhineshbala91@gmail.com

S. K. Ghosh et al. (eds.), *Bioresource Utilization and Bioprocess*,
https://doi.org/10.1007/978-981-15-1607-8_14

1 Introduction

The demand for energy has been increasing in the world due to depletion of fossil energy resources (Karthickeyan 2017, 2018a, b, 2019). The continuous usage of renewable energy resources also causes a big threat to the future generation. In the transportation sector, fossil fuels play a vital role in driving the entire system. This made many researchers to turn on modification of fuel, using naturally derived non-edible seeds. Oils obtained from the non-edible seeds can be used as an alternative fuel in diesel engine, after undergoing a chemical process (Karthickeyan et al. 2017). Diesel engines were used widely from locomotive transport to domestic transport because of its high thermal efficiency and lower costs (Gautam and Kumar 2015). Many researchers have carried out investigations on alternative fuels derived from natural resources like *Karanja*, *Polanga*, *Mahua*, *Rubber* seed, *Cotton*seed, *Croton megalocarpus* (Osawa et al. 2016), *Jojoba*, *Tobacco*, *Neem* (Awolu and Layokun 2013), *Jatropha*, etc. (Ashraful et al. 2014). After chemical treatment, the oils from non-edible seeds have been used as a biodiesel in diesel engine to obtain better performance with lower emission than diesel.

The methods used for production of biodiesel were (1) Transesterification, (2) Micro-emulsion, (3) Pyrolysis and (4) Preheating. Among all, transesterification was widely used for biodiesel production. This process was carried out in the presence of catalyst like KOH/NaOH and allowed to react with alcohols like ethanol/methanol. The last derived component after chemical processes was called as ethyl ester or methyl ester, respectively. The obtained biodiesel was allowed to study the performance and emission characteristics of a single-cylinder four-stroke diesel engine. Indian Government issued an order regarding blending of 20% ethanol with petroleum to drive the vehicle without any engine modifications in January 2006 (Murugesan et al. 2009). Through the blending process, the ethyl or methyl esters can be blended with diesel on volumetric basis under standard atmospheric condition (Subramaniam et al. 2013). The direct use of biodiesel in diesel engine was not advisable because of its high viscosity and low calorific value, which may lead to gum deposition in engine valves, incomplete combustion of fuel mixture (Karthickeyan et al. 2016a). As per ASME standard for diesel fuel, kinematic viscosity was expected to be in the range of 1.9–4.1 (cSt) at 40 °C and cetane number must be 40–55, to run the diesel engine (Tudu et al. 2016). Even though the calorific value of the fuel may be reduced in the blending process, different combustion promoters were used to improve it. Vedharaj et al. used *Kapok* biodiesel with 1,4-dioxane to improve cetane number (Vedharaj et al. 2014). 10 ml of 1,4-dioxane was found to improve the cetane number from 52 to 56. On the other hand, antioxidant additives were used to reduce engine exhaust emissions like NO_x and HC. Narinder Singh et al. investigated the effects of carbon nanotube-emulsified fuel in a single-cylinder water-cooled four-stroke diesel engine (Singh and Bharj 2015). They found that there was an increase in evaporation rate of CNT-emulsified fuel with early ignition and short ignition delay. They observed decline of 48% in NO_x, 9% in HC and 30% in smoke opacity. The use of CNT-emulsified fuel reduced the NO_x emission due to its heat sink effect. They observed decline in CO emission, exhaust gas temperature and increase in CO_2 emission with the increase in CNT concentration in the fuel. Senthil and Silambarasan used *Annona* methyl ester with diesel as fuel in the presence of L-ascorbic acid as an antioxidant and observed 23.38% reduction in NO_x and 29.71% reduction in HC emission at full load (Senthil and Silambarasan 2015). The effort of many researchers towards biodiesel production and utilization in diesel engine was taken up successfully. Many of them attempted to vary the injection pressure, injection timing and bowl geometry in the engine to obtain better performance and emission characteristics with diesel and biodiesel samples. But scanty works were done in the area of variable compression engine mode.

Muralidharan and Vasudevan used single-cylinder four-stroke variable compression ratio diesel engine to analyse the characteristics of waste cooking oil methyl ester and diesel (Muralidharan and Vasudevan 2011). They varied the compression ratio from 18 to 22 with various blended samples. They concluded that the B40 sample with compression ratio of 21 showed an increased BTE with reduced SFC at increasing loads. At full load, maximum mechanical efficiency with B40 was observed. The results showed that the B40 blend exhibited maximum combustion pressure and decreased heat release rate at the beginning of combustion. Amarnath and Prabhakaran investigated the characteristics of the *Karanja* biodiesel samples in VCR engine with compression ratio of 14–18 with different injection pressures of 150, 200 and 250 bar. B95 sample with 18 compression ratio and 250 bars was found to give optimum result (Amarnath and Prabhakaran 2012). Biswajit and Panua used *Jatropha curcas* in VCR engine with 16, 17 and 18 compression ratio and found that B30 sample with 18 compression ratio operations showed a better result than other samples tested (De and Panua 2014). It was also reported from the literature that very few research works were performed in the area of orange oil biodiesel. Some of the research works carried out in this area were reported as follows: Purushothaman and Nagarajan used 300 g of orange skin powder in diesel fuel and varying the injection pressure from 215, 235 and 255 bar, respectively (Purushothaman and Nagarajan 2009a). They concluded that the engine with 235 bar injection pressure showed better results than other injection pressure. Purushothaman and Nagarajan also considered the blend of orange oil (30%) with diesel and same sample with 36 mg/s of diethyl ether (DEE) in diesel engine (Purushothaman and Nagarajan 2009b). The use of DEE with orange oil blend showed better performance and reduced emission characteristics. The above authors also investigated orange oil methyl ester B100 in diesel engine and concluded that except NO_x emission, all the other were lower than those of diesel fuel. Umer Rashid used mandarin orange (*Citrus reticulata*) seed for biodiesel production (Rashid et al. 2013). They used sodium methoxide as a catalyst with methanol and used traditional transesterification method for sample preparation. The obtained results were within the limits of ASTM (D6751) standards. Gokhan Tuccar used *Citrus sinensis* as biodiesel samples (B5, B10, B20) and tested in Mitsubishi Canter four-cylinder direct injection diesel engine (Tüccar et al. 2014). They concluded *Citrus sinensis* can be used as an additive in diesel engine.

Based on the above literature, it was observed that scanty works were carried out in orange oil biodiesel with variable compression ratio. In this present study, different compression ratios were used with orange oil biodiesel sample. The main objectives of the study were as follows:

- To prepare orange oil biodiesel using transesterification process and to study the chemical and physical properties of orange oil biodiesel.
- To investigate the combustion, performance and emission characteristics of B20 sample (20% orange oil methyl ester and 80% diesel) in two different compression ratios (17:1 and 18:1) and to compare with diesel fuel.

2 Materials and Methods

2.1 Production of Orange Oil

Orange oil can be derived from the natural fruit orange (*Citrus sinensis*). The waste products like orange peels and seeds were commonly used for oil production process. The orange oil was produced by three methods, namely (1) Juice extraction, (2) Steam distillation and (3) Cold press (Purushothaman and Nagarajan 2009c). In the local market, three grades of orange oil were available (1) Raw orange oil, (2) Threefold distillation process and (3) Fivefold distillation process (Fold refers to the purity of obtained solution). The raw orange oil was found cheaper in cost when compared to other oils available in the market.

2.2 Preparation of Orange Oil Methyl Ester

The raw orange oil is purchased from the local market located in Coimbatore, Tamil Nadu (Karthickeyan et al. 2016b). To remove the water content, 700 ml of orange oil was heated to about 55–60 °C for a period of 45–60 min. This process was commonly termed as preheating. To prepare sodium methoxide solution, 4 g of sodium hydroxide was used as a catalyst and 200 ml of methanol was used as an alcohol. Both were mixed and stirred continuously for a period of 30–45 min without any heating process. The mixer was connected to a reflux condenser to regain the evaporated methanol. The mixed solution was called as sodium methoxide, and it was a homogeneous mixture. The prepared sodium methoxide solution was mixed with preheated orange oil for a period of 60 min at a temperature range of 60–65 °C. The mixed solution was stirred continuously using magnetic stirrer at a speed of 600–650 rpm. The obtained solution was allowed to settle in a separation funnel. After 24 h, two distinct layers were identified, in which crude methyl ester rises to the top and crude glycerol settle at the bottom. To increase the purity of methyl ester, water washing technique was used. A known quantity of warm distilled water was mixed with obtained crude methyl ester in the separation funnel and mixed vigorously. After

24 h, impurities and water settled down at the bottom and pure methyl ester rises up. This process was repeated thrice to improve the purity of methyl ester. Further, the obtained solution was heated to about 60 °C to remove the traces of water molecules. The solution obtained at the end of heating process was termed as an orange oil methyl ester (OME). Figure 1 represents the production of orange oil methyl ester.

In the present study, 20OME (80% diesel and 20% orange oil methyl ester) was used as biodiesel fuel in diesel engine. The property of mineral diesel and 20 OME samples were investigated under laboratory condition. The obtained results were presented in Table 1 based on ASTM standards. Table 2 shows the presence of fatty acids in the prepared biodiesel sample. Lipid numbers indicates the ratio of number of carbon atoms to number of double bonds. Orange oil methyl ester contains free fatty

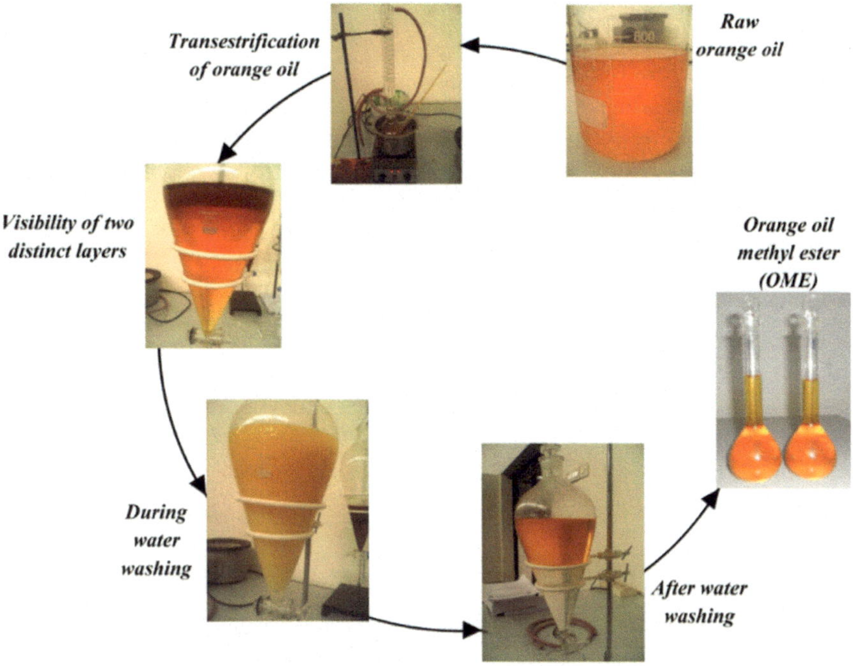

Fig. 1 Production of orange oil methyl ester

Table 1 Properties of prepared biodiesel fuel in comparison with diesel

Property	ASTM Standards	Diesel	20OME
Calorific value (MJ/kg)[a]	ASTM D 240	43.2	40
Kinematic viscosity @ 40 °C (cSt)[a]	ASTM D 445	3.9	3.53
Density (kg/m^3)[a]	ASTM D 1298	823.1	855.6
Flash Point (°C)[a]	ASTM D 93	56	66

[a] All properties were identified based on ASTM standards under laboratory condition

Table 2 Fatty acid composition in OME

Chemical name	Fatty acid	Formula	Lipid number
Octadecadienoic acid	Linoleic acid	$C_{19}H_{34}O_2$	$C_{18:0}$
Hexadeconoic acid	Palmitic acid	$C_{18}H_{36}O_2$	$C_{18:0}$
Tetradecanoic acid	Myristic acid	$C_{15}H_{30}O_2$	$C_{14:0}$

acids namely palmitic acid, myristic acid and linoleic acid. The above mentioned fatty acids composition of the orange oil methyl ester makes it to be a sufficient renewable energy source for the biodiesel production.

2.3 *Experimentation*

Figure 2 shows the schematic diagram of the experimental set-up. The experimentation consists of a single-cylinder multi-fuel four-stroke direct injections (DI) diesel

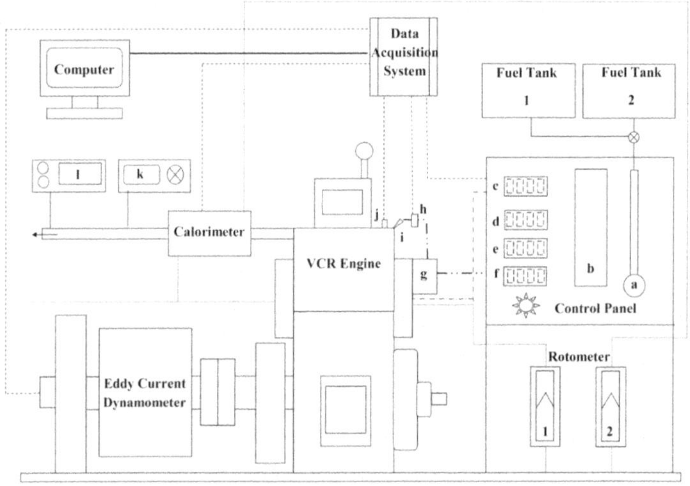

Control panel- a. Fuel measuring burette, b. U- tube manometer, c. Air box with sensors, d. Temperature indicator, e. Speed indicator, f. Load indicator with load adjustor. VCR Engine- g. Fuel filter and pump, h- Fuel line pressure sensor, i- Fuel injector, j- Combustion analyser. Rotometer- 1. Engine water supply, 2. Calorimeter water supply. Engine exhaust line- k. AVL Five gas analyser, l. AVL Smokemeter.

Water line for engine and calorimeter
Electrical line from various sensors to DAQ
Air flow line for engine
Diesel fuel line to engine
Main interface line between DAQ and Computer

Fig. 2 Schematic of the experimental set-up

engine coupled with eddy current dynamometer. Two numbers of piezoelectric sensors were used in the experimentation; one was mounted on the cylinder head for analysing the combustion characteristics, and another was installed in the fuel line to analyse the fuel line pressure. Crank angle encoders were used to analyse the combustion pressure at different crank angles. The piezoelectric sensor was connected to a charge amplifier for pressure measurement. The shaft of the engine was connected to an eddy current dynamometer for loading purpose. The load to the engine was monitored and measured by strain gauge type load cell. Airflow measurements were measured by mass airflow sensor. Fuel flow measurements were noted by graduated burette, and it was connected to a fuel flow sensor. Time taken for fuel consumption was measured using stopwatch. Nearly, six different temperature measurements were noted during engine operation. The salient temperature measurement points were (1) Engine water inlet, (2) Engine water outlet, (3) Engine exhaust before calorimeter, (4) Calorimeter water inlet, (5) Calorimeter water outlet and (6) Engine exhaust after calorimeter. For measuring all the above temperatures, type K thermocouples were used. Heat release rate was measured using calorimeter set-up with simple shell and tube heat exchanger. It was connected directly to the engine exhaust tailpipe. The online data acquisition was done by EngineSoft software with the help of NI data acquisition module and various sensors installed in the engine. Table 3 shows the specification of test engine. AVL 444 Di gas analyser was used to measure the levels of CO, CO_2, HC, O_2 and NO_x emission in the exhaust gas. It works on the principle of NDIR (Non-dispersive infrared) technique for measurement of CO, HC and chemiluminescent for measurement of NO_x. In the prob of analyser, two primary filter elements were placed to prevent the machine from particulate matters and water vapour deposition. AVL 437C free acceleration test equipment was used to measure the opacity of exhaust smoke in Hartridge Smoke Unit (HSU). Figure 3 shows the arrangement of complete experimental set-up.

Table 3 Specification of VCR test engine

Make	Kirloskar
Model	TV 1
Type	Four stroke, compression ignition, water cooled
No. of cylinders	One
Bore	87.5 mm
Stroke	110 mm
Compression ratio	18–15
Rated power	5.2 kW
Rated speed	1500 rpm
Injector nozzle	3 hole
Injection pressure	210 bar
Start of injection	23° BTDC
Dynamometer	Eddy current

Fig. 3 Experimental set-up with DAQ system

2.4 Experimental Procedure

To attain the steady state condition, the engine was started with diesel fuel and observed a rise in engine water outlet temperature to 50 °C (Dhinesh et al. 2017). The processes were carried out in ambient condition to make the system more reliable. During the steady state condition, 50% of load was given to the engine, and the operation prolonged for 30 min (Annamalai et al. 2016). Then, the experiment was started with diesel fuel to record baseline reading. For each load, the engine was allowed to run for 10 min to attain steady state condition, and the last 4 min was used for data recording purpose (Subramani et al. 2018; Dhinesh and Annamalai 2018; Dhinesh et al. 2018). Gas analyser and smoke meter were also used for all loads to analyse the exhaust emissions. All the readings were stored in personal computer (PC) for future data processing and validation process. Then, the compression ratio was varied by adjusting the screw nut provided. After completing the test, diesel fuel was drained out from tank and fuel supply line completely and refilled with prepared sample in fuel tank. Then, the engine was started and allowed to run for 10 min to eliminate the presence of previous diesel fuel during experimentation. The same procedure was repeated for recording the results for the alternative fuel.

3 Results and Discussion

3.1 Combustion Analysis

The aspects such as bowl geometry, injection pressure, operating conditions and fuel properties lead to the pressure variations within the combustion chamber (Vallinayagam et al. 2013). At compression ratio 18, in-cylinder pressure was observed as 71.24 bar and 69.84 bar for diesel and 20OME, respectively. At compression 17, in-cylinder pressure was noticed as 60.3 bar and 58.61 bar for diesel and 20OME, respectively. The ignition occurrence was recorded at 7 °CA, 1 °CA, 8 °CA and 5 °CA for diesel CR 18, 20OME CR 18, diesel CR 17 and 20OME CR 17, respectively. Figure 4 shows the variations of in-cylinder pressure curve for diesel and 20OME with CR 18 and 17 at full load. At both compression ratios, low ignition occurrence was observed with 20OME. The low ignition occurrence was owing to the additional fuel burnt during premixing or uncontrolled combustion of the fuel.

The heat release rate was observed as 58.43 J/°CA, 56.66 J/°CA, 62.39 J/°CA and 56.47 J/°CA for diesel CR18, diesel CR 17, 20OME CR 18 and 20OME CR 17, respectively. It was found that the 20OME sample showed higher heat release rate than diesel at compression ratio 18. This rise in heat release rate was due to the presence of more oxygen content in the prepared biodiesel sample. Figure 5 shows the variation of heat release rate for diesel and 20OME at CR 17 and 18 in full load. The formula used to compute heat release rate was shown as follows,

$$\frac{\partial Q_n}{\partial \theta} = \frac{\gamma}{\gamma - 1} p \frac{\partial V}{\partial \theta} + \frac{1}{\gamma - 1} V + \frac{\partial p}{\partial \theta}$$

where, $\gamma = C_p/C_v$—Ratio of specific heats

Fig. 4 In-cylinder pressure versus crank angle for diesel and 20OME at CR 17 and 18

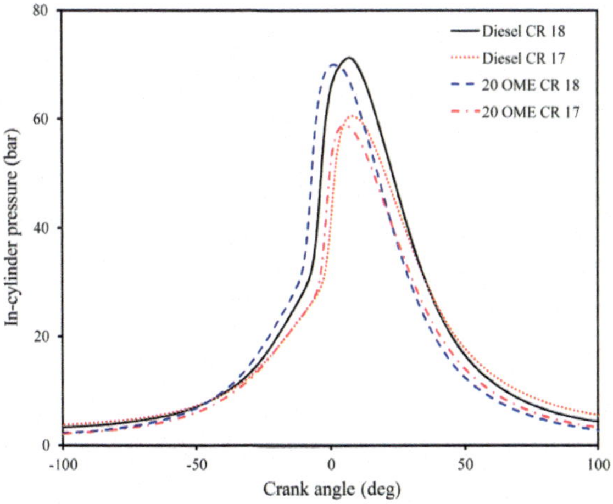

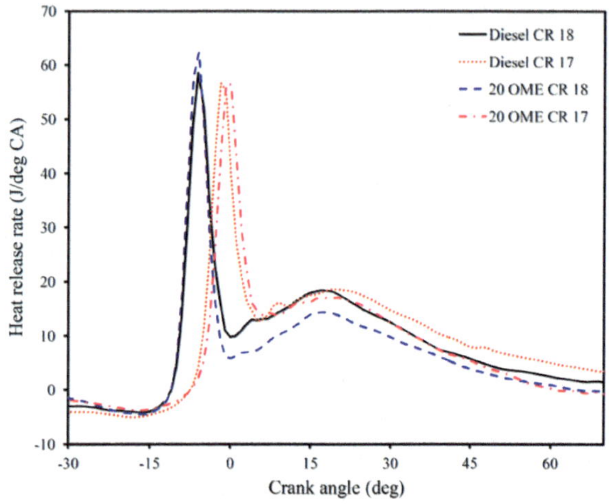

Fig. 5 Heat release rate versus crank angle for diesel and 20OME at CR 17 and 18

$$\frac{\partial Q_n}{\partial \theta} = \text{heat release rate in} \left(\frac{\text{kJ}}{\text{m}^3} \right)$$

$p =$ cylinder pressure in bar
$V =$ cylinder volume in m^3

The maximum rate of pressure rise was noticed as 6.83 bar/°CA, 6.05 bar/°CA, 7 bar/°CA and 6.02 bar/°CA for diesel CR18, diesel CR17, 20OME CR18 and 20OME CR17, respectively. At compression ratio 18, 20OME blend showed higher rate of pressure rise than diesel. Due to high heat release rate, a rise in the rate of pressure was observed during premixed combustion (Sakthivel et al. 2014; Barik and Murugan 2015; Hariharan et al. 2013). If the cylinder volume increased, high rate of pressure rise was observed. Figure 6 shows the variations of rate of pressure rise for diesel and 20OME at CR 17 and 18 at full load.

3.2 *Performance Analysis*

At compression ratio 18, BTE was observed as 31.8% and 30.44% for diesel and 20OME, respectively. At full load, BTE was recorded as 29.61% and 29.29% for diesel and 20OME, respectively, with compression ratio 17. From the results, it was evident that the diesel fuel showed higher BTE than 20OME at both compression ratios. Figure 7 shows the variation of Brake Thermal Efficiency (BTE) for diesel and 20OME at CR 17 and 18. Decrease in BTE for biodiesel sample was due to its low calorific value and improper air–fuel mixture (Wamankar and Murugan 2015).

In the present work, two fuels, namely diesel and orange oil methyl ester (OME), were blended. Each fuel has different density and calorific value which did not provide a consistent factor (Sharma and Murugan 2015; Mani et al. 2009). At compression

Fig. 6 Rate of pressure rise versus crank angle for diesel and 20OME at CR 17 and 18

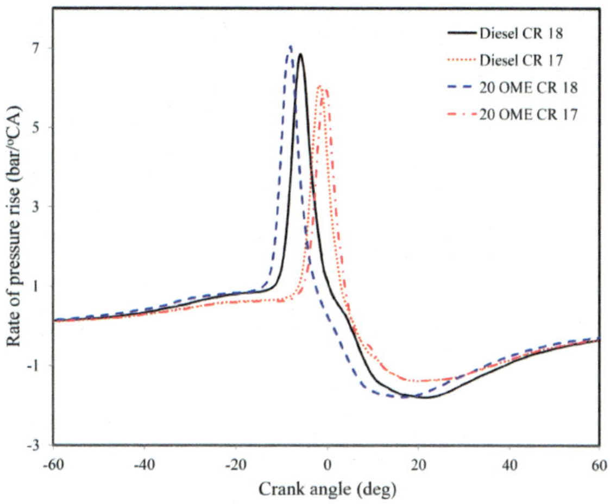

Fig. 7 Effect of compression ratio on BTE for diesel and 20OME

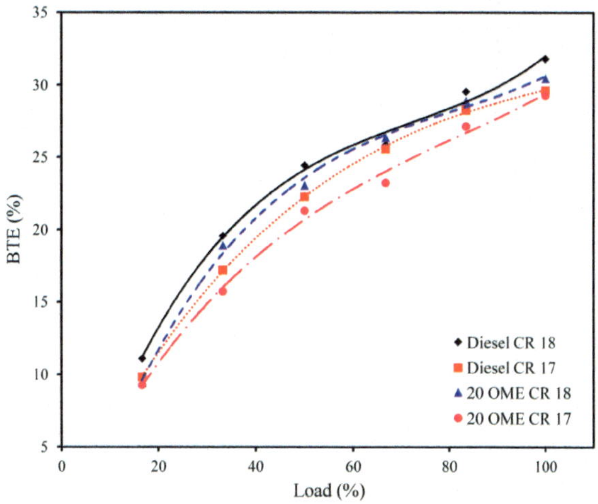

ratio 18, BSEC varies from 33.11 MJ/kWh at low load to 11.61 MJ/kWh at full load for diesel fuel and 37.41 MJ/kWh at low load to 12.47 MJ/kWh at full load for diesel fuel with compression ratio 17. At compression ratio 18, BSEC varies from 38.02 MJ/kWh at low load to 11.98 MJ/kWh at full load for 20OME blend and 38.85 MJ/kWh at low load to 12.39 MJ/kWh at full load for 20OME blend with compression ratio 17. At both compression ratios, 20OME showed higher BSEC than diesel. The rise in BSEC was due to the lower heating value of 20OME than diesel fuel (Murugan et al. 2008). Figure 8 shows the variation of Brake Specific Energy Consumption (BSEC) for diesel and 20OME at CR 17 and 18.

EGT varies from 147.11 to 278.72 °C for diesel with compression ratio 18 and from 169.75 to 298.38 °C with compression ratio 17. The EGT of the prepared 20OME blend varies from 153.9 to 272.31 °C for compression ratio 18 and 181.03

Fig. 8 Effect of compression ratio on BSEC for diesel and 20OME

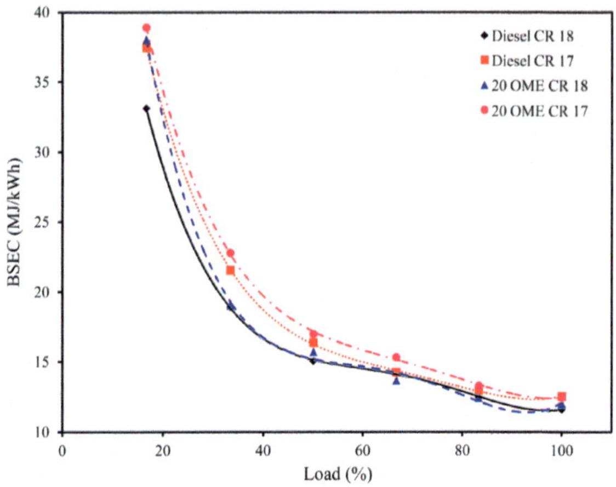

Fig. 9 Effect of compression ratio on EGT for diesel and 20OME

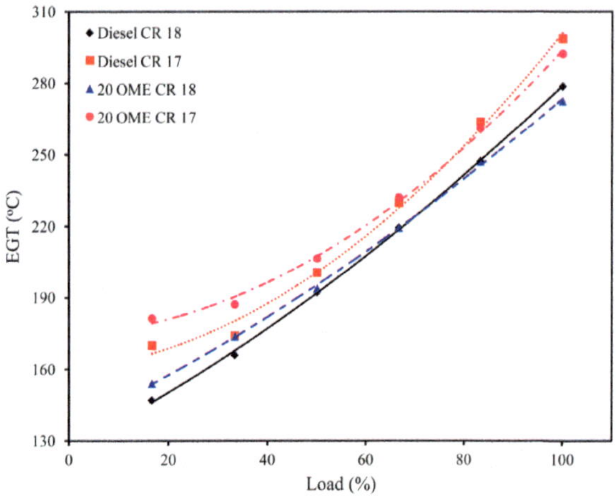

to 291.94 °C for compression ratio 17. Figure 9 shows the variation of exhaust gas temperature (EGT) for diesel and 20OME at CR 17 and 18. At all loads, 20OME exhibited higher EGT than diesel. This increase in EGT was due to the advanced fuel injection (Subbaiah and Gopal 2013) and extended or delayed combustion (Deepanraj et al. 2015).

3.3 *Emission Analysis*

Figure 10 shows the variation of carbon monoxide (CO) emission for diesel and 20OME at CR 17 and 18. CO emission from diesel engine was mainly due to incomplete combustion of fuel. CO emission for diesel with 18 compression ratio varies

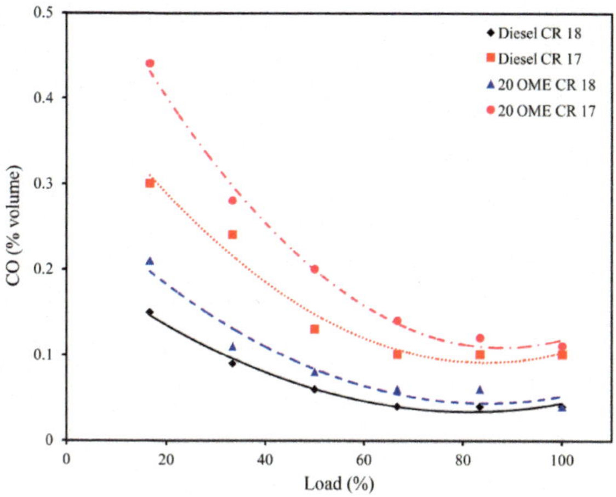

Fig. 10 Effect of compression ratio on CO for diesel and 20OME

from 0.15 to 0.04% and with 17 compression ratio varies from 0.3 to 0.1%. CO emission for prepared 20OME blend with 18 compression ratio varies from 0.21 to 0.04% and with 17 compression ratio varies from 0.44 to 0.11%. At both compression ratios, drastic increment of CO emission at low load and marginal increment of CO emission at high load were observed. Similar results were observed with an increment in CO emission was due to lean air–fuel mixture during the combustion process, which was not enough to provide complete combustion (Karthikeyan and Mahalakshmi 2007).

Figure 11 shows the variation of hydrocarbon (HC) emission for diesel and 20OME at CR 17 and 18. HC emission from diesel engine was due to engine operating condition, working fuel property and fuel spray characteristics (Rizwanul Fattah et al. 2014). HC emission for diesel with compression ratio 18 varies from 38 to 23 ppm and with compression ratio 17 varies from 63 to 39 ppm. For prepared

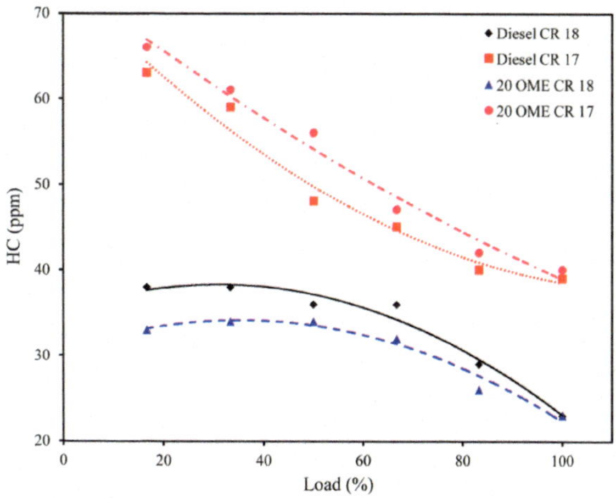

Fig. 11 Effect of compression ratio on HC for diesel and 20OME

20OME blend with the compression ratio 18 varies from 33 to 23 ppm and with compression ratio 17 varies from 66 to 40 ppm. From the results, it was evident that the biodiesel sample 20OME showed low HC emission due to good fuel spray quality.

Figure 12 shows the variation of NO_x emission for diesel and 20OME at CR 17 and 18. NO_x emission for diesel with compression ratio 18 varies from 272 to 1431 ppm and with compression ratio 17 varies from 76 to 1024 ppm. For prepared 20OME blend with compression ratio 18, it varies from 177 to 1481 ppm and with compression ratio 17 varies from 42 to 920 ppm. At compression ratio 17, both diesel and 20OME showed low NO_x emissions at low loads. Conversely, diesel and 20OME showed high NO_x emissions with compression ratio 18 at all loads. NO_x emission highly depends on the following factors such as compression ratio, temperature, in-cylinder pressure and oxygen availability in the fuel (Tudu et al. 2016; Ganesan 2012). Compression ratio plays a vital role in the combustion in-cylinder pressure and temperature (Senthil et al. 2015).

Smoke was nothing but solid soot particles suspended in the engine exhaust gas (Mani et al. 2009). Generally, smoke occurs in the diesel engine due to incomplete combustion of the fuel and partial burning of the carbon content in the liquid fuel during the combustion process (Heywood 1988). Smoke emission for diesel with compression ratio 18 varies from 4.3 HSU to 38.3 HSU and with compression ratio 17 varies from 6.7 HSU to 44.4 HSU. For prepared 20OME blend with compression ratio 18 varies from 3.3 HSU to 38.1 HSU and with compression ratio 17 varies from 6.9 HSU to 41.1 HSU, respectively. At compression ratio 18, both diesel and 20OME showed low smoke emission. High combustion temperature of the both fuels enables attaining better fuel atomization. Figure 13 shows the variation of smoke emission for diesel and 20OME at CR 17 and CR 18.

Fig. 12 Effect of compression ratio on NO_x for diesel and 20OME

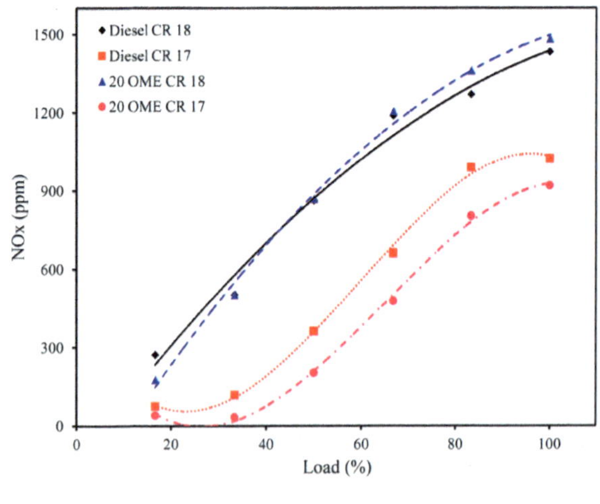

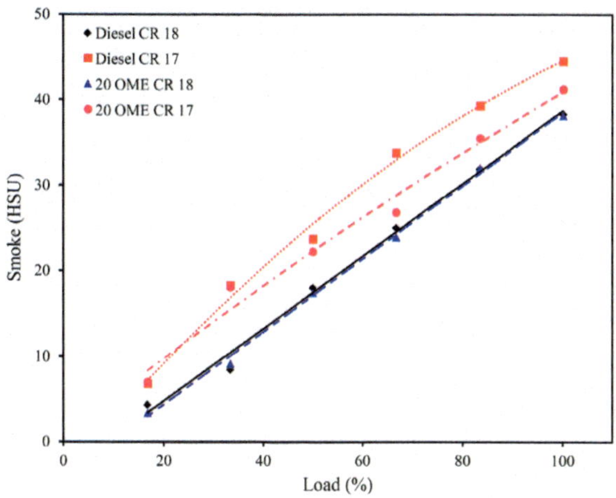

Fig. 13 Effect of compression ratio on Smoke for diesel and 20OME

4 Conclusion

Orange oil methyl ester was prepared from raw orange oil using transesterification process. The prepared sample was blended with neat diesel fuel in the ratio of 20:80 (20% orange oil methyl ester and 80% diesel fuel) which is termed as 20OME. Diesel and 20OME fuel were investigated in two different compression ratios, namely 17:1 and 18:1. A single-cylinder four-stroke direct injection diesel engine with eddy current dynamometer and data acquisition system was used for experimentation. The variation in performance and emission characteristics at full load was tabulated and shown in Table 4. The following conclusions were arrived,

- At compression ratio 18, diesel records peak pressure of 71.24 bar. At compression ratio 17, diesel records peak pressure of 60.3 bar. In both compression ratios, higher in-cylinder pressure was observed with diesel than 20OME which was due to bowl geometry, injection pressure, operating condition and fuel property.

Table 4 Variation of performance and emission characteristics at full load

Characteristics at full load	CR 18		CR 17	
	Diesel	20OME	Diesel	20OME
BTE (%)	1.36 (↑)		0.32 (↑)	
BSEC (MJ/kWh)	0.3757 (↓)			0.071 (↓)
EGT (°C)	6.41 (↑)		6.44 (↑)	
CO (% volume)	–	–	0.01 (↓)	
HC (ppm)	–	–	1 (↓)	
NO_x (ppm)	50 (↓)			104 (↓)
Smoke (HSU)		0.2 (↓)		3.3 (↓)

(↑) increase, (↓) decrease, – no variation observed

- The heat release rate for 20OME CR 18 and 20OME CR 17 was 62.39 J/°CA and 56.47 J/°CA, respectively. This increase in heat release rate at CR 18 was due to the presence of more oxygen content in the prepared biodiesel sample.
- The rate of pressure rise observed for 20OME CR18 and 20OME CR17 was 7 bar/°CA and 6.02 bar/°CA, respectively. Due to high heat release rate, rise in rate of pressure was observed with 20OME CR18 in premixed combustion.
- At both compression ratios, diesel showed higher BTE than 20OME. Decrease in BTE was due to low calorific value and improper air–fuel mixture.
- At all loads, 20OME exhibited higher BSEC than diesel for both compression ratios. This increase in BSEC was due to lower heating value of 20OME than diesel.
- With both compression ratios, 20OME showed higher EGT than diesel at all loads. This rise in exhaust temperature was due to advanced fuel injection and extended or delayed combustion.
- At compression ratio 18, both diesel and 20OME showed similar trend of HC and CO emissions, whereas in compression ratio 17, diesel exhibited lower CO and HC emissions than 20OME.
- At low loads, 20OME showed lower NO_x emission than diesel in compression ratio 17, whereas in CR 18, both fuels exhibited maximum NO_x emission at all loads. This decrease in NO_x emission at CR 17 was due to high viscosity of 20OME.
- Lower smoke emission was observed with 20OME than diesel at both compression ratios. High viscosity of fuel helps in better oxidation than diesel fuel.

From the observations, it can be concluded that 20% orange oil methyl ester with diesel fuel can be a promising alternative fuel in diesel engine without any engine modification.

Acknowledgements The authors would like to express their thanks to University Grants Commission-South Eastern Regional Office, Hyderabad, India, for financial support through Minor research project for teachers with grant number 4-4/2013-14 (MRP-SEM/UGC-SERO).

References

Amarnath, H. K., & Prabhakaran, P. (2012). A study on the thermal performance and emissions of a variable compression ratio diesel engine fuelled with Karanja biodiesel and the optimization of parameters based on experimental data. *International Journal of Green Energy, 9,* 841–863. https://doi.org/10.1080/15435075.2011.647167.

Annamalai, M., Dhinesh, B., Nanthagopal, K., Sivarama Krishnan, P., Lalvani, J. I. J., Parthasarathy, M., et al. (2016). An assessment on performance, combustion and emission behavior of a diesel engine powered by ceria nanoparticle blended emulsified biofuel. *Energy Conversion and Management, 123,* 372–380.

Ashraful, A. M., Masjuki, H. H., Kalam, M. A., Rizwanul Fattah, I. M., Imtenan, S., Shahir, S. A., et al. (2014). Production and comparison of fuel properties, engine performance, and emission characteristics of biodiesel from various non-edible vegetable oils: A review. *Energy Conversion and Management, 80,* 202–228. https://doi.org/10.1016/j.enconman.2014.01.037.

Awolu, O. O., & Layokun, S. K. (2013). Optimization of two-step transesterification production of biodiesel from neem (*Azadirachta indica*) oil. *Journal of Energy and Environmental Engineering, 4,* 39. https://doi.org/10.1186/2251-6832-4-39.

Barik, D., & Murugan, S. (2015). Effects of diethyl ether (DEE) injection on combustion performance and emission characteristics of Karanja methyl ester (KME)–biogas fueled dual fuel diesel engine. *Fuel.* https://doi.org/10.1016/j.fuel.2015.09.094.

De, B., & Panua, R. S. (2014). An experimental study on performance and emission characteristics of vegetable oil blends with diesel in a direct injection variable compression ignition engine. *The Journal of the Brazilian Society of Mechanical Sciences and Engineering, 90,* 431–438. https://doi.org/10.1016/j.proeng.2014.11.873.

Deepanraj, B., Lawrence, P., Sivashankar, R., & Sivasubramanian, V. (2015). Analysis of pre-heated crude palm oil, palm oil methyl ester and its blends as fuel in a diesel engine. *International Journal of Ambient Energy,* 1–6. https://doi.org/10.1080/01430750.2015.1004106.

Dhinesh, B., & Annamalai, M. (2018). A study on performance, combustion and emission behaviour of diesel engine powered by novel nano nerium oleander biofuel. *Journal of Cleaner Production, 196,* 74–83.

Dhinesh, B., Bharathi, R. N., Lalvani, J. I. J., Parthasarathy, M., & Annamalai, K. (2017). An experimental analysis on the influence of fuel borne additives on the single cylinder diesel engine powered by *Cymbopogon flexuosus* biofuel. *Journal of the Energy Institute, 90*(4), 634–645.

Dhinesh, B., Raj, Y. M. A., Kalaiselvan, C., & KrishnaMoorthy, R. (2018). A numerical and experimental assessment of a coated diesel engine powered by high-performance nano biofuel. *Energy Conversion and Management, 171,* 815–824.

Ganesan, V. (2012). *Internal combustion engines*. India: McGraw Hill Education Pvt. Ltd.

Gautam, A., & Kumar, A. (2015). Determination of important biodiesel properties based on fuel temperature correlations for application in a locomotive engine. *Fuel, 142,* 289–302. https://doi.org/10.1016/j.fuel.2014.10.032.

Hariharan, S., Murugan, S., & Nagarajan, G. (2013). Effect of diethyl ether on Tyre pyrolysis oil fueled diesel engine. *Fuel, 104,* 109–115. https://doi.org/10.1016/j.fuel.2012.08.041.

Heywood, J. B. (1988). *Internal combustion engine fundamentals*. McGraw Hill.

Karthickeyan, V. (2017). Impact of partially stabilised zirconia on a single-cylinder diesel engine's performance, using orange oil methyl ester biodiesel. *International Journal of Ambient Energy,* 1–10. https://doi.org/10.1080/01430750.2017.1341429.

Karthickeyan, V. (2018a). Experimental investigation on emission reduction in neem oil biodiesel using selective catalytic reduction and catalytic converter techniques. *Environmental Science and Pollution Research, 25,* 13548–13559. https://doi.org/10.1007/s11356-018-1599-9.

Karthickeyan, V. (2018b). Experimental analysis on thermally coated diesel engine with neem oil methyl ester and its blends. *Heat and Mass Transfer, 54,* 1961–1974. https://doi.org/10.1007/s00231-018-2286-6.

Karthickeyan, V. (2019). Effect of cetane enhancer on *Moringa oleifera* biodiesel in a thermal coated direct injection diesel engine. *Fuel, 235,* 538–550. https://doi.org/10.1016/j.fuel.2018.08.030.

Karthickeyan, V., Balamurugan, P., & Senthil, R. (2016a). Investigation of CI engine fueled with ethanol nano additives blended diesel. In *Proceedings of the First International Conference on Recent Advances in Bioenergy Research* (pp. 121–130). https://doi.org/10.1007/978-81-322-2773-1.

Karthickeyan, V., Balamurugan, P., & Senthil, R. (2016b). Production of orange oil methyl ester and experimental investigation on thermal barrier coated diesel engine. *Asian Journal of Research in Social Sciences and Humanities, 6,* 156–178. https://doi.org/10.5958/2249-7315.2016.00601.8.

Karthickeyan, V., Balamurugan, P., & Senthil, R. (2017). Comparative studies on emission reduction in thermal barrier coated engine using single blend ratio of various non-edible oils. *Journal of the Brazilian Society of Mechanical Sciences and Engineering, 39,* 1823–1833. https://doi.org/10.1007/s40430-016-0645-0.

Karthikeyan, R., & Mahalakshmi, N. V. (2007). Performance and emission characteristics of a turpentine-diesel dual fuel engine. *Energy, 32,* 1202–1209. https://doi.org/10.1016/j.energy.2006.07.021.

Mani, M., Subash, C., & Nagarajan, G. (2009). Performance, emission and combustion characteristics of a DI diesel engine using waste plastic oil. *Applied Thermal Engineering, 29,* 2738–2744. https://doi.org/10.1016/j.applthermaleng.2009.01.007.

Muralidharan, K., & Vasudevan, D. (2011). Performance, emission and combustion characteristics of a variable compression ratio engine using methyl esters of waste cooking oil and diesel blends. *Applied Energy, 88,* 3959–3968. https://doi.org/10.1016/j.apenergy.2011.04.014.

Murugan, S., Ramaswamy, M. C., & Nagarajan, G. (2008). A comparative study on the performance, emission and combustion studies of a DI diesel engine using distilled tyre pyrolysis oil-diesel blends. *Fuel Processing Technology, 87,* 2111–2121. https://doi.org/10.1016/j.fuel.2008.01.008.

Murugesan, A., Umarani, C., Subramanian, R., & Nedunchezhian, N. (2009). Bio-diesel as an alternative fuel for diesel engines—A review. *Renewable and Sustainable Energy Reviews, 13,* 653–662. https://doi.org/10.1016/j.rser.2007.10.007.

Osawa, W. O., Sahoo, P. K., Onyari, J. M., & Mulaa, F. J. (2016). Effects of antioxidants on oxidation and storage stability of *Croton megalocarpus* biodiesel. *International Journal of Energy and Environmental Engineering, 7,* 85–91. https://doi.org/10.1007/s40095-015-0191-z.

Purushothaman, K., & Nagarajan, G. (2009a). Effect of injection pressure on heat release rate and emissions in CI engine using orange skin powder diesel solution. *Energy Conversion and Management, 50,* 962–969. https://doi.org/10.1016/j.enconman.2008.12.030.

Purushothaman, K., & Nagarajan, G. (2009b). Experimental investigation on a C.I. engine using orange oil and orange oil with DEE. *Fuel, 88,* 1732–1740. https://doi.org/10.1016/j.fuel.2009.03.032.

Purushothaman, K., & Nagarajan, G. (2009c). Performance, emission and combustion characteristics of a compression ignition engine operating on neat orange oil. *Renew Energy, 34,* 242–245. https://doi.org/10.1016/j.renene.2008.03.012.

Rashid, U., Ibrahim, M., Yasin, S., Yunus, R., Taufiq-Yap, Y. H., & Knothe, G. (2013). Biodiesel from *Citrus reticulata* (mandarin orange) seed oil, a potential non-food feedstock. *Industrial Crops and Products, 45,* 355–359. https://doi.org/10.1016/j.indcrop.2012.12.039.

Rizwanul Fattah, I. M., Masjuki, H. H., Kalam, M. A., Mofijur, M., & Abedin, M. J. (2014). Effect of antioxidant on the performance and emission characteristics of a diesel engine fueled with palm biodiesel blends. *Energy Conversion and Management, 79,* 265–272. https://doi.org/10.1016/j.enconman.2013.12.024.

Sakthivel, G., Nagarajan, G., Ilangkumaran, M., & Gaikwad, A. B. (2014). Comparative analysis of performance, emission and combustion parameters of diesel engine fuelled with ethyl ester of fish oil and its diesel blends. *Fuel, 132,* 116–124. https://doi.org/10.1016/j.fuel.2014.04.059.

Senthil, R., & Silambarasan, R. (2015). Environmental effect of antioxidant additives on exhaust emission reduction in compression ignition engine fuelled with Annona methyl ester. *Environmental Technology, 20,* 1–7. https://doi.org/10.1080/09593330.2015.1021856.

Senthil, R., Silambarasan, R., & Ravichandiran, N. (2015). Influence of injection timing and compression ratio on performance, emission and combustion characteristics of Annona methyl ester operated diesel engine. *Alexandria Engineering Journal, 54,* 295–302. https://doi.org/10.1016/j.aej.2015.05.008.

Sharma, A., & Murugan, S. (2015). Potential for using a tyre pyrolysis oil-biodiesel blend in a diesel engine at different compression ratios. *Energy Conversion and Management, 93,* 289–297. https://doi.org/10.1016/j.enconman.2015.01.023.

Singh, N., & Bharj, R. (2015). Experimental investigation on the role of indigenous carbon nanotube emulsified fuel in a four-stroke diesel engine. *Proceedings of the Institution of Mechanical Engineers, Part C: Journal of Mechanical Engineering Science,* 1–14. https://doi.org/10.1177/0954406215586021.

Subbaiah, G. V., & Gopal, K. R. (2013). An experimental investigation on the performance and emission characteristics of a diesel engine fuelled with rice bran biodiesel and ethanol blends. *International Journal of Green Energy,* 37–41. https://doi.org/10.1080/15435075.2010.548539.

Subramani, L., Parthasarathy, M., Balasubramanian, D., & Ramalingam, K. (2018). Novel *Garcinia gummi-gutta* methyl ester (GGME) as a potential alternative feedstock for existing unmodified DI diesel engine. *Renewable Energy, 125,* 568–577.

Subramaniam, D., Murugesan, A., Avinash, A., & Kumaravel, A. (2013). Bio-diesel production and its engine characteristics—An expatiate view. *Renewable and Sustainable Energy Reviews, 22,* 361–370. https://doi.org/10.1016/j.rser.2013.02.002.

Tüccar, G., Tosun, E., Özgür, T., & Aydin, K. (2014). Diesel engine emissions and performance from blends of citrus sinensis biodiesel and diesel fuel. *Fuel, 132,* 7–11. https://doi.org/10.1016/j.fuel.2014.04.065.

Tudu, K., Murugan, S., & Patel, S. K. (2016). Effect of diethyl ether in a DI diesel engine run on a tyre derived fuel-diesel blend. *Journal of the Energy Institute, 89,* 525–535. https://doi.org/10.1016/j.joei.2015.07.004.

Vallinayagam, R., Vedharaj, S., Yang, W. M., Lee, P. S., Chua, K. J. E., & Chou, S. K. (2013). Combustion performance and emission characteristics study of pine oil in a diesel engine. *Energy, 57,* 344–351. https://doi.org/10.1016/j.energy.2013.05.061.

Vedharaj, S., Vallinayagam, R., Yang, W. M., Chou, S. K., & Lee, P. S. (2014). Effect of adding 1,4-dioxane with kapok biodiesel on the characteristics of a diesel engine. *Applied Energy, 136,* 1166–1173. https://doi.org/10.1016/j.apenergy.2014.04.012.

Wamankar, A. K., & Murugan, S. (2015). Combustion, performance and emission of a diesel engine fuelled with diesel doped with carbon black. Energy, 1–9. https://doi.org/10.1016/j.energy.2015.04.012.

Experimental Investigation of Performance and Emission Characteristics of Diesel Blended with Palm Methyl Ester Along with Alumina Nano-Additive Using D.I. Diesel Engine

P. V. Inbanaathan, B. Dhinesh and U. Tamilarasan

Abstract In recent years, a crisis in the availability of energy is there with two important reasons, among which a population stretch becomes the major reason and another important reason is continuous consumption of fuels leading to the reduction in fossil fuels. Atmosphere consists of various harmful pollutants in which the burning of fossils fuels becomes a major contributor to it. To reduce these harmful pollutants, severe research for an alternate fuel is made which in turn leads to biodiesel as a good alternate fuel for the usage in CI engine. A biodiesel has healthier properties than those of diesel fuel such as renewable, eco-friendly, non-toxic and basically free of sulphur. In this experimental investigation, a trans-esterified biodiesel, namely palm oil methyl ester is blent with neat diesel with and without addition of nano-alumina additives is prepared, and it is fuelled in D.I. diesel engine. Then, the characteristic of performance, combustion and emission of the tested fuel along with the neat diesel is measured. The trans-esterified palm oil is blended with diesel by the proportion of B10, B10A30, B60 and B60A30 in volume percentages. At various load conditions, the investigational study is carried out by biodiesel blends with and without nano-alumina additives. Brake thermal efficiency is improved with B10A30 sample in comparison with the neat diesel and blends without addition of nanoparticles. The specific fuel consumptions of B10A30 nanoparticle blend are almost similar to that of the normal diesel and better than that of the blend without nanoparticle. With the addition of nanoparticles, the unburned HC emission is decreased and shows a good result for B10A30 sample. At lower conditions, the CO emission seems par against diesel while at the higher load conditions, carbon monoxide emission is getting increased for all the prepared blends in comparison with diesel. NO_x emission has been decreased with B10A30 sample when it is compared with all other blends and diesel fuel. Hence the improvement in emission and performance characteristics, the

P. V. Inbanaathan (✉) · U. Tamilarasan
Department of Production Engineering, Sri Sairam Engineering College, Chennai, Tamil Nadu, India
e-mail: inba29@gmail.com

B. Dhinesh
Department of Mechanical Engineering, Mepco Schlenk Engineering College, Sivakasi, Tamil Nadu, India

S. K. Ghosh et al. (eds.), *Bioresource Utilization and Bioprocess*,
https://doi.org/10.1007/978-981-15-1607-8_15

blend of diesel and biodiesel up to 10% in volume along with the addition of nano-alumina particles is preferred when compared to neat diesel and all other prepared blends.

Keywords Diesel · Palm oil methyl ester · Alumina nanoparticles · Performance · Emission

1 Introduction

In D.I. diesel engine, the characteristics of performance, combustion and emission can be improved by implementing some strategies. The common strategies which many researchers do are making some modifications in the engine, introducing new renewable alternative fuels and also the treatment of the exhaust gas (Shaafi and Velraj 2015). Many researchers have produced researches indicating the importance of biofuels in controlling the engine emissions with a good impact on the performance of the engine. Increase in efficiency of combustion, performance of the engine and reduction in exhaust emissions can be achieved by the addition of nanoparticles as nano-additives in the diesel–biodiesel blend. Metal oxides of alumina are the most common nanoparticles used (Shaafi and Velraj 2015). Under ultrasonic mixing, mass fractions of 30 ppm of carbon-coated aluminium oxide nanoparticles were added. This addition of nanoparticles (B10A30) can reduce BSFC, CO, HC and NO_x emissions in comparison with the B10 blend fuel and neat diesel. Though the particulate number emission is increased, the engine performance is better when comparing to the neat diesel fuel (Wu et al. 2018). Synthesis of nanoparticles is done by using XRD and TEM analysis. Using ultrasonicator, it is then blended at the fraction of 40–120 ppm with biodiesel–diesel blends. An increase in peak in-cylinder pressure is obtained since the combustion duration is earlier and the short duration of ignition delay period (Najafi 2018). B10 and B5 fuel blends were prepared along with the addition of alumina nanoparticles in the proportions following as 90, 60 and 30 ppm. A rise in η_{BTE}, decrease in SFC, CO and HC emissions and rise in NO_x emissions were recorded in considerable percentage at various engine speeds (Hosseini et al. 2017). Here for cerium oxide (CeO_2) nanoparticle blend fuel, a large range of decrease in unburned hydrocarbon, carbon monoxide emission, NO_x emissions and also a negligible reduction in emissions of smoke can be achieved in comparison with standard diesel and emulsion of low-grade oil (LGO) at various power outputs (Annamalai et al. 2016). In this paper, the characteristics of emission and performance of Mahua-based biodiesel blended along with aluminium oxide nanoparticles (APN) were analysed. It is found that the flashpoint of the fuel increases, fuel consumption decreases, and the harmful emissions are reduced considerably (Aalam and Saravanan 2017). R. El-Araby et al. have experimentally arrived at the conclusion that the physical properties of palm oil and blends of biodiesel show no difference in fuel properties when the blend percentage was increased up to 30% volume of biodiesel mixing with the diesel fuel (El-Araby et al. 2018). Hawanam Kim et al. concluded that the use of biodiesel blended fuels always reduces the particulate emissions or smoke emissions from the engine. When comparing the use of biodiesel blends with the use of diesel, diesel fuel causes more particulate emissions which are smaller than 50 nm. These kinds of particulate matters are very harmful to human survival. Since the heating value of biodiesel is lower, the fuel consumption gets increased (Kim and Choi 2010).

Finally, it becomes mandatory to see the effects in adding the nano-alumina particles in the biodiesel–diesel blend to optimize the engine characteristics.

2 Materials and Methods

2.1 Blends

In recent days of retail diesel marketplace, conventional diesel and blends of biodiesel are the products which are commonly distributed for use. "B" is the letter which denotes the quantity of biodiesel in the mixture of biodiesel with the diesel fuel in many parts of the world. For example, B100 means 100% biodiesel. While B20 means 20% biodiesel and the remaining 80% is diesel fuel. In comparison with the various oilseeds, the most potential biodiesel feedstock is palm oil. Today in the world market, with respect to the availability of vegetable oils, palm oil refined with impurities is at the top. The removal of impurities is done in the refineries. After the removal of impurities, the palm refined oil is easily converted into methyl ester, which acts as the biodiesel. The obtained refined palm oil can be blended with the diesel. Envo diesel can be prepared from blending a certain quantity of palm oil with the petroleum diesel. The prepared biodiesel from the palm oil has more stability in oxidation. This tends to lower the engine HC and CO emissions apart from the slight reduction in high NO_x emission.

2.2 Nanoparticles as Additives to Biodiesel

Nowadays, mixing additives in biodiesel has shown better engine characteristics. The various additives used are oxygenates, antioxidants and nanoparticles. By using nanoparticles as additive, clean combustion and reduced exhaust emission can be achieved (Balasubramanian et al. 2018; Dhinesh et al. 2017a; Parthasarathy et al. 2016; Lalvani et al. 2016; Lalvani et al. 2015). A technique of fuel formulation which best achieves in emission and performance characteristics has become the main focus by many researchers in recent days (Dhinesh and Annamalai 2018; Dhinesh et al. 2016, 2017b, 2018; Subramani et al. 2018). Among various additives to the biodiesel blended fuel, nanoparticle as an additive material has shown a major contribution in attaining enhanced performance characteristics, combustion and exhaust emissions (Vigneswaran et al. 2018; Ramalingam et al. 2018).

2.3 Advantages of Nano-Additives in Biodiesel

- Substantial enhancement in BTE.
- At the higher load condition, there is a drop in ignition delay period which quickens the prior commencement of combustion resulting in the reduced in-cylinder pressure and HRR (Jayakar et al. 2018; Nanthagopal et al. 2019).
- Enhancement in oxidation and reduction processes with the nanoparticles leads to the simultaneous decrease in CO, HC and smoke opacity.

2.4 Test Sample Preparation

Test sample preparation is of two stages:

- Diesel–biodiesel blend preparation
- Addition of nanoparticles to the biodiesel blend.

2.4.1 Preparation of Diesel–Biodiesel Blend

Although many types of blends with a varied composition are made, tests were carried out with B10 (10% palm methyl ester 90% diesel by volume) and B60 blends. The diesel fuel is designated as B0 while the neat biodiesel is meant as B100. The fuels were prepared at the optimum room temperature of 28 °C on volume basis. Mixing of fuels to create blends is usually done on the basis of volume at the normal room temperature. Because of this, the blending fraction in terms of volumetric means has been selected as the primary work (Fig. 1).

2.4.2 Mixing of Nanoparticles to Biodiesel–Diesel Blend

The mixing of nanoparticles with the prepared fuel blends is done by using ultrasonicator in which the nanoparticles dispersion was carried out for 45 min to the prepared test fuels. The test fuels types which are prepared for experimental analysis are designated as BxAy where y denotes ppm and x denotes the fraction of volume. The test fuels prepared are B10 (which contains 90% of neat diesel and 10% of biodiesel), B10A30 (which contains 90% of neat diesel, 10% of biodiesel with 30 ppm of aluminium oxide nanoparticle), B60 (contains 60% biodiesel and 40% of diesel) and B60A30 (contains 60% of biodiesel and 40% of diesel with 30 ppm of aluminium oxide nanoparticle) (Figs. 2 and 3).

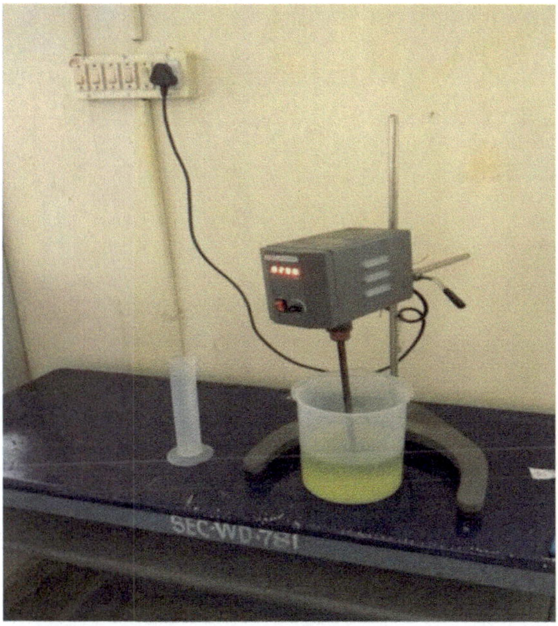

Fig. 1 Diesel–biodiesel blend preparation

Fig. 2 Biodiesel blend without nanoparticles

3 Biofuel Properties

There are various standards to test the physical and chemical properties of biodiesel. These standards decide the quality of the prepared biodiesel. Some notable important properties in biodiesel which affect the quality of combustion inside the engine are viscosity, density, flashpoint and calorific value. The viscosity of biodiesel is high which can be controlled by mixing it with the neat diesel in varying proportions. This

Fig. 3 Biodiesel blend with nanoparticles

Table 1 Property comparison of diesel and biodiesel blend

Property	Fuel		
	Diesel	B10	B60
Viscosity @ 35 °C in CST	3.97	3.70	2.93
Flash point (°C)	52–96	51	62
Fire point (°C)	62–106	58	69
Gross calorific value in kJ/kg	45,600	41,402.94	39,070.38
Density in kg/m^3	832	856	881

enhances the combustion inside the engine. Density is also an important parameter in the preparation of biodiesel as the storage and transportation of biodiesel are necessary in terms of the daily usage. The density of biodiesel is higher in comparison with the neat diesel. Flashpoint is an expressive characteristic to describe the liquids which are not used as fuels. The calorific value of the biodiesel is lower in comparison with the neat diesel which intends to increase the specific fuel consumption of the biodiesel (Table 1).

4 Experimental Procedure

A four-stroke, single cylinder and water-cooled D.I. diesel engine, attached along an electrical dynamometre is used for the study. The investigational set-up is presented in Fig. 4.

The engine specifications are given in Table 2. The computing devices attached are expressed in Table 3.

Fig. 4 Investigational set-up

Table 2 Engine specifications

Make	Kirloskar engine (TV1)
Type	Four-stroke, single cylinder and water-cooled direct injection CI engine
Power rating in kW	5.20
Rated speed in rpm	1500
Stroke length (L) in mm	110
Compression ratio	17.5:1
Diameter of bore (D) in mm	87.5
Diameter of orifice (d) in mm	20
Discharge coefficient	0.60
Cooling type	Water cooled

Table 3 Measuring instruments used

S. No.	Measuring parameter	Instruments used
1	CO, HC, NO_x, CO_2, O_2	Five gas analyser (AVL DI GAS 444)
2	Smoke	AVL 437C smoke metre
3	Cylinder pressure	AVL pressure transducer
4	Air flow rate	Orifice and manometre

The quantity of intake is measured by using an anti-pulsating drum which is attached to the test engine suction side. The air is allowed inside the engine through a 20 mm orifice plate. A thermocouple is used to measure the difference in pressure created due to the flow air into the orifice. To study the combustion characteristics, a combustion analyser is used. To measure the various types of emissions, several devices like five gas analysers, smoke metre, exhaust gas thermocouple and piezoelectric transducer are used.

The experimental set-up also consists of two tanks full of test fuel fitted along with a burette and usually coupled through a three-way valve. The total fuel consumption of the engine is measured by using a fuel gauging device. The output shaft of engine is connected to an electrical dynamometre followed by a load rheostat. The arrangement is provided to measure the voltage, current and power with a digital voltmetre, ammetre and wattmetre, respectively, for various loads. The heat release rate, combustion pressure and crank angles for a stipulated number of cycles are measured by using an 8-bit data acquisition system which is connected to the engine. The system sends the signal, and this in turn is connected to a computer to measure the various combustion parameters.

5 Performance and Emission Characteristics

It is witnessed that from a four-stroke, single cylinder, D.I. diesel engine, the various biodiesel and diesel blend's performance and emission characteristics were studied at injection timing at 22° BTDC and injection pressure at around 210 bar. There is no modification made to the diesel engine.

5.1 Parameters of Performance Characteristics

The parameter which governs the performance of the fuel is given by the following:

- η_{mech} (Efficiency—Mechanical)
- η_{BTE} (Efficiency—Brake Thermal)
- η_{ITE} (Efficiency—Indicated Thermal)
- SFC (Specific Fuel Consumption).

5.1.1 SFC (Specific Fuel Consumption)

The variations in SFC in order to the various engine loads for diesel, blends (B10 and B60) without the addition of nanoparticles and blends with the addition of nanoparticles (B10A30 and B60A30) are shown in Fig. 5. When compared to diesel fuel, SFC for B10 and B60 is higher. The reason behind is the low heating value of biodiesel. Rise in SFC is directly proportional to the amount of biodiesel blended with the diesel. Al_2O_3 nanoparticles addition to the blend decreases SFC because of better

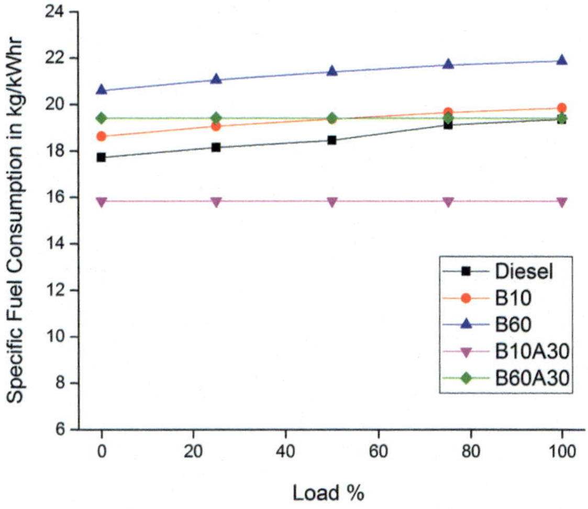

Fig. 5 SFC versus load graph

atomization of nanoparticles added which results in good combustion in comparison with neat diesel and blends without nanoparticles.

The specific fuel consumptions of B10A30 nanoparticle blend are better than the SFC of the B60A30 nanoparticle blend, normal diesel and blends without nanoparticle. The specific fuel consumption is higher for the B60 biodiesel blend without nanoparticles.

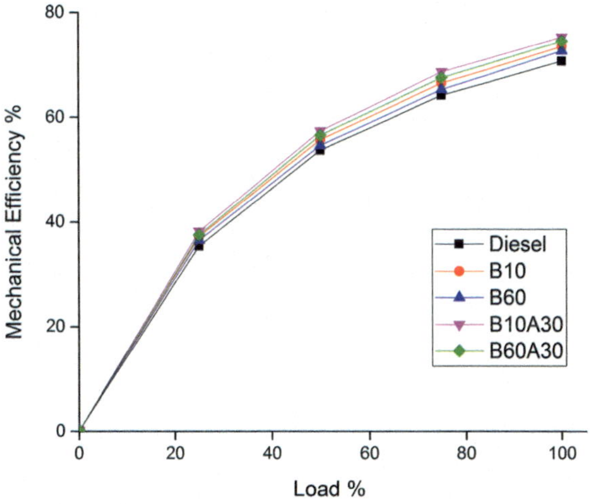

Fig. 6 Mechanical efficiency versus load graph

5.1.2 η_{mech} (Mechanical Efficiency)

The mechanical efficiency versus the load graph is presented in Fig. 6. Mechanical efficiency of B10A30 is greater than the diesel and other blends, while the diesel fuel has the least mechanical efficiency. The mechanical efficiency is improved by 2.8% and 3.96% for B60 and B10 blends, respectively and 5.41% and 6.58% for B60A30 and B10A30 blends, respectively. From the above discussion, it is clear that the B10A30 blend performed well when it is compared with the all the other prepared blends and also with the neat diesel.

5.1.3 η_{BTE} (Brake Thermal Efficiency)

From Fig. 7, BTE of B10A30 is higher than the diesel and other blends in comparison with the diesel fuel which has the least BTE. η_{BTE} increases due to the use of biodiesel blends and also for the blends with the addition of alumina nanoparticle. Alumina nanoparticles present here as oxygen enhancer in the combustion of the test fuel. This attributes to the enhanced combustion of the alumina nanoparticle added with the blend. More presence of highly active surfaces, the alumina nanoparticle acting as the catalyst has improved its activity. The brake thermal efficiency is improved by 0.92% and 2% for B60 and B10 blends and 9.37% and 12.2% for B60A30 and B10A30 blends, respectively. From the above discussion, it is clear that the B10A30 blend performed well when it is compared with all the other prepared blends and also with the neat diesel.

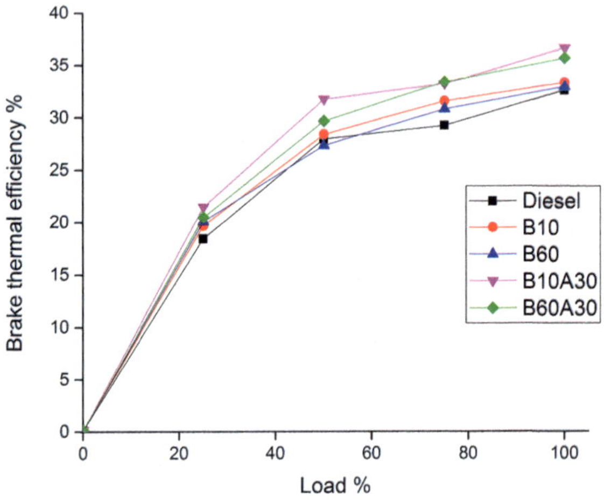

Fig. 7 Brake thermal efficiency versus load graph

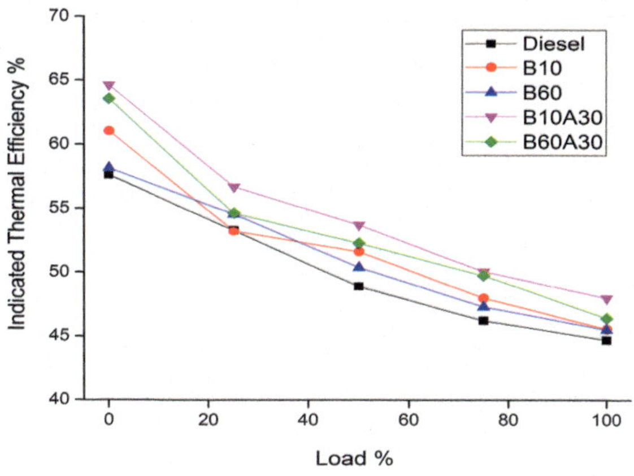

Fig. 8 Indicated thermal efficiency versus load graph

5.1.4 η_{ITE} (Indicated Thermal Efficiency)

η_{ITE} of B10A30 is greater than diesel and other blends, while the diesel fuel has the least indicated thermal efficiency. From Fig. 8, the same trend is followed in comparison with the graph that has been shown for the η_{BTE}. η_{ITE} is enhanced by 0.90% and 5.87% for B60 and B10 blends and 10.26% and 12.1% for B60A30 and B10A30 blends, respectively. From the above discussion, it is clear that the B10A30 blend performed well when it is compared with the all the other prepared blends and also with the neat diesel.

5.2 Parameters of Emission Characteristics

From the emission control point of view, measurement of emissions from the engine is very important. The exhaust of the engine may contain the following gases:

- HC—Hydrocarbon emission
- CO—Carbon monoxide emission
- NO_x—Nitrous oxide emission
- O_2—Oxygen emission.

5.2.1 CO Emission

From Fig. 9, it is evident that the emission of CO is slightly decreased in all the blends of biodiesel. The reason behind this is carbon content in the diesel is lesser when compared to the carbon content in the biodiesel fuels. At low load conditions, there is no substantial difference in CO emission for all the prepared blends in comparison with the diesel fuel. But at higher load conditions, CO emission increases for all

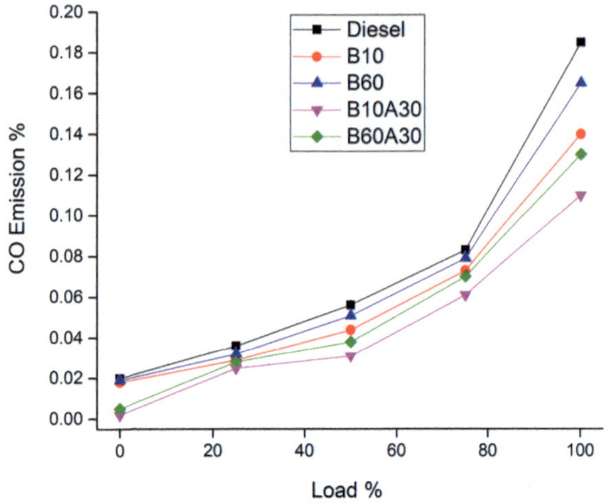

Fig. 9 CO emission versus load graph

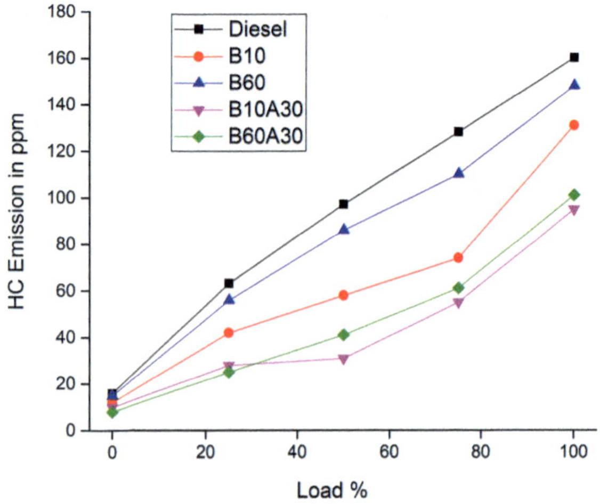

Fig. 10 HC emission versus load graph

the fuels. Comparatively, the addition of alumina nanoparticles has reduced the CO emission level even at higher load conditions. The ignition delay period shortening happened due to the surface contact area of nanoparticles. The fuel–air mixing, sudden evaporation and enhanced spray atomization of fuel have improved with the addition of nanoparticles in the biodiesel–diesel blend.

5.2.2 HC Emission

Figure 10 depicts the deviation of unburned hydrocarbon emission with increase in load. Hydrocarbon emission upturns as the load rises for all the cases. In fuel-rich region, due to the less amount of oxygen molecule, HC emissions are formed.

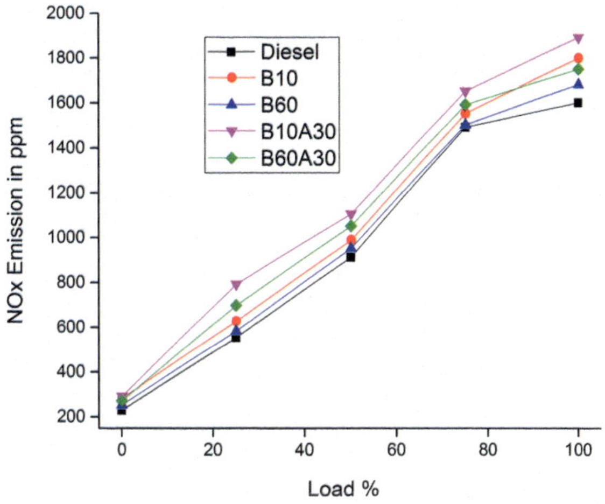

Fig. 11 NO_x emission versus load graph

Biodiesel blends usage reduces the combustion delay period. Further, the addition of nanoparticles supplied the O_2 molecules for the oxidation of HC during the combustion process. The fuel B10A30 shows better results in HC emission at higher load conditions while the fuel B60A30 showed good results at lower load conditions.

5.2.3 NO_x Emission

The variation of NO_x emissions is depicted in Fig. 11. The figure clearly shows the NO_x emission rises due to the usage of biodiesel. Rise in NO_x emission is due to double bond molecules of biodiesel causing greater adiabatic flame temperature. Due to the addition of nanoparticles, the NO_x emission seems to be increasing still because of higher peak pressure, greater HRR and also since alumina nanoparticles present here as an oxygen donating catalyst.

5.2.4 Smoke Opacity Emission

The smoke opacity versus the engine load graph is depicted in Fig. 12. Owing to partial combustion of fuel in diffusive combustion phase, smoke emission rises as the load rises for all cases. When comparing the biodiesel blends with the neat diesel, a substantial decrease in emission of smoke is obtained. The reason behind is the disposal of more O_2 molecules in the biodiesel blend and also in the alumina nanoparticles.

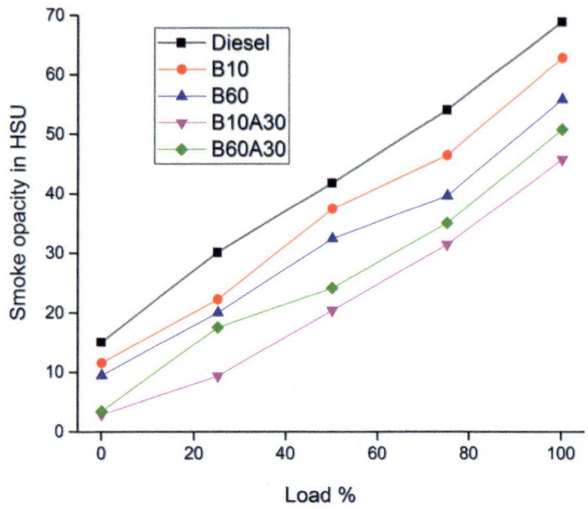

Fig. 12 Smoke emission versus load graph

6 Conclusion

The palm methyl ester (with and without nano-alumina additives) used here as a biofuel is compared with the diesel, and it results that palm methyl ester with nano-additives is having a potential to be used in diesel engines. The trans-esterified palm oil is blended with diesel by the proportion of B10, B10A30, B60 and B60A30. The performance and emission characteristics were carried out on Kirloskar TV1 engine. The obtained results are related with that of diesel on same engine.

From the results, B10A30 blend shows lesser SFC, higher η_{mech}, higher η_{ITE} and higher η_{BTE} by load variations than that of conventional diesel. While at the conditions of 100% load, brake thermal efficiency (BTE) of the other blends was found to be B10 33.19%, B60 32.84%, B60A30 35.59%, and B10A30 blend is 36.51% which is better when compared to 32.54% of the normal diesel.

The emission test was carried out on a 437C smoke metre and DIGAS 444N analyser, and the results have been recorded. It is clear that the biodiesel blend has slightly increased NO_x emission percentage than diesel. This can be seen from the full load condition results of diesel, B10, B60, B60A30 and B10A30 which shows 1601, 1798, 1681, 1750 and 1890 ppm, respectively. The NO_x emission results show a considerable increase when the nano-alumina particles are added, and this is due to better fuel combustion. At higher load conditions, CO emissions are diesel (0.185%), B60 (0.165%), B10 (0.14%), B60A30 (0.13%) and B10A30 (0.11%) respectively. Addition of alumina nanoparticle reduced the CO emissions. The unburned HC content in diesel at full load condition was found to be 160 ppm, and it was found to be decreased in B60 blend to 148 ppm and almost similar in B10 blend which showed 131 ppm. Here also, the blends with nanoparticles showed a lesser amount of HC content as B10A30 showed 95 ppm and B60A30 101 ppm.

Therefore, the palm methyl ester-based biodiesel blended with aluminium oxide nanoparticles B10A30 can reduce pollutant emission like CO and HC to a considerable extent as well as improve the performance of the engine considerably by increasing the brake thermal efficiency.

References

Aalam, C. S., & Saravanan, C. G. (2017). Article on effects of nano metal oxide blended Mahua bio-diesel on CRDI engine. *Ain Shams Engineering Journal., 8*(11), 689–696.

Annamalai, M., Dhinesh, B., Nanthagopal, K., Sivarama Krishnan, P., Lalvani, J. I. J., Parthasarathy, M., et al. (2016). An assessment on performance, combustion and emission behavior of a diesel engine powered by ceria nanoparticle blended emulsified biofuel. *Energy Conversion and Management, 123*(1), 372–380.

Balasubramanian, D., Arumugam, S. R. S., Subramani, L., Chellakumar, I. J. L. J. S., & Mani, A. (2018). A numerical study on the effect of various combustion bowl parameters on the performance, combustion, and emission behavior on a single cylinder diesel engine. *Environmental Science and Pollution Research, 25*(3), 2273–2284.

Dhinesh, B., & Annamalai, M. (2018). A study on performance, combustion and emission behavior of diesel engine powered by novel nano nerium oleander biofuel. *Journal of Cleaner Production, 196*(20), 74–83.

Dhinesh, B., Annamalai, M., Lalvani, J. I. J., & Annamalai, K. (2017a). Studies on the influence of combustion bowl modification for the operation of *Cymbopogon flexuosus* biofuel-based diesel blends in a DI diesel engine. *Applied Thermal Engineering, 112,* 627–637.

Dhinesh, B., Lalvani, J. I. J., Parthasarathy, M., & Annamalai, K. (2016). An assessment on performance, emission and combustion characteristics of single cylinder diesel engine powered by *Cymbopogon flexuosus* biofuel. *Energy Conversion and Management, 117,* 466–474.

Dhinesh, B., Lalvani, J. I. J., Parthasarathy, M., & Annamalai, K. (2017b). An experimental analysis on the influence of fuel-borne additives on the single cylinder diesel engine powered by *Cymbopogon flexuosus* biofuel. *Journal of the Energy Institute, 90*(4), 634–645.

Dhinesh, B., Raj, Y. M. A., Kalaiselvan, C., & Krishna Moorthy, R. (2018). A numerical and experimental assessment of a coated diesel engine powered by high-performance nano biofuel. *Energy Conversion and Management, 171,* 815–824.

El-Araby, R., Ashraf Amin, A. K., Morsi, El, El-Ibiari, N. N., & El-Diwani, G. I. (2018). Study on the characteristics of palm oil–bio-diesel–diesel fuel blend. *Egyptian Journal of Petroleum, 27*(2), 187–194.

Hosseini, S. H., Taghizadeh-Alisaraei, A., Ghobadian, B., & Abbaszadeh-Mayvan, A. (2017). Effect of added alumina as nano-catalyst to diesel-bio-diesel blends on performance and emission characteristics of CI engine. *Energy, 124*(1), 543–552.

Jayakar, J., Elumalai, P. V., Annamalai, K., Mailainathan, M., Dhinesh, B., & Harikrishnan, G. (2018). Macroscopic characteristics of palm oil and palm oil methyl ester using dimensionless analysis. *Journal of Palm oil research, 30*(4), 666–673.

Kim, H., & Choi, B. (2010). The effect of bio-diesel and bioethanol blended diesel fuel on nanoparticles and exhaust emissions from CRDI diesel engine. *Renewable Energy, 35*(1), 157–163.

Lalvani, J. I. J., Parthasarathy, M., Dhinesh, B., & Annamalai, K. (2015). Experimental investigation of combustion, performance and emission characteristics of a modified piston. *Journal of Mechanical Science and Technology, 29*(10), 4519–4525.

Lalvani, J. I. J., Parthasarathy, M., Dhinesh, B., & Annamalai, K. (2016). Pooled Effect of injection pressure and turbulence inducer piston on combustion, performance and emission characteristics of a D.I Diesel engine powered with bio-diesel blend. *Ecotoxicology and Environmental Safety, 134*(2), 336–343.

Najafi, G. (2018). Diesel engine combustion characteristics using nano-particles in bio-diesel-diesel blends. *Fuel, 212*(15), 668–678.

Nanthagopal, K., Ashok, B., Garnepudia, R. S., Tarun, K. R., & Dhinesh, B. (2019). Investigation on diethyl ether as an additive with calophyllum inophyllum bio-diesel for CI engine application. *Energy Conversion and Management, 179*(1), 104–113.

Parthasarathy, M., Lalvani, J. I. J., Dhinesh, B., & Annamalai, K. (2016). Effect of hydrogen on ethanol-bio-diesel blend on performance and emission characteristics of a direct injection diesel engine. *Ecotoxicology and Environmental Safety, 134*(2), 433–439.

Ramalingam, K., Kandasamy, A., Subramani, L., Balasubramanian, D., & Thadhani, J. P. J. (2018). An assessment of combustion, performance characteristics and emission control strategy by adding an anti-oxidant additive in emulsified fuel. *Atmospheric Pollution Research, 9*(5), 959–967.

Shaafi, T., & Velraj, R. (2015). Influence of alumina nano-particles, ethanol and isopropanol blend as an additive with diesel–soybean bio-diesel blend fuel: Combustion, engine performance, and emissions. *Renewable Energy, 80,* 655–663.

Subramani, L., Parthasarathy, M., Balasubramanian, D., & Ramalingam, K. (2018). Novel garcinia gummi-gutta methyl ester (GGME) as a potential alternative feedstock for existing unmodified DI diesel engine. *Renewable Energy, 125,* 568–577.

Vigneswaran, R., Annamalai, K., Dhinesh, B., & Krishnamoorthy, R. (2018). Experimental investigation of unmodified diesel engine performance, combustion, and emission with multipurpose additive along with water-in-diesel emulsion fuel. *Energy Conversion and Management, 172,* 370–380.

Wu, Q., Xi, X., Wang, Y., & Roskilly, T. (2018). Effect of carbon coated aluminum nano-particles as an additive to bio-diesel-diesel blends on performance and emission characteristics of the diesel engine. *Applied Energy, 221*(1), 597–604.

Efficacy of Metal Extraction from Discarded Printed Circuit Board Using *Aspergillus tubingensis*

Amber Trivedi and Subrata Hait

Abstract The immense growth in the usage of electrical and electronic equipment (EEE) and their shorter lifespan are accelerating their discarding rate and the generation of waste EEE (WEEE) or electronic waste (e-waste). To conserve natural resources and to reduce adverse environmental impact, recycling of e-waste is getting wider attention worldwide. Printed circuit board (PCB) is an integral part of EEE. Recycling of e-waste especially PCB has become a major concern owing to its high metallic content including base, precious, and toxic metals. As the metals are generally embedded in polymer matrix in the PCB, the bioleaching efficacy of selected abundant metals, viz. Cu, Zn, and Ni from the PCB of discarded computer has been attempted in the study using *Aspergillus tubingensis*, a polymer-degrading fungal species. The bioleaching experiments were performed using the pulverized PCB with a particle size ranging from 0.038 to 1 mm at e-waste pulp density varying from 2.5 to 10 g/L for duration of 33 days. Results from the study revealed that the bioleaching efficacy of selected metals was decreased with the increase in e-waste pulp density. However, the bioleaching efficiency of Zn was found to be the maximum among the selected metals considered in the present study. Thus, the present study highlights the practical feasibility of bioleaching of metals from e-waste using *A. tubingensis*.

Keywords Printed circuit board · Metals · Bioleaching · Polymer-degrading · *Aspergillus tubingensis*

1 Introduction

The advancement and usage of electrical and electronic equipment (EEE) have increased drastically in recent years. Due to the shorter life period, the obsolescence rate of EEE has increased and thereby resulting into the increased generation of waste EEE (WEEE) or electronic waste (e-waste). E-waste generation is estimated

A. Trivedi · S. Hait (✉)
Department of Civil and of Environmental Engineering, Indian Institute of Technology Patna, Patna, Bihar 801106, India
e-mail: shait@iitp.ac.in

S. K. Ghosh et al. (eds.), *Bioresource Utilization and Bioprocess*,
https://doi.org/10.1007/978-981-15-1607-8_16

to grow from 44.7 Mt in 2016 to 52.2 Mt in 2021 at a growth rate of 3–4% per annum globally (Baldé et al. 2017). Asia with 18.2 Mt of e-waste generation tops the list and India is the third-largest contributor in Asia with 1.7 Mt of e-waste generation (Baldé et al. 2017). Printed circuit board (PCB) is one of the central parts of any EEE. The PCB matrix is complex and diverse in nature that contains metallic and non-metallic fractions of 28–30% and 70–72%, respectively (Li et al. 2004; Zhou and Qiu 2010). Apart from precious metals in the range of 0.3–0.4%, the metallic portion of PCB mainly comprises of Cu, Pb, and Ni of around 15%, 2.5%, and 1.5%, respectively (Huang et al. 2009; Li et al. 2007; Wang et al. 2005; Zhao et al. 2004). Because of higher metallic content than the natural ores makes waste PCBs a potential secondary metallic reservoir. Further, the presence of toxic metals makes waste PCBs hazardous to the environment upon disposal. Therefore, the e-waste recycling for metal extraction from the PCBs is required not only to minimize environmental contamination but metal recycling too for sustainable development. However, conventional metal extraction techniques, i.e., pyrometallurgy and hydrometallurgy are associated with high energy requirements and secondary environmental pollution.

The application of biotechnology with the hydrometallurgical process by using various microorganisms is known as biometallurgy. Metal recycling from e-waste using biometallurgy or bioleaching has gained wider attention as this is an eco-friendly, efficient, and cost-effective alternative to conventional recycling techniques (Lundgren et al. 1986; Priya and Hait 2017). Various autotrophic bacteria like *Acidithiobacillus ferrooxidans* and *Acidithiobacillus thiooxidans* and heterotrophic fungi like *Aspergillus niger* and *Penicillium simplicissimum* have been used for bioleaching of metals from e-waste (Atlas and Bartha 1997; Beolchini et al. 2012; Brandl et al. 2001; Narayanasamy et al. 2018; Priya and Hait 2017, 2018; Wang et al. 2009; Xia et al. 2018; Xin et al. 2012). During bioleaching, various metabolic byproducts and biomass of microorganisms facilitate the metal dissolution from the PCB (Lundgren et al. 1986). It is evident from literature that most of the bioleaching studies for metal recycling from e-waste have employed bacterial culture. Fungal bioleaching of metals from e-waste is comparatively less explored area. Fungi have more ability to tolerate toxic materials and have a shorter lag phase and a faster leaching rate. Further, they grow in alkaline and acid-consuming materials. The fungi can grow at high pH solutions which makes fungi more efficient for bioleaching or extraction of alkaline metals (Burgstaller and Schinner 1993). Moreover, the metabolites like organic acids excreted by fungi in shorter lag phase help leach the metals (Wu and Ting 2006). Thus, there is ample scope to explore the fungal bioleaching technique for metal recycling from e-waste. The specific attribute of fungal species like polymer-degrading ability of *Aspergillus tubingensis* is yet to be explored for fungal bioleaching of metals from e-waste as metals are embedded in polymer/complex matrices of PCB. In this context, the fungal bioleaching of selected metals such as Cu, Zn, and Ni from the PCB of obsolete computer was investigated employing *A. tubingensis* species. Further, the effect of e-waste pulp density in the range of 2.5–10 g/L on the efficacy of fungal bioleaching was also assessed in this study.

Fig. 1 Typical PCB of obsolete computer used in the present study

2 Materials and Methods

2.1 *Obsolete PCB Collection and Sample Preparation*

PCBs of obsolete computer were obtained from electrical equipment repairing shops in Bihta, Patna, Bihar, India. Representative image of PCB used in the study is presented in Fig. 1. Prior to pulverization, the mounted parts were removed manually from the PCBs. The PCBs were then subjected to manual chopping into smaller parts. Subsequently, the chopped PCB parts were milled for pulverization using a cutting mill (SM200, Retsch, Germany) and fractionated to the particle size of range between 0.038–1 mm. Upon homogenization, the pulverized PCB sample was divided into three subparts for morphological characterization, metallic content quantification, and bioleaching experiments, respectively.

2.2 *Microorganism*

The fungal strain used in the study was *A. tubingensis* (ATCC-76608), a polymer-degrading species. The pure culture of the fungal strain was procured from the American Type Culture Collection (ATCC), USA. The malt extract (ME) medium as prescribed by the ATCC, USA was used for fungal growth and culture. To culture the fungal strain, 500 mL of malt extract (ME) media in 1000mL conical flasks were inoculated. The flasks were kept in a shaker incubator (SIF 5000R, Jeio Tech, South Korea) at 170 rpm and 30°C for 48 h for growth and culture.

2.3 Morphological Characterization and Metal Quantification

A subpart of pulverized and homogenized PCB sample was observed under a scanning electron microscope (SEM) (EVO 18, Zeiss, Germany) for morphological characterization. For the quantification of selected metals, another subpart of pulverized and homogenized PCB sample was acid digested in a microwave digester (Ethos Easy, Milestone, Italy) following the USEPA method 3052 (USEPA 1995). The digested samples were subsequently filtered using 0.22 μm filter. The filtered samples were then analyzed using inductively coupled plasma mass spectrometry (ICP-MS) (7800, Agilent, USA) for metal quantification.

2.4 Bioleaching Experiments

All bioleaching experiments were carried out in 250mL conical flasks containing 100 mL of malt extract (ME) medium. Pulverized PCB sample in the size range of 0.038 mm–1 mm was added at four different pulp densities of 2.5, 5, 7.5, and 10 g/L into the conical flasks containing malt extract (ME) medium. The conical flasks were then sterilized using an autoclave at 121°C for 20 min. The sterilized flasks were then inoculated with 1mL spore suspension of *A. tubingensis* and incubated in a shaker incubator (SIF 5000R, Jeio Tech, South Korea) at 170 rpm and 30°C. Apart from pH, the dissolution of selected metals, viz. Cu, Zn, and Ni in the bioleaching reactors along with the control reactors was monitored at three days interval by analyzing for metal concentration in bioleached liquor using ICP-MS (7800, Agilent, USA). All experiments were conducted in triplicate and average values were reported.

2.5 Analytical Measurements

Morphological characterization of the pulverized and homogenized PCB sample was performed using a scanning electron microscope (SEM) (EVO 18, Zeiss, Germany). Acid digestion was performed using a microwave digester (Ethos Easy, Milestone, Italy) for the quantification of selected metals in the PCB following the USEPA method 3052 (USEPA 1995). The pH of the bioleached liquor was measured using a pH meter (Hach, India). ICP-MS (7800, Agilent, USA) was used for the determination of metals. All analyses were carried out in triplicate.

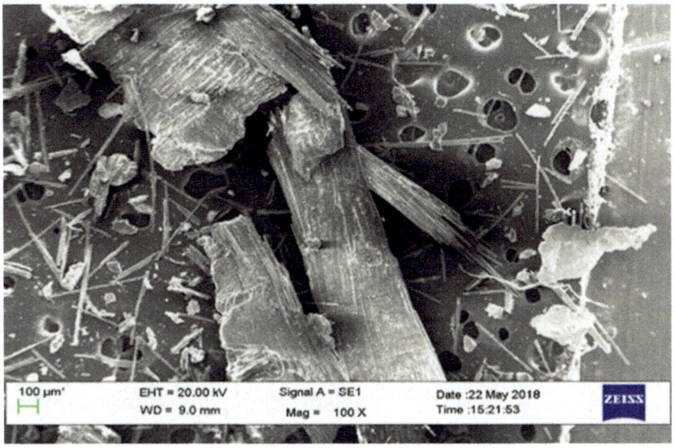

Fig. 2 SEM micrograph of pulverized PCB of obsolete computer

3 Results and Discussion

3.1 *Morphological Characterization and Metallic Content of PCB*

The morphological characterization of pulverized PCB performed under the SEM revealed the heterogeneity in particle sizes and shapes (Fig. 2). It is clear from Fig. 2 that the comminuted PCB of obsolete computer in the size range of 0.038–1 mm consisted of predominantly rods and lumps of various metals indicative of various forces under which PCBs were grounded. The observed variation in particles shapes and sizes can be attributed to the variety of metals, polymeric substances, glass fibers, etc. present in the PCB (Murugan et al. 2008).

The metal quantification revealed that the major metallic constituents present in the PCB of the obsolete computer include Cu, Ni, and Zn. The most abundant metal in the waste computer PCB was found to be Cu with the content of 21.50%, followed by the Zn and Ni with the content of about 0.10% and 0.08%, respectively. Considering the abundant presence of the base metals like Cu, Ni, and Zn, the PCB of obsolete computer can be considered as a high-grade secondary metal reservoir.

3.2 *Bioleaching of Selected Metals*

The pH of the bioleaching medium is one of the most significant parameters regulating microbial growth and metabolism in addition to the major role in the dissolution of metals from the solid substrate. With the progression of fungal bioleaching, the variation in pH of the bioleaching medium at different pulp densities is presented in Fig. 3. The variation in pH trend demonstrated a sharp increase in pH with the increase in e-waste pulp density initially followed by a subsequent steady decrease with the progression of fungal bioleaching. The increase in pH during the initial bioleaching stage can be attributed to the alkaline nature of PCB (Brandl et al. 2001). The production of various organic acids like citric acid and oxalic acid with the growth and metabolic activity of fungi can be ascribed for the subsequent steady decrease in pH with the bioleaching time (Kim et al. 2016).

Metal bioleaching results indicated that the bioleaching efficiency of the selected metals from pulverized PCB of obsolete computer using *A. tubingensis* was observed to vary considerably at different e-waste pulp densities ranging from 2.5 to 10 g/L with better dissolution at 2.5 g/L (Fig. 4). It is evident from Fig. 4 that the metal bioleaching was observed to increase gradually during initial period following which an asymptotic increasing trend was observed beyond 21 days. At 2.5 g/L of e-waste pulp density, the maximum bioleaching of 54% Zn, 34% Cu, and 8% Ni was observed in the present study. However, the bioleaching efficiency was observed to decrease to 28%, 5%, and 1% for Zn, Cu, and Ni, respectively, with the increase in e-waste pulp density to 10 g/L. Lower metal biodissolution at higher e-waste pulp density can be attributed to higher toxicity imparted by the PCB and thereby inhibiting microbial metabolism and growth leading to lesser production of organic acids as lixiviant (Zhou et al. 2008). The metal bioleaching efficacy of the fungal species, i.e., *A. tubingensis* can be ascribed to its ability to produce organic acids which further keep the pH of the medium low and thereby facilitating metal biodissolution from the PCB (Kim et al. 2016). It has been shown that citric and oxalic acids are the predominant organic acids produced by *Aspergillus* species (Aung and Ting 2005; Horeh et al. 2016; Santhiya and Ting 2005).

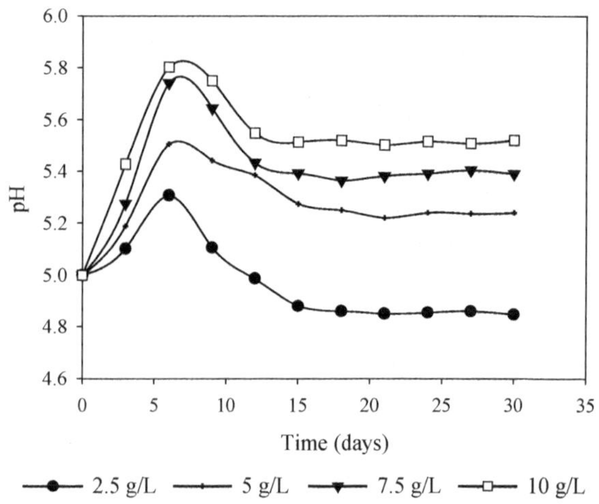

Fig. 3 Variation in pH with time at different e-waste pulp densities during fungal bioleaching

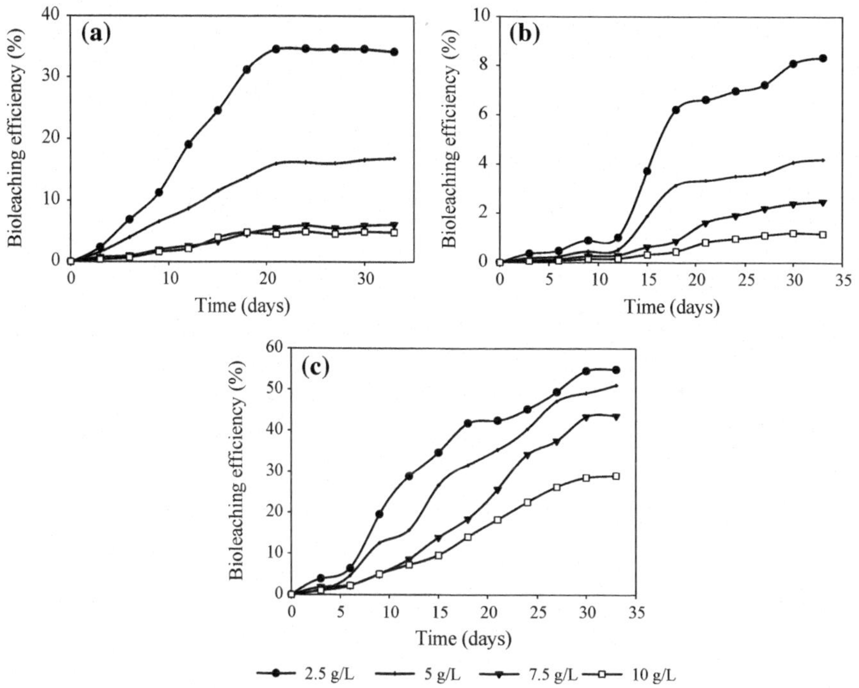

Fig. 4 Bioleaching efficacy of *Aspergillus tubingensis* for selected metals, (**a**) Cu, (**b**) Ni, and (**c**) Zn from PCB of obsolete computer at different e-waste pulp densities

3.3 Future Implications

The present study has shown the efficacy of fungal bioleaching of metals from waste PCB employing the pure culture of *A. tubingensis*, a polymer-degrading species at different e-waste pulp densities ranging from 2.5 to 10 g/L in 33 days. This highlights the practical feasibility of fungal bioleaching of metals from e-waste using *A. tubingensis*. Therefore, it is obligatory to improvise the bioleaching efficiency of metals from e-waste using mixed culture comprising of *Aspergillus* spp. and other complimentary fungal strain from a commercial point of view. The mixed culture of two or more different fungal strains can be employed for proficient metal extraction from waste PCB. Acclimatized fungal species with prior exposure to the e-waste environment may lead to enhanced metal bioleaching. Further, the metal extraction rate can be increased by optimizing the various factors governing the fungal bioleaching process like pH, inoculum size, continuous supply of nutrients, etc.

4 Conclusions

The feasibility of using *A. tubingensis*, a polymer-degrading fungal species in bioleaching the abundant metals present in the PCB of obsolete computer at different e-waste pulp density has been attempted in the present study. Results revealed that there was a decreasing trend in metal bioleaching efficiency employing *A. tubingensis* as the e-waste pulp density was increased from 2.5 to 10 g/L. Maximum bioleaching of 54% Zn, 34% Cu, and 8% Ni at 2.5 g/L of e-waste pulp density was achieved in 33 days in the present study. However, the metal extraction rate and the time required for bioleaching can be further enhanced by employing a consortium of different fungal strains and optimizing various biotic and abiotic parameters controlling the process.

References

Atlas, R. M., & Bartha, R. (1997). *Microbial ecology: Fundamentals and applications*. USA: Benjamin Cummings.

Aung, K. M. M., & Ting, Y. P. (2005). Bioleaching of spent fluid catalytic cracking catalyst using *Aspergillus niger*. *Journal of Biotechnology, 116*(2), 159–170.

Baldé, C. P., Forti, V., Gray, V., Kuehr, R., & Stegmann, P. (2017). *The global e-waste monitor 2017: Quantities, flows and resources*. United Nations University, International Telecommunication Union, International Solid Waste Association.

Beolchini, F., Fonti, V., Dell'Anno, A., Rocchetti, L., & Vegliò, F. (2012). Assessment of biotechnological strategies for the valorization of metal bearing wastes. *Waste Management, 32*(5), 949–956.

Brandl, H., Bosshard, R., & Wegmann, M. (2001). Computer-munching microbes: Metal leaching from electronic scrap by bacteria and fungi. *Hydrometallurgy, 59*(2–3), 319–326.

Burgstaller, W., & Schinner, F. (1993). Leaching of metals with fungi. *Journal of Biotechnology, 27*(2), 91–116.

Horeh, N. B., Mousavi, S. M., & Shojaosadati, S. A. (2016). Bioleaching of valuable metals from spent lithium-ion mobile phone batteries using *Aspergillus niger*. *Journal of Power Sources, 320*, 257–266.

Huang, K., Guo, J., & Xu, Z. (2009). Recycling of waste printed circuit boards: A review of current technologies and treatment status in China. *Journal of Hazardous Materials, 164*(2–3), 399–408.

Kim, M. J., Seo, J. Y., Choi, Y. S., & Kim, G. H. (2016). Bioleaching of spent Zn–Mn or Ni–Cd batteries by *Aspergillus* species. *Waste Management, 51*, 168–173.

Li, J., Lu, H., Guo, J., Xu, Z., & Zhou, Y. (2007). Recycle technology for recovering resources and products from waste printed circuit boards. *Environmental Science and Technology, 41*(6), 1995–2000.

Li, J., Shrivastava, P., Gao, Z., & Zhang, H. C. (2004). Printed circuit board recycling: A state-of-the-art survey. *IEEE Transactions on Electronics Packaging Manufacturing, 27*(1), 33–42.

Lundgren, D. G., Valcova-Valchanova, M., & Reed, R. (1986). Chemical reactions important in bioleaching and bioaccumulations. In H. L. Ehrlich & D. S. Holmes (Eds.), *Biotechnology for the mining, metal-refining, and fossil fuel processing industries* (pp. 7–22). New York: Wiley.

Murugan, R. V., Bharat, S., Deshpande, A. P., Varughese, S., & Haridoss, P. (2008). Milling and separation of the multi-component printed circuit board materials and the analysis of elutriation based on a single particle model. *Powder Technology, 183*(2), 169–176.

Narayanasamy, M., Dhanasekaran, D., Vinothini, G., & Thajuddin, N. (2018). Extraction and recovery of precious metals from electronic waste printed circuit boards by bioleaching acidophilic fungi. *International Journal of Environmental Science and Technology, 15*(1), 119–132.

Priya, A., & Hait, S. (2017). Comparative assessment of metallurgical recovery of metals from electronic waste with special emphasis on bioleaching. *Environmental Science and Pollution Research, 24*(8), 6989–7008.

Priya, A., & Hait, S. (2018). Feasibility of bioleaching of selected metals from electronic waste by *Acidiphilium acidophilum*. *Waste Biomass Valorization, 9,* 871–877.

Santhiya, D., & Ting, Y. P. (2005). Bioleaching of spent refinery processing catalyst using *Aspergillus niger* with high-yield oxalic acid. *Journal of Biotechnology, 116*(2), 171–184.

USEPA. (1995). *Microwave assisted acid digestion of siliceous and organically based matrices USEPA method 3052* (3rd ed.). Washington, DC: United States Environmental Protection Agency.

Wang, J., Bai, J., Xu, J., & Liang, B. (2009). Bioleaching of metals from printed wire boards by *Acidithiobacillus ferrooxidans* and *Acidithiobacillus thiooxidans* and their mixture. *Journal of Hazardous Materials, 172*(2–3), 1100–1105.

Wang, H., Gu, G. H., & Qi, Y. F. (2005). Crushing performance and resource characteristic of printed circuit board scrap. *Journal of Central South University of Technology, 12*(5), 552–555.

Wu, H. Y., & Ting, Y. P. (2006). Metal extraction from municipal solid waste (MSW) incinerator fly ash—Chemical leaching and fungal bioleaching. *Enyzme and Microbial Technology, 38*(6), 839–847.

Xia, M., Bao, P., Liu, A., Wang, M., Shen, L., Yu, R., et al. (2018). Bioleaching of low-grade waste printed circuit boards by mixed fungal culture and its community structure analysis. *Resources, Conservation and Recycling, 136,* 267–275.

Xin, B., Jiang, W., Aslam, H., Zhang, K., Liu, C., Wang, R., et al. (2012). Bioleaching of zinc and manganese from spent Zn–Mn batteries and mechanism exploration. *Bioresource Technology, 106,* 147–153.

Zhao, Y., Wen, X., Li, B., & Tao, D. (2004). Recovery of copper from waste printed circuit boards. *Mining, Metallurgy & Exploration, 21*(2), 99–102.

Zhou, H. B., Liu, F. F., Zou, Y. Q., Zeng, X. X., & Qiu, G. Z. (2008). Bioleaching of marmatite using moderately thermophilic bacteria. *Journal of Central South University of Technology, 15*(5), 650–655.

Zhou, Y., & Qiu, K. (2010). A new technology for recycling materials from waste printed circuit boards. *Journal of Hazardous Materials, 175*(1–3), 823–828.

Effect of CR on the Performance, Emission and Heat Release Rate of a DI Diesel Engine Run by B20 Blend of Waste Cooking Biodiesel in Diesel Fuel

Santosh Kumar Dash

Abstract This study experimentally investigates the performance, combustion and emission characteristics of a single-cylinder DI diesel engine at different compression ratio (CR16–CR18) run by popular B20 blend of waste cooking biodiesel (WCB20). The brake thermal efficiency improves, whereas the exhaust gas temperature is reduced for higher CR. The start of combustion advances for higher CR operation, which was ensured from HRR curves. At full load, peak HRR reduced as the CR increased as a result of lower premixed combustion due to the lower ignition delay at higher CR. Maximum HRR is observed for CR16. With increase in CR, the CO and HC emission drastically reduced irrespective of the loading. However, with ascending of load level the CO emission increased as a result of both lesser oxygen content and more fuel supplied at higher load. This study supports the use of WCB20 at higher CR (preferably CR18), which improves the performance as well as emissions.

Keywords Waste cooking biodiesel · Performance · Emission · Heat release rate · Diesel engine

1 Introduction

Present status of the world petroleum reserves and environmental concerns has emerged as potential threats to the humankind and other living species. To curb the pollutant emissions, most of the countries have collectively started imposing stringent emission norms for the on-road vehicles. In this regard, biodiesel made from vegetable oils and fats seems appropriate, which serves the purpose without sacrificing power (Demirbas and Demirbas 2016). Biodiesel prepared from vegetable oils is a sustainable energy resource. In the developed countries, the vegetable oils employed for the production of BD are edible in nature. However, it is not feasible in developing or underdeveloped countries, where edible oils costs are too high and it

S. K. Dash (✉)
Department of Mechanical Engineering, Aditya Engineering College, Surampalem 533437, Andhra Pradesh, India
e-mail: 90nash@gmail.com

S. K. Ghosh et al. (eds.), *Bioresource Utilization and Bioprocess*,
https://doi.org/10.1007/978-981-15-1607-8_17

intensifies the food versus fuel issues (Dash and Lingfa 2017). National biofuel policy (Government of India 2009) is in accord with the idea and encourages achieving a target of 20% blending by the end of 2017 (Government of India-Ministry of New and Renewable Energy 2009). However, as of 2018, even the implementation of 10% biodiesel blending has not achieved in reality. Several biodiesel production plants remain non-operational throughout the year owing to the paucity of Jatropha feedstock (Bandyopadhyay 2015). Hence, focus must be shifted to other suitable sources. The present updated policy (National policy on Biofuels 2018) stressed on utilizing non-edible feedstock and waste-based feedstock for the production of biodiesel (Government of India-Ministry of Petroleum and Natural Gas 2018). Numerous works have been carried out by several researchers for using various indigenously available non-edible feedstocks for the production of biodiesel and its utilization in diesel engines (Dash et al. 2018; Mofijur et al. 2013; No 2011; Dewangan et al. 2018; Dash and Lingfa 2018a). However, comparatively lesser number of articles reports the application of waste oil-based biodiesel in a diesel engine. This is the motivation behind selecting waste cooking biodiesel (WCB) for the present study. Simple modifications of engine hardware like change in compression ratio, injection timing and injection pressure have resulted in some improvement in combustion, which led to achieve both improved performance and emission characteristics as far as earlier literature concerned (Barik and Murugan 2014). Out of these simple modifications, effect of compression ratio (CR) has a significant role on the combustion characteristics, which ultimately affect performance and emission characteristics of test engine. Nagaraja et al. (2012) reported lower specific fuel consumption and improvement in brake thermal efficiency for higher compression ratio. Sharma and Murugan (2015) experimentally investigated tire pyrolysis oil—Jatropha biodiesel blend in a diesel engine at different CR. They found a significant drop of HC at higher CR and improvement of thermal efficiency. Different feedstock possesses dissimilar fuel characteristics (Dash and Lingfa 2018b). Hence, study for a particular feedstock should not be ignored. This study deliberates the experimental investigation of a popular B20 blend of waste cooking biodiesel (20% waste cooking biodiesel mix with 80% high-speed diesel)-powered DI VCR research engine at different compression ratios (16, 17 and 18) and diesel at standard compression ratio of 17.5. The properties of WCB, WCB20 and diesel are listed in Table 1.

2 Experimental

The experimental work was carried out on a single-cylinder, four-stroke DI diesel engine with a power output of 5.2 kW at 1500 rpm. Air box, fuel tank and other flow measurement devices were mounted on the appropriate places of control panel box. Software package "EngineSoft" was employed for online monitoring of engine performance and combustion parameters. AVL DiGas 444 exhaust gas analyzer is employed to measure engine emissions such as carbon monoxide (CO) and unburnt

Table 1 Properties of test fuels

Property	WCB	WCB20	Diesel	ASTM method
Acid value (mg KOH/g)	0.21	0.06	–	D664
Calorific value (MJ/kg)	38.357	41,368	42.137	D4809
Viscosity (cSt@40 °C)	4.7	2.72	2.8	D445
Flash point (°C)	148	81	67	D93
Pour point (°C)	3.5	−5.7	−8.5	D97
Density (g/cc)	0.876	0.840	0.833	D1298
Cetane number	50.6	48.73	48.24	D613

hydrocarbon (HC). AVL 437 smoke meter (accuracy of $\pm 1\%$) was used for the smoke opacity (%) measurement.

All the tests pertaining to diesel fuel are carried out at standard engine settings, i.e., at 17.5 compression ratio, 23° BTDC injection timing and 210 bar injection pressure. For blended fuel, changing of CR was achieved by tilting the cylinder head with respect to the reference mark provided by the manufacturers. A leak check was performed for AVL DiGas analyzer, before collecting any readings. For a particular result, at least three time average value is calculated and presented for minimizing undesired fluctuation in result value. The uncertainties of some of the important parameters involved in this study are calculated by the method described by Holman (2011). Overall uncertainties of the experiments are found to be within $\pm 3.1\%$ for performance and combustion. An uncertainty for emission study is found within $\pm 4\%$.

3 Result and Discussion

3.1 Brake Thermal Efficiency

From Fig. 1, it can be observed that the brake thermal efficiency increases with an increase in load as expected. This is due to the increase in brake power to sustain the applied load. At full load and CR18, the EGT for the blend has been reduced by 5.57% and 6.8% as compared to diesel at CR17.5 and blend at CR16 respectively. Better atomization and spray formation owing to the high pressure and temperature in the combustion chamber at 18 CR have resulted in an improved performance.

Fig. 1 Variation of BTE with load at different CRs

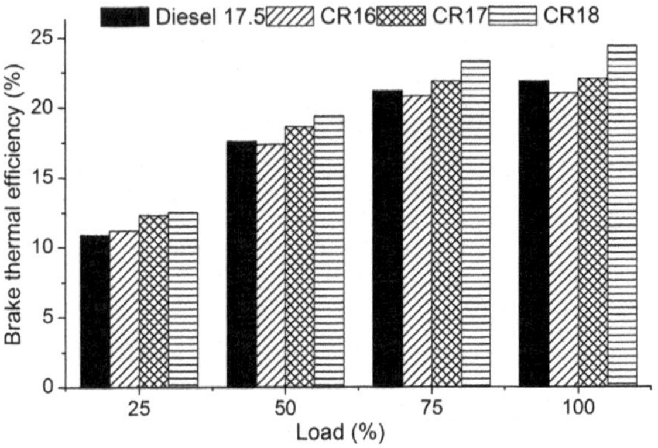

3.2 Exhaust Gas Temperature

Figure 2 shows the exhaust gas temperature (EGT) which increases with the increase in load. More amount of fuel burned as the load increases, which is the primary reason for the increase in EGT. Slight increase in EGT is observed for CR16, and significant decrease in EGT is observed for CR18. This may be attributed to the fact that at lower CR, the expansion could not take place properly, which resulted in loss of heat to the exhaust pipe, whereas at higher CR more expansion resulted in decrease in heat loss as a consequence yields lower exhaust temperature. The lower exhaust gas temperature for the blend at 18 CR is also evident from the higher thermal efficiency at 18 CR. At full load, for 18 CR 5.57% and 6.8% lower EGT is observed for the blend compared to diesel at normal setting and blend at 16 CR respectively.

Fig. 2 Variation of EGT with load for different CRs

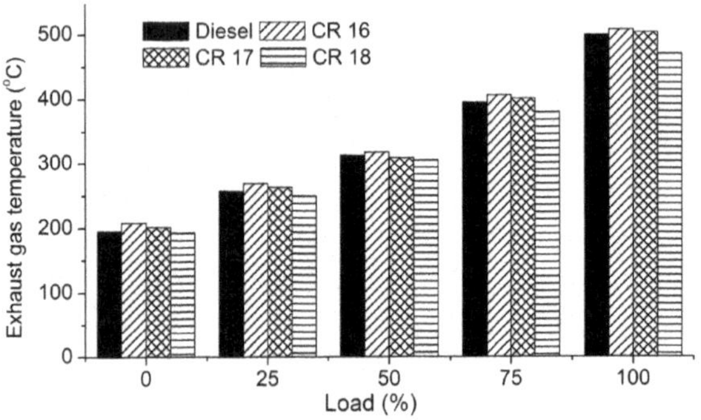

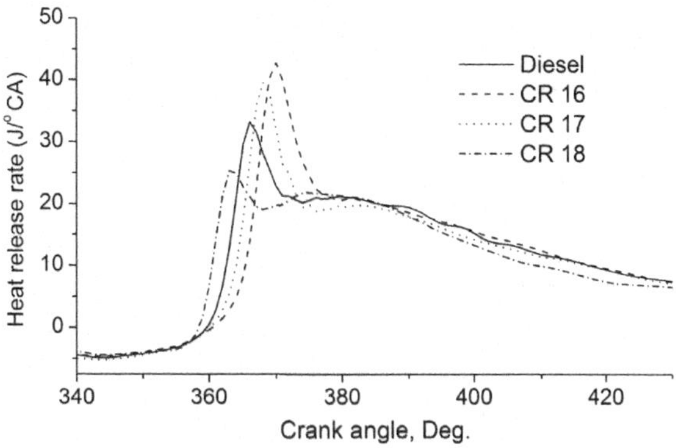

Fig. 3 Variation of HRR with crank angle for different CRs

3.3 Heat Release Rate

The pressure crank angle data were analyzed for the estimation of heat release rate. An equation derived from the first law of thermodynamics is employed for evaluating heat release rate and is given below.

$$\frac{\mathrm{d}Q}{\mathrm{d}\theta} = p\frac{\gamma}{\gamma - 1}\left(\frac{\mathrm{d}v}{\mathrm{d}\theta}\right) + \frac{1}{\gamma - 1}V\frac{\mathrm{d}p}{\mathrm{d}\theta} + Q_w$$

where $\frac{\mathrm{d}Q}{\mathrm{d}\theta}$ is the heat release rate, P and V are the instantaneous cylinder pressure and volume, respectively, and γ is the ratio of specific heats, $\frac{Cp}{Cv}$. The wall heat transfer is neglected in the present study. The plot of heat release rate at different CR with respect to crank angle is shown in Fig. 3. It can be seen from the figure that at lower CR peak heat release rate increased significantly for the blend. This is due to the higher ignition delay at lower CR resulted in a rapid burning of more fuel particles once ignition initiates, which is responsible for significant surge in premixed heat release rate. The peak HRR is obtained as 33.23 J/°CA at 366 °CA, 42.67 J/°CA at 370 °CA, 39.58 J/°CA at 368 °CA and 25.37 °CA at 363 °CA for diesel, blend at 16 CR, 17 CR and 18 CR, respectively.

3.4 CO Emission

The variation of CO emission with load at different CR is presented in Fig. 4. It can be observed that with the increase in load from 25 to 75%, the CO emission increases very slowly for both the diesel and the blend at all operating conditions. However, it is drastically increased at full load condition. This is due to the insufficient oxygen availability at higher load as a result of lower air fuel ratio, which resulted in an incomplete combustion. At full load, the blend at 18 CR shows 39, 38 and 30% lower CO emission compared to diesel at normal setting, blend at 16 CR and blend

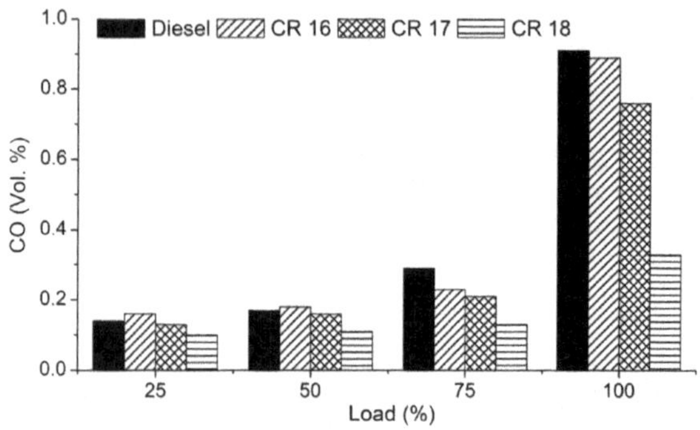

Fig. 4 Variation of CO emission with load for different CRs

at 17 CR, respectively. This may be due to the complete combustion of the blend at 18 CR owing to the improved atomization of the fuel injected to the combustion chamber at high temperature and pressure.

3.5 HC Emission

The effect of CR and load on HC emission of test fuels is depicted in Fig. 5. At part load, minimum HC emission observed due to the combined influence of improved air–fuel mixture and combustion temperature. At full load, unburned hydrocarbon (HC) for blend 18 CR decreased by 23.44, 28.66 and 26.35% compared to diesel fuel, blend at 16 CR and blend at 17 CR, respectively. The drop in HC and CO emissions for 18 CR is attributed to the improved temperature and pressure which resulted in the formation of fine spray of fuel particles before combustion.

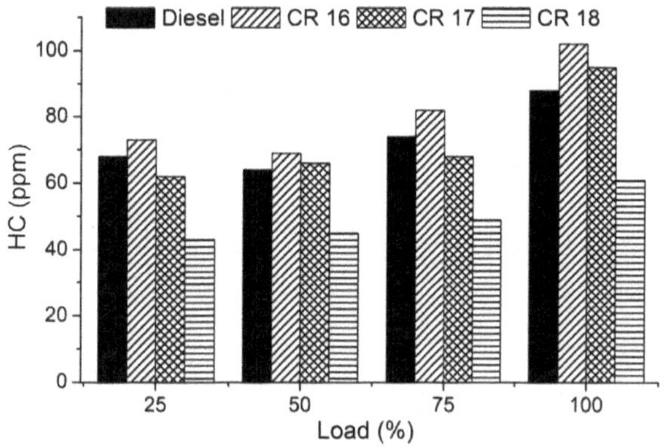

Fig. 5 Variation of HC emission with load for different CRs

4 Conclusion

B20 blend of waste cooking biodiesel has been experimentally investigated at different loads and compression ratios in a direct injection, single-cylinder and four-stroke variable compression ratio diesel engine. The conclusion drawn from the performance, emission and combustion characteristics is given below.

- At maximum CR, the brake thermal efficiency increased (by 10.4%) significantly compared to baseline diesel fuel. Owing to lower heat loss for 18 CR, the exhaust gas temperature dropped.
- At full load, both CO and HC emission drastically dropped by 39% and 23.44%, respectively, for 18 CR.
- Owing to the earlier combustion of test fuel at higher CR, premixed phase of combustion reduced significantly and better mixing controlled phase of combustion are observed for 18 CR.
- Keeping in view all the experimental results, it is concluded that possible higher CR is desirable for better performance and emission characteristics of B20 blend of waste cooking biodiesel-powered diesel engine.

References

Bandyopadhyay, K. R. (2015). Biofuel Promotion in India for Transport: Exploring the Grey Areas. Policy Brief, The Energy and Resources Institute, February 2015. http://www.teriin.org/policybrief/docs/biofuel.pdf. Last accessed April 26, 2017.

Barik, D., & Murugan, S. (2014). Simultaneous reduction of NO_x and smoke in a dual fuel DI diesel engine. *Energy Conversion and Management, 84,* 217–226.

Dash, S. K., & Lingfa, P. (2017). A review on production of biodiesel using catalyzed transesterification. *AIP Conference Proceedings, 1859*, 020100. https://doi.org/10.1063/1.4990253. http://dx.doi.org/10.1063/1.4990253.

Dash, S. K., & Lingfa, P. (2018a). An overview of biodiesel production and its utilization in diesel engines. *IOP Conference series: Material Science and Engineering, 377,* 012006. https://doi.org/10.1088/1757-899X/377/1/012006.

Dash, S. K., & Lingfa, P. (2018b). Performance evaluation of Nahar oil-diesel blends in a single cylinder direct injection diesel engine. *International Journal of Green Energy, 15,* 400–405.

Dash, S. K., Lingfa, P., & Chavan, S. B. (2018). An experimental investigation on the application potential of heterogeneous catalyzed Nahar biodiesel and its diesel blends as diesel engine fuels. *Energy Sources, Part A: Recovery, Utilization, and Environmental Effects, 40,* 2923–2932. https://doi.org/10.1080/15567036.2018.1514433.

Demirbas, A., & Demirbas, B. (2016). Sustainable rural bioenergy production for developing countries. *Energy Sources, Part A: Recovery, Utilization, and Environmental Effects, 38,* 3578–3585.

Dewangan, A., Yadav, A. K., & Mallick, A. (2018). Current scenario of biodiesel development in India: prospects and challenges. *Energy Sources, Part A: Recovery, Utilization, and Environmental Effects, 40,* 2494–2501. https://doi.org/10.1080/15567036.2018.1502849.

Government of India-Ministry of New and Renewable Energy. (2009). National Policy on Biofuels, New Delhi. http://mnre.gov.in/file-manager/UserFiles/biofuel_policy.pdf.

Government of India-Ministry of Petroleum and Natural Gas. (2018). National Policy on Biofuels, New Delhi. http://petroleum.nic.in/sites/default/files/biofuelpolicy2018_1.pdf. Last accessed on 20/07/2018.

Holman, J. P. (2011). *Experimental methods for engineers* (8th ed.). New Delhi: Tata McGraw Hill.

Mofijur, M., Atabani, A. E., Masjuki, H. H., Kalam, M. A., & Masum, B. M. (2013). A study on the effects of promising edible and non-edible biodiesel feedstocks on engine performance and emissions production: A comparative evaluation. *Renewable and Sustainable Energy Reviews, 23,* 391–404.

Nagaraja, S., Shakthivel, M., & Sudhakaran, R. (2012). comparative study of the combustion, performance, and emission characteristics of a variable compression ratio engine fuelled with diesel, corn oil methyl ester, and palm oil methyl ester. *Journal of Renewable and Sustainable Energy, 4,* 063122.

No, S. Y. (2011). Inedible vegetable oils and their derivatives for alternative diesel fuels in CI engines: A review. *Renewable and Sustainable Reviews, 15,* 131–149.

Sharma, A., & Murugan, S. (2015). Potential for using a tyre pyrolysis oil-biodiesel blend in a diesel engine at different compression ratios. *Energy Conversion and Management, 93,* 289–297.

Leaf Litter Biogas Digester Residue—A Nutrient Supplement for Mushroom Cultivation

Sreesha Malayil and Hoysall N. Chanakya

Abstract Partially digested biogas digester residues (BDR) arising from leaf litter/agro-residues fed biogas plants (typically used as fertilizer) were tried and tested as a substrate for mushroom cultivation. Such digested residues are outputs of emerging MSW and peri-urban biomethanation plants and can greatly create value addition and increase economic sustainability of the overall MSW treatment systems. Low-value agro-residues such as sugarcane trash, groundnut husk, and sorghum stalk were substituted with 30% BDR and used as substrate for cultivation of *Pleurotus florida* and *Pleurotus djamor*. The biological efficiency, yield, and nutritional value achieved with these substrates were studied. The most suitable substrate that produced higher yields and biological efficiency was sorghum stalk mixed with BDR (2 kg/kg substrate). Addition of BDR with agro-residues could increase mushroom yield by 17–20% and reduce growth stages by 3–5d. Anaerobic digestion of biomass removes >60% of carbohydrates as gas while conserving the NPK in the BDR. BDR rich in N was also increased the protein content of both *P. florida* and *P. djamor* by 7–12%.

Keywords Biogas digester residue · Pleurotus cultivation · Enhanced mushroom yeilds · VAP from biogas plants

1 Introduction

Demand for alternatives to petroleum is increasing the production of biofuels from municipal solid wastes (MSW), agro-residues, and leaf litter. This has resulted in design and implementation of large-scale biogas plants (BGP) with MSW, crop residues and leaf litter as feedstock. Unlike food waste, anaerobic digestion (AD) of MSW (with street sweepings and leaf litter), crop residues/leaf litter is difficult

S. Malayil (✉) · H. N. Chanakya
Centre for Sustainable Technologies (Formerly ASTRA), Indian Institute of Science, Bengaluru 560012, India
e-mail: srisham@iisc.ac.in

H. N. Chanakya
e-mail: chanakya@iisc.ac.in

S. K. Ghosh et al. (eds.), *Bioresource Utilization and Bioprocess*,
https://doi.org/10.1007/978-981-15-1607-8_18

to achieve due to the recalcitrance of lignin in biomass. Therefore, AD of crop residue/leaf litter generates large quantities (40–50%) of undigested residue (BDR) which is rich in essential nutrients (N, P, K) and trace elements. However, nutrients are present in inorganic plant available forms at a markedly higher extent in BDR, compared to untreated waste, due to the large input of organic nutrients that are mineralized during the digestion process (Gerardi 2003). For instance, BDR contains 25% more accessible ammonium (NH_4^+-N) than untreated liquid manure (Monnet 2003). Consistent with these findings, several trials show that BDR enhances crop yield (Tiwari et al. 2000). Additionally, BDR inhibits plant diseases and induces of plant pathogen resistance (Zhao et al. 2011) and has a direct effect on soil-borne diseases and indirect effect on stimulation of biological activity (Odlare et al. 2008). While conventional dung-based BGP gives two outputs, biogas and slurry in conventional, the plug flow digester designed by ASTRA-IISc of AD of agro-residues gives three outputs (1) biogas, (2) digester liquid (BDL), and (3) partially digested material (biogas digester residue, BDR) because there is a phase separation between the outputs as the material is not pulverized when fed into the reactor (Chanakya and Malayil 2012; Chanakya and Sreesha 2012).

Recovery of nutrients from anaerobic compost that can be enhanced by increasing the incorporation efficiency of nitrogen in the form of protein, by growing mushrooms on agricultural residues which are supplemented with BDR as nitrogen-rich source. BDR as a single feedstock for cultivation of *Pleurotus* spp. has been tried and found inefficient due to the collapse of the feedstock bed after 14d spawn run (Gangulli and Chanakya 1994). This was infested by insects and lead to very low yield in the first run and also inefficient aeration of the biomass bed due to compaction as the material is too soft. It was reported that BDR on its own is an inefficient feedstock for mushroom cultivation (Gangulli and Chanakya 1994). BDR amended with paddy straw (0, 50, and 100% mix) have been tested by Ganguli and Chanakya (1994) for the cultivation of *Pleurotus* and found efficient in doubling the yield. Also, BDR high in lignin content can only be digested by Basidiomycetes which have the ability to release extracellular lignin degrading enzymes like laccase and manganese peroxidase (Mishra and Kumar 2007). BDR with high N-content if applied directly to soil as fertilizer leads to the loss of nitrogen by nitrification and denitrification, whereas if combined if agro-residue for mushroom cultivation leads to effective use and recycling of N and other nutrients (Chanakya and Sreesha 2012).

The aim of this study is to study the yield of *Pleurotus florida* and *Pleurotus djamor* on feedstocks like sugarcane trash, groundnut husk, and sorghum stalk combined with anaerobically digested biomass (BDR).

2 Materials and Methods

The agro-residues, sugarcane trash, sorghum stalk, and groundnut husk were obtained from Hassan District, Karnataka, India. The primary inocula of *P. djamor* and *P.*

florida were obtained from Indian Institute of Horticultural Science (IIHR), Bangalore. BDR was obtained from a plug flow digester designed by ASTRA-CST, fed with banana leaf biomass (30d SRT). The digested material was pulled out from the digester, sun-dried and solarized for 7 days to prevent contamination and later oven-dried at 90 °C for 8 h. The agro-residue for mushroom cultivation (groundnut husk; GN, sorghum stalk; SS, and sugarcane trash; ST) was dried in oven at 90 °C for 24 h. Sorghum stalk and sugarcane trash were chopped into 5–7 cm length pieces. All the three substrates were then filled in jute bags and soaked in tap water for 12 h. Excess water was drained off to maintain moisture content of 60%, and the substrates were pasteurized by dipping in hot water at 75 °C for 30 min. BDR was sun-dried and solarized for seven days. Steam sterilization of BDR was conducted in order to prevent disintegration of the material during pasteurization in hot water (Gangulli and Chanakya 1994). The mixing ratio of BDR and substrate was 1:1—a ratio chosen based on previous study (Gangulli and Chanakya 1994). Later the moisture content of the single substrate and admixtures was adjusted to 50%, and spawning (5% of TS) was carried out, 3 replicates of each admixture and single substrate feedstocks were made. After spawning, mixtures and single biomass feedstocks were filled in perforated polythene bags (Bano et al. 1962). The cultivation was carried out in a humid chamber covered with jute cloth in laboratory conditions. The poly-bag cultures were frequently sprayed with water everyday till the stage of fruiting body initiation in order to maintain adequate moisture content and high humidity (Gangulli and Chanakya 1994).

A total weight of all the fruiting bodies harvested from the three subsequent pickings were measured and reported as the total yield of mushroom. The biological efficiency (yield of mushroom per kg dry substrate) was calculated using the formula given by Chang and Hayes (1978). The statistical analysis was carried out using RStudio [version 2.15.2 (2012-10-26)] and other worksheet packages.

3 Results and Discussion

Pleurotus florida was cultivated on ST, GN, and combination of this with BDR, and *P. djamor* was cultivated on SS, ST, and in combination with BDR. The aim of the experiment was to check the effect of BDR on mushroom in terms of total yield, biological efficiency, and time required to reach stages of pinhead formation, fruit body emergence, duration required for the first harvest, and subsequent harvests, etc. The addition of BDR to agro-residues was not only expected to increase the yields but also hasten the process of mushroom production (Gangulli and Chanakya 1994). The highest yield of *Pleurotus* reported so far was on paddy straw and wheat straw (0.5–0.1 kg/kg, Ragunathan et al. 1996; Kalmis and Sargin 2004; Liang et al. 2009), whereas in the current experiment we have chosen low economic value agro-residues such sugarcane trash (ST), groundnut shell (GN), and sorghum stalk (SS) and their combinations with BDR. Figure 1 shows the various species of mushroom grown on various agro-residues.

Fig. 1 **a** *P. florida* grown on sugarcane trash amended with BDR, **b**, **c** *P. djamor* grown on sorghum stalk amended with BDR, **d** mycelial network formation of *P. florida* on groundnut shell amended with BDR

3.1 Pinhead and Fruiting Body Formation

Addition of BDR to agro-residue for both the *Pleurotus* spp. showed different responses depending on the type of feedstock mixed and the species used (Table 1). Pinhead formation for *P. florida* using ST as feedstock required 20d, and this was reduced to 17d when mixed with BDR. In case of GN, 17d was required for pinhead formation and this was reduced to 15d when mixed with BDR for *P. florida*. Many authors report primordia initiation to take place around 24–30 days with single feedstocks like PS, CP, maize stover, sugarcane bagasse, wheat straw, while using various *Pleurotes* species such as *P. ostreatus*, *P. sajor-cajo*, and *P. citrinopileatus* (Ragunathan et al. 1996; Kalmis and Sargin 2004; Liang et al. 2009). This suggests that addition of BDR to agro-residue for cultivation of *P. florida* could potentially reduce the pinhead formation time and hasten fruiting formation. On the other hand, addition of BDR to ST for cultivation of *P. djamor* increased pinhead formation time from 20d (ST as single feedstock) to 21d which was different from other feedstocks studied in this laboratory earlier (paddy straw, coir pith, and their combination with BDR, Chanakya et al. 2015), and this needs further investigation. In case of SS and

Table 1 Time taken for pinhead formation, fruiting body formation, and total yield obtained for various substrates

Substrate	Pinhead (d)	Fruiting body (d)	Yield (kg/kg)	BE%	Substrate	Pinhead (d)	Fruiting body (d)	Yield (kg/kg)	BE%
P. florida					*P. djamor*				
ST	20 ± 0.80	26 ± 0.55	1.21 ± 2.41	121.08 ± 0.80	ST	20 ± 0.45	26 ± 0.48	0.95 ± 1.25	95.11 ± 0.88
ST + BDR	17 ± 0.68	23 ± 0.69	1.57 ± 1.92	156.98 ± 0.88	ST + BDR	21 ± 0.26	23 ± 0.43	1.19 ± 2.10	119.21 ± 0.90
GN	17 ± 1.07	23 ± 1.18	1.07 ± 0.97	107.53 ± 0.92	SS	15 ± 0.65	18 ± 0.39	1.67 ± 2.51	167.44 ± 0.81
GN + BDR	15 ± 0.96	19 ± 0.88	1.34 ± 1.54	134.35 ± 0.87	SS + BDR	15 ± 0.99	18 ± 0.54	2.04 ± 1.87	203.97 ± 0.85

its combination with BDR, there was no reduction in the time taken for pinhead formation and this was achieved in 15d which was the least time taken when comparing both the species of *Pleurotus* and type of agro-residue used.

Fruiting body formation for the three substrates and their combination with BDR for both the species of *Pleurotus* understudy occurred between 18 and 26d. For *P. florida* and ST as single feedstock, fruiting body formation occurred at 26d and addition of BDR to this reduced this time to 23d. In case of *P. djamor* also this pattern of 3d reduction in fruiting body formation from 26 to 23d by addition of BDR was seen to be repeated. With *P. florida* and GN as feedstock, the fruiting body formation occurred at 23d for single feedstock and addition of BDR reduced it to 19d. This reduction of 3d in fruiting body formation repeated with *P. djamor* when grown on ST with BDR. But in case of *P. djamor* and SS as feedstock, there was no noticeable change in reduction of fruiting body formation (18d for both single feedstock and with BDR) with addition of BDR.

3.2 Effect of BDR Mix with Reference to Yield

The yield of mushroom discussed here is the total yield of three flushes (80d) and reported as fresh weight of mushroom harvested per unit fresh weight of feedstock. In case of the three feedstocks understudy SS with BDR gave the highest yield (2.03 kg/kg substrate) with *P. djamor* (Table 1). Previous study in this laboratory showed that *P. djamor* gave a maximum yield of 1.25 kg/kg when grown on a mixture of PS + BDR (50% each; unpublished data, with weed feedstock derived BDR at 30% substitution) and *P. florida* for PS + BDR with banana leaf derived BDR (2.32 kg/kg, with 30% BDR substitution, Chanakya et al. 2015). All the three feedstocks gave higher yield (18–23% higher than single feedstocks) with the addition of BDR. Addition of BDR to ST gave 22.8% higher yield for *P. florida* and 20.2% for *P. djamor*. In case of GN, 19.9% higher yield was achieved for *P. florida* with addition of BDR and SS gave 17.9% higher yield for *P. djamor* with addition of BDR. When comparing the two species of *Pleurotus*, *P. djamor* gave higher yield for SS + BDR (2.03 kg/kg substrate, BE of 20.397%, Table 1), but for ST + BDR *P. florida* gave better yield of 1.56 kg/kg substrate (BE of 156.98%, Table 1). Earlier studies with BDR substitution showed higher yields for *P. florida* in case of paddy straw when compared to *P. flabellatus* (2.32 kg/kg which is the highest yield reported with BDR substitution, Chanakya et al. 2015). Further, it is observed that in this study, firstly addition of BDR increased the productivity compared to single feedstocks. Secondly 30% substitution was found to be more efficient compared to earlier studies Finally, among all the feedstocks studied in this laboratory PS + BDR gave the highest yield followed by SS + BDR and among the species studied *P. florida* gave higher yields for PS + BDR and *P. djamor* for SS + BDR (Ganguli and Chanakya 1994; Chanakya et al. 2015).

Addition of BDR to agro-residue for both the *Pleurotus* spp. showed different responses depending on the type of feedstock mixed and the species used (Table 1).

Pinhead formation for *P. florida* using ST as feedstock required 20d, and this was reduced to 17d when mixed with BDR. In case of GN, 17d was required for pinhead formation and this was reduced to 15d when mixed with BDR for *P. florida*. Many authors report primordia initiation to take place around 24–30 days with single feedstocks like PS, CP, maize stover, sugarcane bagasse, wheat straw while using various *Pleurotes* species such as *P. ostreatus, P. sajor-cajo,* and *P. citrinopileatus* (Ragunathan and Swaminathan 2003; Kalmis and Sargin 2004; Liang et al. 2009). This suggests that addition of BDR to agro-residue for cultivation of *P. florida* could potentially reduce the pinhead formation time and hasten fruiting formation. On the other hand, addition of BDR to ST for cultivation of *P. djamor* increased pinhead formation time from 20d (ST as single feedstock) to 21d which was even different from other feedstocks studied in this lad earlier (paddy straw, coir pith, and their combination with BDR, Chanakya et al. 2015), and this needs further investigation. In case of SS and its combination with BDR, there was no reduction in the time taken for pinhead formation and this was achieved in 15d which was the least time taken when comparing both the species of *Pleurotus* and type of agro-residue used.

Fruiting body formation for the three substrates and their combination with BDR for both the species of *Pleurotus* understudy occurred between 18 and 26d. For *P. florida* and ST as single feedstock fruiting body formation occurred at 26d and addition of BDR to this reduced this time to 23d. In case of *P. djamor*, also this pattern of 3d reduction in fruiting body formation from 26 to 23d by addition of BDR repeated. With *P. florida* and GN as feedstock, the fruition body formation occurred at 23d for single feedstock and addition of BDR reduced it to 19d. This reduction of 3d in fruiting body formation repeated with *P. djamor* grown on ST with BDR. But in case of *P. djamor* and SS as feedstock, there was no noticeable change in reduction of fruiting body formation (18d for both single feedstock and with BDR) with addition of BDR (Table 2).

Table 2 Moisture and protein content of mushroom

Substrates	Moisture (%)	Crude protein (%)
P florida		
ST	88.56 ± 0.10	22.50 ± 0.08
ST + BDR	89.12 ± 0.06	23.10 ± 0.05
GN	88.68 ± 1.12	20.25 ± 0.10
GN + BDR	89.74 ± 0.98	23.04 ± 0.09
P djamor		
ST	92.17 ± 0.07	22.50 ± 0.09
ST + BDR	92.59 ± 0.08	23.67 ± 0.08
SS	91.13 ± 0.07	19.32 ± 0.07
SS + BDR	91.45 ± 0.09	19.89 ± 0.10

3.3 Protein and Moisture Content of Mushroom

Addition of BDR to agro-residues was found to increase the moisture content of *P. florida* for both ST and GN. In case of *P. djamor*, the moisture content was not increased with BDR addition in both ST and SS and the reason for this need to be reached further. The maximum moisture content reported so far for *Pleurotus* cultivated on various agro-residues was found to be 88–93% (Bonatti et al. 2004; Ragunathan and Swaminathan 2003), and in the current experiment addition of BDR even though enhanced some moisture content, it did not exceed the range mentioned in literature. Enhanced yields were thus not merely increase in moisture content of the mushrooms.

Addition of BDR was found to increase the protein content of mushroom (both in case of *P. florida* and *P. djamor*) for all agro-residues tried and tested. BDR is found to have higher locked N since AD of biomass removes a lot more CHO as gas (CH_4 and CO_2) generally leaving a greater extent of N behind in the BDR when compared to aerobic compost (Chanakya and Malayil 2012). With a lot of easy to decompose C removed, and yet with a higher N-content, this feedstock appears a good supplement to mushroom cultivation without potential to suffer contamination emerging from N supplementation as reported above (Chanakya et al. 2015). The higher and faster pick of N from BDR by the *Pleurotus* spp. has resulted in higher yields and protein content.

4 Conclusion

Biogas digester residue (BDR) generated from a biogas plant fed with leaf litter/agro-residue was used as supplement to grow edible mushroom (*P. florida* and *P. djamor*). Low-value agro-residues such as sugarcane trash, groundnut husk and sorghum stalk which cannot be used as fodder was supplemented with 30% BDR and tested for mushroom cultivation. BDR rich in nutrients and microelement not only enhanced yields (17–20% higher) but also reduced the time for growth stages (pinhead formation, fruiting body formation, and first flush) by 3–5d. Addition of BDR was also found to enhance the protein content of both *Pleurotus* spp. studied for various agro-residues chosen. These results indicate that BDR which otherwise would have been used merely as compost can be now be used in the mushroom cultivation industry for higher process efficiency and nutrient capture and greater value addition to MSW and biogas plant operation.

References

Bano, Z., Srinivasan, K. S., & Srivastava, H. C. (1962). Amino acid composition of the protein from a mushroom (*Pleurotus* sp.). Central Food Technological Research Institute, Mysore, India.

Bonatti, M., Karnopp, P., Soares, H. M., & Furlan, S. A. (2004). Evaluation of *Pleurotus ostreatus* and *Pleurotus sajor-caju* nutritional characteristics when cultivated in different lignocellulosic wastes. *Food Chemistry, 88*(3), 425–428. https://doi.org/10.1016/j.foodchem.2004.01.050.

Chanakya, H. N., & Malayil, S. (2012). Anaerobic digestion for bioenergy from agro-residues and other solid wastes—An overview of science, technology and sustainability. *Journal of the Indian Institute of Science, 92*(1), 111–144. Retrieved from http://journal.library.iisc.ernet.in/index.php/iisc/article/view/25.

Chanakya, H. N., Malayil, S., & Vijayalakshmi, C. (2015). Cultivation of *Pleurotus* spp. on a combination of anaerobically digested plant material and various agro-residues. *Energy for Sustainable Development, 27,* 84–92. https://doi.org/10.1016/j.esd.2015.04.007.

Chanakya, H. N., & Sreesha, M. (2012). Anaerobic retting of banana and arecanut wastes in a plug flow digester for recovery of fiber, biogas and compost. *Energy for Sustainable Development, 16*(2), 231–235. https://doi.org/10.1016/j.esd.2012.01.003.

Chang, S. T., Hayes, W. A. (1978). *The biology and cultivation of ediblemushrooms. Academic Press, New York.*

Gangulli, N. K., & Chanakya, H. N. (1994). Mushroom cultivation on spent biomass from biogas plants. *Current Science, 66*(1), 70–73.

Gerardi, M. H. (2003). The microbiology of anaerobic digesters. Online ISBN: 9780471468967.

Kalmiş, E., & Sargin, S. (2004). Cultivation of two *Pleurotus* species on wheat straw substrates containing olive mill waste water. *International Biodeterioration and Biodegradation, 53*(1), 43–47. https://doi.org/10.1016/j.ibiod.2003.08.002.

Liang, Z. C., Wu, C. Y., Shieh, Z. L., & Cheng, S. L. (2009). Utilization of grass plants for cultivation of *Pleurotus citrinopileatus*. *International Biodeterioration and Biodegradation, 63*(4), 509–514. https://doi.org/10.1016/j.ibiod.2008.12.006.

Mishra, A., & Kumar, S. (2007). Cyanobacterial biomass as N-supplement to agro-waste for hyper-production of laccase from *Pleurotus ostreatus* in solid state fermentation. *Process Biochemistry, 42*(4), 681–685.

Monnet, F. (2003). An introduction to anaerobic digestion of organic wastes. In Remade Scotland; Final Report Biogas.

Odlare, M., Pell, M., & Svensson, K. (2008). Changes in soil chemical and microbiological properties during 4 years of application of various organic residues. *Waste Management, 28,* 1246–1253.

Ragunathan, R., Gurusamy, R., Palaniswamy, M., & Swaminathan, K. (1996). Cultivation of *Pleurotus* spp. on various agro-residues. *Food Chemistry, 55*(2), 139–144. https://doi.org/10.1016/0308-8146(95)00079-8.

Ragunathan, R., & Swaminathan, K. (2003). Nutritional status of *Pleurotus* spp. grown on various agro-wastes. *Food Chemistry, 80*(3), 371–375. https://doi.org/10.1016/S0308-8146(02)00275-3.

Tiwari, V. N., Tiwari, K. N., & Upadhyay, R. M. (2000). Effect of crop residues and biogas slurry incorporation in wheat on yield and soil fertility. *Journal of the Indian Society of Soil Science, 48,* pp. 515–520.

Zhao, Q., Dong, C., Yang, X., Mei, X., Ran, W., Shen, Q., & Xu, Y. (2011). Biocontrol of Fusarium wilt disease for Cucumis melo melon using bio-organic fertilizer. *Applied Soil Ecology, 47,* pp. 67–75.

Using Solid Waste Biomass for Dye Adsorption in Water Treatment

Mohan Rao Tamtam and Basava Rao Vudata Venkata

Abstract The abundantly available biomass *Ficus religiosa* leaf powder (FRLP) was used as an adsorbent for the removal of pollutants like Congo red (CR) from aqueous solution. The feasibility and capacity of removal were determined from the equilibrium and thermodynamic studies. The results exhibited that FRLP is a competitive and effective adsorbent compared to those commercially available adsorbents with maximum uptake of 227.7 mg per gram of adsorbent. The studies also indicated that the removal process is spontaneous and exothermic.

Keywords *Ficus religiosa* · Congo red · Adsorption · Isotherms · Thermodynamics

1 Introduction

The residual dyes from various industries like paper and pulp, textile, tannery, dye and dye intermediates, pharmaceutical, and kraft bleaching industries, etc., are considered to be the most important pollutants. The effluents from these industries initiate the pollution of natural water resources. The synthetic dyes in effluent act as eco-toxic hazards, affecting photosynthesis of aquatic plants and causing depletion of oxygen concentration in water and suffocation of aquatic flora and fauna (Wang et al. 2010). The most common method that has been employed for the removal of these effluents appears to be adsorption.

Adsorption is a simple and efficient process with easy recovery and reusability of the adsorbent. Activated carbon is extensively used as an adsorbent in many industries because of its high adsorption capacity and wide applicability. However, activated carbon being expensive and difficult to regenerate, the researchers have been

M. R. Tamtam (✉)
Department of Chemical Engineering, Bapatla Engineering College, Bapatla, AP, India
e-mail: mohanrtam@gmail.com

B. R. Vudata Venkata
Department of Chemical Engineering, University College of Technology, OU, Hyderabad, Telangana, India
e-mail: profbasavarao_1964@yahoo.com

S. K. Ghosh et al. (eds.), *Bioresource Utilization and Bioprocess*,
https://doi.org/10.1007/978-981-15-1607-8_19

investigating the potential of alternative low-cost and readily available adsorbents (Kumar et al. 2007). In this regard, several natural materials like zeolites (Wang et al. 2009; Alver and Metin 2012), clays (Errais et al. 2011; Basava Rao and Mohan Rao 2015; Akar and Uysal 2010), coal (Ting-Chu 2008; Rusu et al. 2014), and waste biomass (RaoT and Basava Rao 2015; Mohan Rao and Basava Rao 2016; Mao et al. 2008; Kumar et al. 2010; Mehmood et al. 2015; Lafi et al. 2014; Gaikwad and Kinldy 2009) have been studied out of which the use of biomass for removal of dye has gathered attention because of its cost-effectiveness as well as environment- and eco-friendly features (Aksu 2005).

Congo red (CR) in industrial effluent is considered to be highly toxic as it causes hypersensitive effect in biological systems leading to cytotoxicity, genotoxicity, hematotoxicity, neurotoxicity, carcinogenicity, and mutagenicity (Han et al. 2008). Conventional methods of separation are expensive (Ozcan et al. 2007). Hence, cost-effective removal of CR from the effluent streams has gained enormous significance in recent times.

In the current investigation, we researched the biomass *Ficus religiosa* (FR) as an effective adsorbent for the removal of (CR). This genus belongs to Ficus family and is known by more than 150 names worldwide. The FR leaf powder contains different functional groups which are able to interact with pollutants like heavy metals and organic matters (Chitra Gupta 2012). Influence of various parameters like temperature, pH, etc., was investigated. Mass transfer mechanism and energy changes were estimated from the kinetic and thermodynamic studies. Maximum biosorption capacity of FR was determined from isothermal studies.

2 Materials and Methods

2.1 Preparation of Anionic Dye Solutions

Congo red (CI = 22,120) was supplied by Merck (Mumbai, India). Initially, a solution of 1000 mg/L was prepared and then diluted as per the requirement.

2.2 Preparation of the Adsorbent

Ficus religosa leaves were collected directly from the trees in the local area, rinsed with water to remove the debris, and dried in an oven at 80 °C to a constant weight. The dried leaves were made to powder of required size and finally packed in desiccators.

2.3 Biosorption Experiments

Batch studies were carried out by preparing a stock solution of the dye with a concentration of 1000 mg L^{-1} with Millipore water and is diluted to have various concentrations from 25 to 1000 mg/L. Precalculated amount of adsorbent (0.01–1 g) was measured accurately with an analytical balance (SHIMADZU AX200) and added to 50 ml of feed solution. The adsorbent solution was agitated with Remi make Temperature-Controlled Orbital Shaker (REMI—CIS 24 BL). At the end, the samples were collected and centrifuged to remove the suspended solid particles using REMI C 24 centrifuge. The clear liquid was collected and analyzed using UV–VIS spectrophotometer (SYSTRONICS-117) at a wavelength of 498 nm.

3 Results and Discussion

3.1 Adsorption Isotherms Study

Adsorption isotherms relate equilibrium concentrations of two phases at constant temperature. Several adsorption equilibrium models have been widely accepted to describe inter phase adsorption behavior. In general, the Langmuir and Freundlich isotherm equations are broadly used for interpretation of adsorption data obtained. The characteristics and the interaction between two phases can be inferred from the shape of the equilibrium isotherm (WongY et al. 2003; Duong 1998).

3.1.1 Langmuir Isotherm Equation

This is the simplest and the most widely used expression based on the assumption of homogeneous adsorption sites with each site accommodating one molecule or one atom of zero interaction with each other; the adsorption is assumed to be a monolayer coverage. The Langmuir isotherm is represented by the equation:

$$q_e = \frac{q_{max} K_L C_e}{1 + K_L C_e} \tag{1}$$

Its linear expression is

$$\frac{1}{q_e} = \frac{1}{q_{max}} + \frac{1}{K_L q_{max} C_e} \tag{2}$$

where q_e is the amount of dye adsorbed per unit mass of adsorbent (mg/g); C_e is the equilibrium dye concentration (mg/L), respectively; q_{max} and K_L are the Langmuir constants representing the monolayer adsorption capacity (mg/g) and the energy of adsorption (L/mg), respectively. q_{max} and b are calculated from the slopes and intercepts of the straight lines of a plot of $1/q_e$ versus $1/C_e$.

3.1.2 Freundlich Isotherm Model

The Freundlich model is an empirical relation between the concentrations of solute on the adsorbent surface to that in the liquid phase. It represents the sorption on heterogeneous surface, which is not restricted to a monolayer. It is mathematically expressed as (Duong 1998):

$$q_e = K_f C_e^{1/n} \tag{3}$$

where K_f and n are the Freundlich constants, representing adsorption capacity and adsorption intensity, respectively. The linearized form of Freundlich isotherm can be written as:

$$\ln q_e = \ln K_f + 1/n \ln C_e \tag{4}$$

The value of K_f and n can be calculated by plotting ln q_e versus ln C_e. Units of K_f are mg/g and can be used as an alternative measure of adsorptive capacity, while $1/n$ (1/mg) determines the adsorption intensity. If the value of n is equal to unity, the adsorption is linear. If it is below unity, it implies that the adsorption process is chemical, while values above unity indicate that adsorption is favorable and is a physical process.

Figures 1 and 2 display linear plots of Langmuir and Freundlich isotherms for CR removal by FRLP. The parameters estimated from the slope and intercept of these plots have been presented in Table 1. The values of correlation coefficients (R^2) (0.98, 0.927) indicate the applicability of both isotherms. From the Langmuir parameters (Table 1), the maximum adsorption capacity observed is 227.778 mg/g. It is higher compared to other biosorbents like *Azadirachta indica* leaf powder and

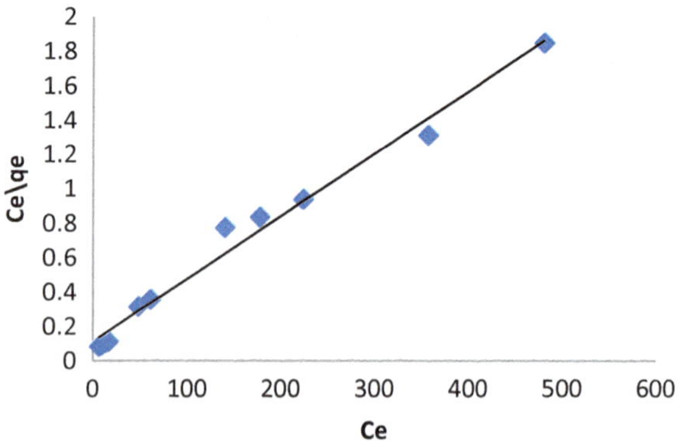

Fig. 1 Langmuir isotherm for biosorption of CR using FRLP

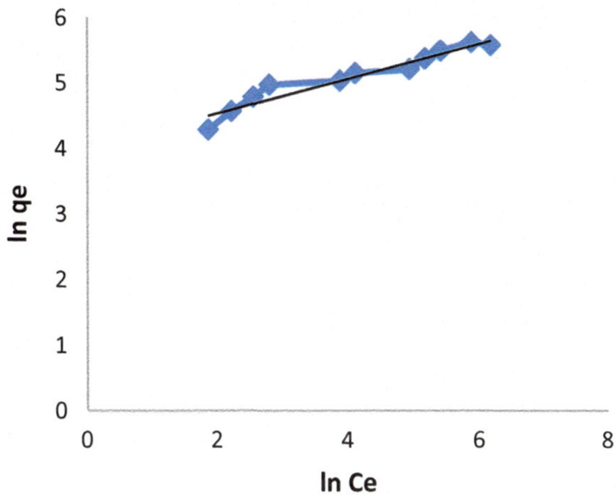

Fig. 2 Friendlich isotherm for biosorption of CR using FRLP

Table 1 Langmuir and Freundlich adsorption constant for FRL powder

Langmuir adsorption isotherm	$q_{\max}$ (mg/g)	K_L (L/mg)	R^2	R_L
	277.778	0.02317	0.98	0.079
Freundlich adsorption isotherm	K_F (mg/g)	$1/n$ (1/mg)	R^2	n (mg)
	54	0.264	0.927	3.7879

waste orange peel (Ozcan et al. 2007; Webi and Chakravorti 1974). Hence, FRLP is an effective biosorbent for the removal of CR from aqueous solutions.

Freundlich parameter 'n' value is 3.7878 (Table 1). It implies that adsorption is favorable and is physical in nature, involving weak binding forces (Hall et al. 1966).

Separation factor (R_L) is an important parameter that can be evaluated from Langmuir parameters. It is used to assess the feasibility of the process, which is calculated from the following equation:

$$R_L = \frac{1}{1 + K_L C_o} \tag{5}$$

where K_L is the Langmuir constant and C_o is the highest initial dye concentration. The values of R_L were calculated (using the above equation) for different initial concentrations, and the results of the same have been displayed in Table 2. All the R_L values (Table 2) are between 0 and 1. This indicates that CR biosorption on FRLP is favorable.

Table 2 R_L values for different concentrations for biosorption of CR on FRLP

C_o (mg/L)	150	200	250	300	350	400	500	600	700	900	1000
R_L	0.172	0.135	0.111	0.094	0.082	0.072	0.059	0.049	0.043	0.033	0.030

3.2 Adsorption Thermodynamics Study

Thermodynamic parameters such as the Gibbs free energy (ΔG°), enthalpy (ΔH°), and entropy (ΔS°) can be determined by using the following equations:

$$K_c = \frac{C_A}{C_s} \tag{6}$$

$$\Delta G^\circ = -RT \ln K_c \tag{7}$$

$$\ln K_c = -\frac{\Delta G^\circ}{RT} = \frac{\Delta H^\circ}{RT} + \frac{\Delta S^\circ}{R} \tag{8}$$

where K_c is the equilibrium constant and C_A and C_s are the equilibrium concentrations (mg/g) of dye in solid and liquid phases. T is the temperature in Kelvin, and R is the gas constant. In any adsorption process, both energy and entropy considerations should be taken into account in order to determine thermodynamic feasibility of the process (Purkait et al. 2007).

Based on Eq. 8, a plot of ln K_c as a function of $1/T$ was generated from the experimental data conducted at three different temperatures (30, 40, 50 °C) with five solutions of different initial concentrations (25, 50, 75, 100, and 150 mg/L). Thermodynamic parameters were calculated from the slope and intercept of the straight lines, displayed in Fig. 3 and the results were presented in Tables 3 and 4.

Table 3 gives the values of ΔH° and ΔS° for different solutions. ΔH° values are −36.862, −24.069, −34.015, −22.515, and −23.480 kJ/mol at the five concentrations, respectively. These negative values indicate the exothermic nature of the process. An adsorption process is considered to be physical when ΔH° is <84 kJ/mol (Yu et al. 2001).

Fig. 3 Plot of ln K_c versus $1/T$ for biosorption of CR using FRLP

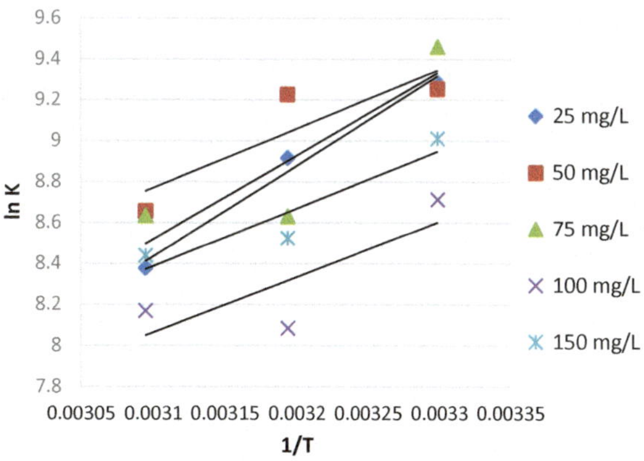

Table 3 Thermodynamic parameters for the adsorption of CR on FRLP

Conc. (mg/L)	$\Delta H°$ (kJ/mol)	$\Delta S°$ (J/mol K)
25	−36.862	−44.1532
50	−24.069	−1.70936
75	−34.015	−34.627
100	−22.516	−2.76191
150	−23.480	−3.06787

Table 4 Gibbs free energy data for CR biosorption

Temperature (°C)	30	40	50
Concentration (mg/L)	ΔG° (kJ/mol)	ΔG° (kJ/mol)	ΔG° (kJ/mol)
25	−23.3959	−23.2099	−22.5017
50	−23.3236	−24.0202	−23.2584
75	−23.846	−22.4724	−23.1981
100	−21.9618	−21.0463	−21.9451
150	−22.7069	−22.1866	−22.6668

$\Delta S°$ values are −44.1532, −1.70936, −34.627, −2.76191, and −3.06787 J/mol. The negative values suggest a decrease in the randomness at interface during the adsorption process.

Table 4 gives the values of $\Delta G°$ at all the temperatures and dye concentration values. The negative values indicate the spontaneous and feasible nature of CR adsorption on FRLP.

4 Conclusions

In conclusion, FRLP is an efficient biosorbent for the removal of CR from aqueous solutions. It is competitive with other commercially available adsorbents with maximum uptake of 227.7 mg/g. Equilibrium studies confirm that the adsorption is a favorable process, and it is a monolayer covering the surface. Thermodynamic studies conclude that the adsorption process is physical, apart from being spontaneous and exothermic.

References

Akar, S. T., & Uysal, R. (2010). Untreated clay with high adsorption capacity for effective removal of CI Acid Red 88 from aqueous solutions: Batch and dynamic flow mode studies. *Chemical Engineering Journal, 162,* 591–598.

Aksu, Z. (2005). Application of biosorption for the removal of organic pollutants: A review. *Process Biochemistry, 40,* 997–1026.

Alver, E., & Metin, A. U. (2012). Anionic dye removal from aqueous solutions using modified zeolite: Adsorption kinetics and isotherm studies. *Chemical Engineering Journal, 200,* 59–67.

Basava Rao, V. V., & Mohan Rao, T. (2015). Thermodynamic and isothermal studies of congo red adsorption onto modified bentonite. *Korean Chemical Engineering Research, 53*(6), 1–6.

Chitra Gupta, S. S. (2012). Taxonomy, phytochemical composition and pharmacological prospectus of *Ficus religiosa* Linn. (*Moraceae*)—A review. *The Journal of Phytopharmacology, 1*(1), 57–70.

Duong, D. (1998). Pure component adsorption in microporous solids. *Adsorption analysis: Equilibria and kinetics* (pp. 188–190)., Series on Chemical Engineering London: Imperial College Press.

Errais, E., Duplay, J., Darragi, F., Rabet, I. M., Aubert, A., Huber, F., et al. (2011). Efficient anionic dye adsorption on natural untreated clay: Kinetic study and thermodynamic parameters. *Desalination, 275,* 74–81.

Gaikwad, R. W., & Kinldy, S. A. M. (2009). Studies on auramine dye adsorption on psidium guava leaves. *Korean Journal of Chemical Engineering, 26,* 102–107.

Hall, K., Eagleton, L., & Acrivos, A. (1966). Pore-and solid-diffusion kinetics in fixed-bed adsorption under constant-pattern conditions. *Industrial and Engineering Chemistry Fundamentals, 5*(2), 212–223.

Han, R., Ding, D., Xu, Y., Zou, W., Wang, Y., Li, Y., et al. (2008). Use of rice husk for the adsorption of congo red from aqueous solution in column mode. *Bioresource Technology, 99,* 2938–2946.

Hsuand, T.-C. (2008). Adsorption of an acid dye onto coal fly ash. *Fuel, 87,* 3040–3045.

Kumar, A., Kumar, S., & Gupta, D. V. (2007). Adsorption of phenol and 4-nitrophenol on granular activated carbon in basal salt medium: equilibrium and kinetics. *Journal of Hazardous Materials, 147,* 155–166.

Kumar, G. V., Ramalingam, P., Kim, M. J., Yoo, C. K., & Kumar, M. D. (2010). Removal of acid dye (violet 54) and adsorption kinetics model of using musa spp. waste: A low-cost natural sorbent material. *Korean Journal of Chemical Engineering, 27,* 1469–1475.

Lafi, R., Fradj, A., Hafiane, A., & Hameed, B. H. (2014). Coffee waste as potential adsorbent for the removal of basic dyes from aqueous solution. *Korean Journal of Chemical Engineering, 312,* 198–2206.

Mao, J., Won, S. W., Min, J., & Yun, Y. S. (2008). Removal of Basic Blue 3 from aqueous solution by Corynebacterium glutamicum biomass: Biosorption and precipitation mechanisms. *Korean Journal of Chemical Engineering, 25,* 1060–1064.

Mehmood, A., Bano, S., Fahim, A., Parveen, R., & Khurshid, S. (2015). Efficient removal of crystal violet and eosin B from aqueous solution using *Syzygium cumini* leaves: A comparative study of acidic and basic dyes on a single adsorbent. *Korean Journal of Chemical Engineering, 32,* 882–895.

Mohan Rao, T., & Basava Rao, V. V. (2015). Biosorption of Congo Red using Allium Sativum skin powder. *Research Journal of Chemistry and Environment, 19*(11), 35–44.

Mohan Rao, T., & Basava Rao, V. V. (2016). Biosorption of Congo Red from aqueous solution by crab shell residue: A comprehensive study. *SpringerPlus, 5,* 537.

Ozcan, A. S., Erdem, B., & Ozcan, A. (2007). Adsorption of Acid Blue 193 from aqueous solutions onto BTMA-bentonite. *Colloids and Surfaces A: Physicochemical and Engineering Aspects, 266*(1–3), 73–81.

Purkait, M. K., Maiti, A., DasGupta, S., & De, S. (2007). Removal of congo red using activated carbon and its regeneration. *Journal of Hazardous Materials, 145,* 287–295.

Rusu, L., Harja, M., Simion, A. I., Suteu, D., Ciobanu, G., & Favier, L. (2014). Removal of Astrazone Blue from aqueous solutions onto brown peat. Equilibrium and kinetics studies. *Korean Journal of Chemical Engineering, 31*(6), 1008–1015.

Wang, C., Li, J., Wang, L., Sun, X., & Huang, J. (2009). Adsorption of dye from wastewater by zeolites synthesized from fly ash: Kinetic and equilibrium studies. *Chinese Journal of Chemical Engineering, 17,* 513–521.

Wang, L., Li, J., Wang, Y., Zhao, L., & Jiang, Q. (2010). Adsorption capability for Congo red on nanocrystalline MFe_2O_4 (M = Mn, Fe Co, Ni) spinel ferrites. *Chemical Engineering Journal, 72,* 181–182.

Webi, T. W., & Chakravorti, R. K. (1974). Pore and solid diffusion models for fixed-bed adsorbents. *AIChE Journal, 20,* 228–238.

Wong, Y. C., Szeto, Y. S., Cheung, W. H., & McKay, G. (2003). Equilibrium studies for acid dye adsorption onto chitosan. *Langmuir, 19,* 7888–7894.

Yu, Y., Zhuang, Y. Y., & Wang, Z. H. (2001). Adsorption of water-soluble dye onto functionalized resin. *Journal of Colloid and Interface Science, 242*(2), 288–293.

Bioconversion of Mango Industrial Waste into Vermicompost: Evaluation of Nutritional Status of Vermicompost

Mannarapu Mastan

Abstract The cellulose substances bagasse, vegetable waste, leaf litter, and mango industrial waste were mixed with cow dung and treated with *Eudrilus eugenia*, *Eisenia fetida*, *Perionix excavates,* and *Mega scolex*. The vermicompost originated was chemically characterized for their total N, P, K, Fe, Zn, Cu, Mn, S, Ca, and Mg. The yield of vermicastings was determined on every month up to six months. The highest yield of vermicastings was obtained when the compost was treated with *Perionix excavatus*, the same compost exhibited higher concentration of NPK and other plant nutrients. The seedlings of *Albizia lebbeck*, *Azadarachta indica*, *Pterocarpus santalinus,* and *Tamarindus indica* exhibited significantly ($p < 0.05$) high growth and biomass in vermicompost obtained from *Perionix excavatus* compare to control and used as potting medium.

Keywords Vermicompost (Mango industrial waste) · Earthworm species · Growth responses · Vermiculture

1 Introduction

Human and animal habitations generate large quantities of waste matter; some of these wastes such as metals, glass, plastics, and paper can be collected and recycled. The rest is mostly organic and non-toxic and is referred to as organics. Though several technologies are harnessed to dispose organics, electrical and chemical resources but incurred expenditure is high to dispose the above waste through these technologies. The disposal techniques popularly referred to as treatment technique are costly to install and operate and do not generate net income. Hence, these are not willingly adopted the resource-deficient developing countries.

Earthworms play a key role in the management of organics in soil. Vermiculture eco-technology ensures effective bioconversion of organic residues into vermicastings or biosoil with plant nutrients and other growth factors, which promote the growth of plants. Conventional composting, i.e., microbial processing without the

M. Mastan (✉)
Department of Biotechnology, Dravidian University, Kuppam 517426, India
e-mail: mastanmadhav@gmail.com

S. K. Ghosh et al. (eds.), *Bioresource Utilization and Bioprocess*,
https://doi.org/10.1007/978-981-15-1607-8_20

presence of earthworm invariably becomes acidic and may loss nutrients through acidic leachates. Earthworms, however, produce their excrete (vermicastings) with 2–3 times plant nutrients than the conventional compost (Tomati and Galli 1995). Because there are known losses due to denitrification and leaching.

Potting medium plays a vital role in determining the growth of healthy fibrous root system. The medium physically supports growing seedlings to supply the nutrients, water, and air to the root systems. The better the media, the better will be the development of a healthy fibrous root system and subsequently, a better quality seedling is produced which survives out planting.

Mango (*Mangifera indica* L.) is the most important tropical fruit and national fruit of India. It is also known as "The king of fruits", which belongs to the family Anacardiaceae.

The number of mango pulp industries established in Chittoor district and it contributes prime area of mango yield. In the fruit industries, 30–40% of fruit goes in the waste form. These wastes are dumping on the open areas during the monsoon season. Due to heavy rain and cloudy weather in these days, results start fermentation of waste with bad odor. This leads to serious environmental pollution as well as showing adverse effect like livestock extermination due to feeding of this waste (Andhra Jyothi Newspaper 2012). So, management of this waste became a big challenge for mango fruit processing industrialists.

Vermicompost is a process in which the microorganism and earthworms are used for waste conversion to produce a good end product. Generally, the earthworms fed the organic waste materials and these pass through via digestive tract and let out in a casting form. These are known as vermicastings and also called vermicompost. These vermicomposts will strengthen the soil nutrient status as well as strength. The enzymes secreted by the saliva and digestive tract of earthworms break down soil and organic matter, so the casting contains highly nutritional values and more powerful growth-promoting agents over the conventional composts. The vermicompost protects the soil status when compared to the chemical fertilizers.

Vermicompost contains beneficial microorganisms, plant growth hormones like auxins, cytokinins, etc. and also a good amount of enzymes. In addition to this, vermicompost is rich in micro-nutrients like Zn, Mo, B, Fe, Cu and Mn as well as NPK. The above all macro- and micro-nutrients, plant hormones, and vitamins status will depend on the stock feed supply to the earthworms.

In the present work, four different species of earthworms viz., *Eudrilus eugenia*, *Eisenia foetida*, *Perionyx excavates,* and *Mega scolex* were used to evaluate the

(i) The efficiency of the species used
(ii) Rate of production of vermicastings
(iii) Status of micro- and macro-nutrients, and
(iv) Growth response of some tree species to vermicompost amended potting medium.

2 Materials and Methods

The organic waste processing high content of cellulose was selected for composting. The sugarcane bagasse was obtained from local sugarcane factory, air-dried and chopped into small pieces. The air-dried leaf litter was collected from the Biotechnology Research Center, Tirupati. The mixed vegetable waste obtained from the local market was chopped and air-dried. Partially, degraded mango fruit processing industrial waste was chopped and air-dried. Twenty kgs of mango dried waste, 5 kgs of cow dung, and 5 kgs of leaf litter and sugarcane bagasse were added and mixed well. The mixture was filled up in a rectangular tub and the surface was plastered with 2.0-cm-thick soil paste with holes of 2-cm diameter at 30-cm apart to provide needed aeration and was left for two weeks to eliminate noxious gasses like CO_2, CH, CO, H_2, etc. This evolves during anaerobic fermentation of the compost and is toxic to the other organisms. The selected species of earthworms i.e., *Eudrilus eugenia, Eisenia foetida, Perionyx excavates* and *Mega* scolex were obtained from Gandhi Krishi Vignana Kendra, Bangalore. These were maintained in an earthworm rearing box at room temperature (30 °C) for further experiments.

Fifty worms of each species were introduced separately in each tub. The mouth of the tubs was tied with a wet muslin cloth to avoid the escaping of worms and was placed in 50% shade created by using agronet in nursery for six months.

The black granular excreta known as vermicastings produced as an end product of this complex processing was collected with the help of thick plastic sheet carefully to retain the granular structure and the yield was determined on every month. At the end of the experiment, the vermicastings obtained in each month were mixed, shade dried and sieved form a 2-mm sieve to obtain uniform granules and were chemically characterized for their pH, electrical conductivity, status of macro-nutrients, and micro-nutrients. The nitrogen content was estimated by micro-kjeldahl method. Phosphorous was estimated by calorimetric method. Potassium by Flame photometer. Calcium and magnesium by titration methods. Zinc, copper, manganese, and iron were estimated by atomic absorption spectrophotometer.

To study the effect of vermicompost on plant growth, three different volumes of (50 cc, 350 cc, 165 cc) root trainers were filled with red earth, farmyard manure, and vermicompost (1:1:1) ratio. The seeds of *Albizia lebbeck*, *Azadirachta indica*, *Pterocarpus santalinus,* and *Tamarindus indica* were surface sterilized (except *A. indica*) with 0.1% $HgCl_2$ solution for 15 min. Four to five seeds of each species were sown in the root trainers of each capacity. After germination, the seedlings were thinned to one wedding and watering was done periodically.

After 120 days of growth, the seedlings were uprooted and growth characters like shoot length and total biomass (shoot and root system) were estimated. The seedlings were separated into root and shoot systems. Roots were washed with tap water to eliminate soil particles adhered to the root system. The washed samples were air-dried and subsequently ovened to get constant dry biomass. The controls were maintained using red earth and farmyard manure (1:1) without vermicompost. The data obtained was subjected to analysis of variance (ANOVA) for statistical analysis to test the significance level.

3 Results and Discussion

Four different species of earthworms were cultured in plastic tubs containing prefermented medium as non-fermented medium was found toxic to the earthworms. The worms fed and grew on the fermented aerated composted and voided black colloidal casts (vermicastings). The day of initiation of casting production was found different in all the species. At the end of the experiment, maximum yield of castings was obtained from the *Perionyx excavatus* followed *Eisenia fetida*, *E. eugenia,* and *M. scolex*. Zajonc and Sidor (1990) reported the variation in the yield of vermicastings produced from the cellulose-rich substrates. The pH of the castings produced from different species of earthworm was found varied between 6.63 and 6.9 and electrical conductivity between 3.86 and 8.79 m mhos. The production of vermicompost from manure resulted in a pH shift toward neutral, a reduction in electrical conductivity, and a large increase in oxidation potential. The amount of total nitrogen was found maximum in the castings obtained by *E. fetida* followed by *E. eugenia*, *M. scolex,* and *P. excavatus*. The content of available phosphorous was found varied from one species to another being maximum in *E. eugenia* and minimum in *E. fetida*. Similarly, the amount of potassium was maximum in the compost obtained by *P.excavatus* and minimum in *E. fetida* treated compost (Table 1). The content of Fe, Zn, Cu, Mn, Ca, and Mg was also found varied from one species to others (Table 1). These observations support the earlier findings (Kalem Basha 1996; Singh et al. 1997; Mitchel and Edwards 1997).

The shoot length and dry biomass of all the tree species differed depending on the sizes of the root trainers. The *A. lebbeck* responded well when compared to other plants. There was a significant ($p < 0.05$) increase in shoot length and dry biomass of all tree species over control (Table 2). There are reports of increase in biomass of *Acacia mearnsii*, *Eucalyptus grandis*, and *Pinus patula* (Donald and Visser 1989) in vermicompost supplemented potting media. Vermiculture is eco-friendly and conventionally fit for any type of crop cultivation to boost the yield. In the present experiment, it is proved that the use of vermicompost had a positive effect on the growth of seedlings. Hence, vermiculture technology can be well exploited to increase the quality of the seedlings in forest nurseries to undertake national developmental programs of afforestation, social forestry, and agroforestry.

Nowadays, organic waste is dumping more through the fruit industries. Every day the amount is increasing and can cause more problems to the environment. The decomposition of organic waste is highly difficult through conventional methods of composting. But, the vermicomposting technique is a novel technique of converting decomposable organic wastes into valuable vermicastings by introduction

Table 1 Physical and chemical characterization of vermicompost obtained from four different species of earthworms

Species	pH	E.C.	N (%)	P (%)	K (%)	Fe (ppm)	Zn (ppm)	Cu (ppm)	Mn (ppm)	S (ppm)	Ca (ppm)	Mg (ppm)
Perionyx excavates	6.33 ± 0.11	7.68 ± 0.17	1.42 ± 0.01	1.45 ± 0.02	0.44 ± 0.013	2784 ± 2.07	89 ± 0.19	18 ± 0.04	237 ± 0.22	477 ± 0.41	28.4 ± 0.17	35.6 ± 0.14
Mega scolex	6.09 ± 0.03	5.62 ± 0.13	1.26 ± 0.02	1.33 ± 0.01	0.24 ± 0.014	2810 ± 2.07	77 ± 0.13	22 ± 0.06	144 ± 0.11	246 ± 0.21	25. 4 ± 0.14	22.6 ± 0.11
Eudrillus eugenia	6.24 ± 0.01	4.22 ± 0.04	1.33 ± 0.01	1.74 ± 0.06	0.38 ± 0.014	2985 ± 2.074	81 ± 0.14	17 ± 0.03	127 ± 0.17	338 ± 0.27	23.4 ± 0.12	26.4 ± 0.15
Eisenia fetida	6.42 ± 0.06	8.15 ± 0.09	1.69 ± 0.05	1.37 ± 0.02	0.44 ± 0.018	2901 ± 2.08	67 ± 0.15	19 ± 0.04	211 ± 0.21	282 ± 0.17	26.5 ± 0.15	26.4 ± 0.14

Values are average of six sets of separate experiments (Mean ± SD)

Table 2 Growth response of some tree species to vermicompost application

Species	Shoot length (cm)						Dry biomass (gm/plant)					
	A	Con	*B*	Con	*C*	Con	*A*	Con	*B*	Con	*C*	Con
Albiazia lebbeck	28.4 ± 0.17	18 ± 0.02	26.5 ± 0.12	20.2 ± 0.09	15.5 ± 0.06	11.5 ± 0.01	8.5 ± 0.04	5.2 ± 0.14	7.4 ± 0.01	5.2 ± 0.02	5.5 ± 0.12	2.4 ± 0.04
Azadirachta indica	23.2 ± 0.12	17.5 ± 0.03	18.2 ± 0.07	11.4 ± 0.04	11.4 ± 0.04	4.4 ± 0.01	1.4 ± 0.01	0.8 ± 0.02	1.1 ± 0.01	0.5 ± 0.02	0.52 ± 0.04	0.44 ± 0.02
Pterocarpus santalinus	17.1 ± 0.07	14.2 ± 0.01	16.2 ± 0.04	9.6 ± 0.01	14.2 ± 0.06	6.8 ± 0.02	2.8 ± 0.02	1.6 ± 0.04	2.2 ± 0.02	2.4 ± 0.06	2.7 ± 0.06	1.65 ± 0.02
Tamarindus indica	26.5 ± 0.15	22.4 ± 0.11	27.1 ± 0.13	22.4 ± 0.06	22.6 ± 0.09	12.5 ± 0.04	2.1 ± 0.06	1.0 ± 0.01	1.46 ± 0.01	1.2 ± 0.02	0.9 ± 0.01	0.24 ± 0.01

Values are average of six sets of separate experiments (Mean ± SD)
A 500 CC
B 350 CC
C 165 CC
Con Control

earthworms within a short span of time which is cost-effective and environmental-friendly for agriculture and also improves soil nutritional status as well as increases water holding capacity to the soil.

Vermicomposting is a biological process which may be a future technology for the management of agro-industrial waste. This study was undertaken to produce vermicompost from mango industrial waste by mixing with a combination of different potting media. This vermicompost with potting media inoculated with different species of earthworms allowed to vermicomposting for 40–60 days.

In our India, small-scale production of vermicompost is possible by converting kitchen waste into high-quality soil modifications, where space is limited. The earthworms can decompose organic matter without the influence of human physical effort. But the large-scale vermicomposting is practiced in Canada, Italy, Japan, Malaysia, Philippines, and the USA (Asha et al. 2008).

The earthworms helped break down organic matter and aerate the soil. For plowed fields, this is adequate, as the aerated soil is not allowed to build bulk density. Recently, earthworms are being used for bioremediation of degraded soils (Satchell 1983; Lee 1985). Vermicompost is an excellent water-soluble nutrient-rich organic fertilizer and soil conditioner.

References

Andhra Jyothi News Paper. (2012, August). District edition; Chittoor Dist. Andhra Pradesh, India.

Asha, A. A. K., Tripathi, & Soni, P. (2008). Vermicomposting: A better option for organic solid waste management. *Journal of Human Ecology, 24*(1), 59–64.

Donald, D. G. M., & Visser, L. B. (1989). The use of vermicompost in sustainable agriculture: Impact on plant growth and soil fertility. *Applied Plant Science, 3*(2), 110–113.

Kalem Basha, D. (1996). *Organic waste and the protection and productivity of agrocenosis* (Vol. **437**, pp. 244–252).

Lee, K. E. (1985). *Earthworms, their ecology and relationships with soils and land use*. New York: Academic Press.

Mitchel, A., & Edwards, C. A. (1997). Vermitechnology based bioprocessing of agricultural wastes and testing the efficiency of various vermin based products. *Soil-Biology and Biochemistry, 29*(3–4), 763–766.

Satchell, J. E. (1983). *Earthworm ecology, from Darwin to vermiculture*. London: Chapman and Hall.

Singh, L., Bhonde, S. R., & Mishra, V. K. (1997). Effect of different organic manures and inorganic fertilizers on yield and quality of rabi onion. *News-letter, 17*(3), 1–3.

Tomati, U., & Galli, E. (1995). Earthworms, soil fertility and plant productivity. *Acta Zoologica Fennica, 196,* 11–14.

Zajonc, I., & Sidor, V. (1990). Use of some wastes for vermicompost preparation and their influence on growth and reproduction of the earthworm *Eisenia fedita. Research Institute of Animal Production, 36*(8), 742–752.

Enhancement of Compost Rate by Adding Microbes: A Comparative Study on Nutritional Values

J. Vimal Amaladoss, M. K. Sutharsan, S. Sunil Gowthamraj, S. Gowri Sankar, Nyanthung Y. Humtsoe and V. Kirubakaran

Abstract Solid waste management is a critical issue for huge country like India. Even though several technologies have been developed and demonstrated, the collection to a centralized place makes the process not economical. The open dumping is the traditional method for the treatment of solid waste. The compost is the easiest method and produces value-added fertilizer. The main problem in the compost is the duration and thus occupies more volume. This paper analyzes the possibility of fast degradation of solid waste using the biodigester effluent as well as external microbes. The degradation kinetics of normal compost and the biodigester-treated compost has been evaluated and presented. The nutritional value of the final compost has been estimated and reported.

Keywords Solid waste management · Critical · Fertilizer · Compost · Effluent · Microbes · Nutritional value

---

J. Vimal Amaladoss · M. K. Sutharsan · S. Sunil Gowthamraj · S. Gowri Sankar · N. Y. Humtsoe · V. Kirubakaran (✉)
Rural Energy Centre, The Gandhigram Rural Institute—Deemed University, Dindigul, Tamil Nadu 624302, India
e-mail: kirbakaran@yahoo.com

J. Vimal Amaladoss
e-mail: Johnsonvimal82@gmail.com

M. K. Sutharsan
e-mail: sumonsuthar@gmail.com

S. Sunil Gowthamraj
e-mail: sunilgautamchetty@gmail.com

S. Gowri Sankar
e-mail: gowrishankar1510@gmail.com

N. Y. Humtsoe
e-mail: humtsoenyanthung@yahoo.com

S. K. Ghosh et al. (eds.), *Bioresource Utilization and Bioprocess*,
https://doi.org/10.1007/978-981-15-1607-8_21

1 Introduction

Our country produces an amount of 960 million tons of solid waste in a year out of which are biodegradable and the remaining are non-biodegradable. Several technologies have been introduced in processing the solid waste generated daily; yet, the means of segregating and managing the waste with respect to energy conversion point of view has been lacking behind. Composting of a biodegradable waste has been a common practice and an easy technique to all concern citizens. Some of the common methods practiced in our country are Bangalore method, Coimbatore method, Indoor method, NADEP method, Vermicomposting and some unique systems developed by several institutions. Such common methods require months for proper compost product. However, the method to composting the biodegradable waste using microbes in order to speed up the reaction time and generate suitable quality manures have not been well known to all. Therefore, focusing on the necessity and the awareness for the best of all well-being and the community, this study has been carried out on a comparative study between the common composting method and the microbe added method so as in the reduction and recycling of the degradable waste also in the replacement of chemical fertilizers with the by-products obtained containing more nutritional value.

2 Literature Survey

*Nakasakiet al. (*1993*)*

Composting of municipal solid waste was conducted by using laboratory-scale reactor under well-controlled experimental conditions and the effects of PH value control were quantitatively analyzed. Lime is added to composting to maintain the pH decreasing below 7, nitrogen loss is enhanced by the control of pH value, to maintain pH growth rate and the degradation activity of proteins of the microorganisms was in the range of 7–8. Therefore, the decomposition of glucose proceeded rapidly at an early stage in a pH from 6 to 9.

*Pattnaik and Vikram Reddy (*2009*)*

The nutrient and compost of the organic carbon, C/N and C/P ratios decreased as the composting processed from 0 to 60 days. The nutrient value of vermicomposting of all earthworm species produced from both the wastes was more than that of the compost. The MW has high nutrient values than the FW. Vermicomposting of vegetable market waste and floral waste were assessed across different periods in relation to their respective innovative substrates. The general parameters like temperature, moisture, pH, nutrients has also been measured and reported.

Wei et al. (2007)

Municipal solid waste compost contains an amount of humic substances, the compost of residual MSW with the metal, plastic and glass removed. To enhance degradation processes and the degree of composting humification and inoculated microorganisms has been employed. During the MSW composting, humic acid was extracted and purified. The elemental analysis, UV, FTIR and fluorescence spectroscopy has been used to evaluate the elements of humic substances. Added inoculums with microbes led to a greater degree of aromatization of control process, during composting gave a greater degree of HA aromatization than inoculation with complex microorganisms or lingo-cellulolytic alone. But comparing with the HA of soil, the HA of, MSW compost revealed a lower degree of aromatization.

Jared et al. (2018)

In chemical fertilizers are not well suitable for poor and low of agricultural production system. So the alternative source to make nutrients by using of enhance sustainable agriculture and also promote organic farming. Descriptive statistics on result show that the percentage concentration of nitrogen, phosphorous, potassium was high in bio-slurry as compared to slurry compost and farmyard manure. This study will be used by agricultural laboratories in Kenya which results of getting bio-slurry and improves agricultural production and soil structure.

Haile and Ayalew (2018)

Kale is one of the nutritious vegetables with a high nitrogen fertilizer requirement. However, soil fertility is declining progressively due to the imbalanced use of inorganic fertilizer. Discharging bio-slurry as waste will cause environmental pollution and disposing will also lead to costs because of its large pollution load. Consequently, replacing chemical fertilizers with bio-slurry can not only achieve efficient resource utilization, but also reduce the amount of chemical fertilizer used and the resulting environmental pollution. Results revealed that the treatment had significant effect on growth and yield attributes of kale. The highest (455.10 g) leaf fresh weight and fresh biomass (814.86 g) were obtained when 100% sole application of liquid bio-slurry was used.

3 Composting

The Composting is the process of converting an organic matter into a soil fertilizer with the use of inoculums. Depending on the type of inoculums used and the system procedure, it varies. Composting is a times a difficult task. If no proper aeration and water are supplied, the system will lead into an anaerobic process which will produce methane and hydrogen sulfide, causing unpleasant aroma. Also, if the temperature rises above 71 °C during the process, spontaneous combustion can result, which might lead to the outbreak of fire. Having improper ratios of compost can also lead to inefficiency (i.e. too much green waste, too little brown waste to catalyze reactions) or pollution (i.e. too much brown waste, which can cause ammonia formation and lead to odor or run-off pollution) (https://en.wikipedia.org/wiki/Compost).

In India, composting is an alternative to landfills, for both economic and environmental reasons alike. Villages that compost instead of landfill their domestic waste earn their daily wages by using the organic manure to produce vegetables for their sustenance and sell the remaining. The actual reason for this particular technique is that it not only develops the vegetative production through the use of organic manure from the compost and reduces the composting duration but it can also reduce the degradable waste produced everyday leading to cleaner and better environment also minimizing the use of chemical fertilizers in agriculture and domestic purposes.

4 Experimental Setup

For the experiment involves the analysis of the decomposition rate of the biodegradable waste. Two different pit has been made to study the difference in the decomposition rate of the biodegradable waste, one with the addition of the digested slurry and the other for non-addition of the slurry. The experimental outcome has a comparative analysis with the experimental model and the real-time implementation. Both the setup has their own measured values and they are explained, respectively.

4.1 Biodegradable Waste

The waste that gets degraded with the help of biological means is termed as biodegradable waste. Through the process of biological degradation, they will release some amount of gas and chemical substrates, and these can be employed for different constructive processes. These also include the biomass but the degradation rate is very slow.

4.2 Biomass

The biomass is an organic material that is obtained from the plants and animals. The biomass that is with liquid components such as cow dung is subjected to biogas production because the cow dung contains less than 10% solid matter. Certain biomass can not be employed for the biogas production such biomass can be subjected for different energy extraction processes (https://en.wikipedia.org/wiki/Biomass).

4.3 Degradation of Biological Components

The biological matter is made up of several biological components that serves as food resources for the microbial organisms. If the biological is found dead, the growth of the microorganisms is found more over the dead biological matter. The microorganisms help in the process of degradation by consuming the dead matter, i.e. breaking down the higher sugar molecules into smaller or simple sugar complex, and during this process, evaluation of some gas like methane is found. It is also found to be that there is a reduction in weight of the dead biological mater and this process is found out to be the degradation of the biological matter.

4.4 Composting Pit

The biomass are the main source for the composting pit, and the main aim for composting is for the development of high nutrient content composition that can be employed as the fertilizer for agriculture. The composting pit is made of with a layer of mud and the biomass to be employed is applied above the layer of mud. The biomass filled will be made as a layer at a height for about 2–4 in inches and then mud is made as a layer over the biomass. This setup is subjected to the external atmospheric conditions for the microbial activities, so that the biomass is converted into a rich in nutrient content. The nutrient content of the composting pit is analyzed with variable test methods to find the nutrient composition (Fig. 1).

4.5 Slurry

The anaerobically digested product is identified to be the slurry which is found to have high nutrient content in abundant but the nutrient content depends upon the type of waste given to the biogas plant. The digested slurry will be of semisolid or liquid form, and the pH value of the slurry is found to be around 6.5–7.5 (https://en.wikipedia.org/wiki/Slurry) (Fig. 2).

5 Experimental Setup and Procedure

5.1 Experimental Model in Real Time

This model includes three pit, each of at a depth of 10 in. The pits one and two are loaded with a layer of leaves (biomass) for up to 3 in. and then it is then increased to

Fig. 1 Composting pit

Fig. 2 Digested slurry

a height of about 7 by adding up the slurry. Then, the pit is added up with a layer of mud for one inches and is left open to the atmosphere (Fig. 3).

The third pit is made up by adding the same biomass used in other two pits, then the pit is left over to the atmospheric conditions by adding up a layer of mud over the biomass (Fig. 4).

Fig. 3 Composting pit with slurry

Fig. 4 Composting pit

5.2 *Experimental Model Using Conical Flask*

Table 1 readings are obtained with the help of two conical flasks that is filled with the biodegradable waste along with slurry and without slurry. The slurry that is added is the residue from the bio digester.

The flask one contains 60 g of biodegradable waste with the addition of slurry of about 20 g and is left over to the optimum atmospheric conditions. The flask two contains the same biodegradable waste of 60 g and is left over to the controlled atmospheric conditions.

Table 1 Weight reduction versus days

Day's	With slurry (60 + 20) (g)	Without slurry (60) (gm)
1	80	60
10	57	55
15	43	54
20	30	49
25	22	45
30	16	41
35	9.9	37.3

The weight of the setup is monitored in a regular basis interval of 5 days; the reduction in the weight of the sample gives the decomposition rate of the waste. From the outcome of the sample, change in the weights is obtained (Figs. 5 and 6).

Fig. 5 Sample with and without slurry in day 1

Fig. 6 Sample with and without slurry in day 35

Graph 1 Days versus with slurry

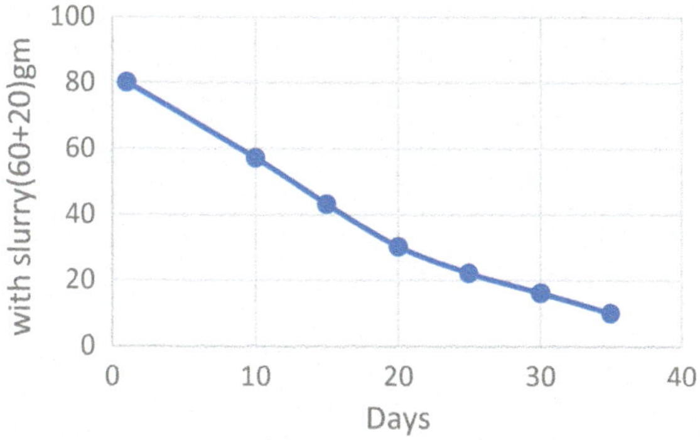

Graph 2 Days versus without slurry

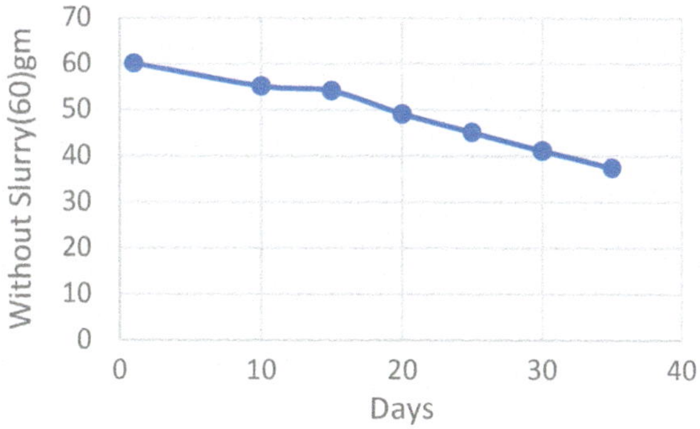

6 Result and Discussion

This paper deals with increase in decomposition of the biomass with the help of adding the slurry to the composting pit. The process is observed as successful and gives a greater loss in weight than the normal composting method is also been observed. The nutrient content of the composite pit resultant is yet to be studied for the available of the nutrients for using it as fertilizer and shown in Graphs 1 and 2.

7 Conclusion

The need for the fertilizer for the purpose of agriculture is to improve the yield. The natural means of fertilizer is developed from the composite pit; but, the days taken for the process of the composting of the biomass takes too long thereby employing the slurry to the composite pit has resulted in faster degradation process which is observed in through the weight monitoring of the sample, and thus, the system out is found out to be successful and more nutritious than the normal method of composting.

References

Biomass, https://en.wikipedia.org/wiki/Biomass.

Compost basics, types and methods, an overview. https://en.wikipedia.org/wiki/Compost.

Haile, A., & Ayalew, T. (2018). *Comparative study on the effect of bio-slurry and inorganic N-Fertilized on the growth and yield of Kale*. Accepted 13 February 2018. ISSN 1996-0824 Copyright © 2018 Author(s) retain the copyright of this article. http://www.academicjournals.rg/AJPS.

Jared, N., Erastus, G., Christopher, N., & Steve, A. (2018, September 10). *Evaluation of biogas slurry as an alternative organic fertilizer*. https://www.researchgate.net/publication/308892484_Evaluation_of_Biogas_Slurry_as_An_Alternative_Organic_Fertilizer_A_Case_Study_in_Kenya.

Nakasaki, K., Yaguchi, H., Sasaki, Y., & Kubota, H. (1993 March 1). Effect of pH control on composting of garbage. https://doi.org/10.1177/0734242X9301100204.

Pattnaik, S., & Vikram Reddy, M. (2009, September 21). *Nutrient status of vermicompost of urban green waste processed by three earthworm species—Eisenia fetida, Eudrilus eugeniae* and *Perioyx excavatus*. Department of Ecology and Environmental Science, Pondicherry University, Puducherry 605014, India.

Slurry, https://en.wikipedia.org/wiki/Slurry.

Wei, Z., Xi, B., Zhao, Y., Wang, S., Liu, H., & Jiang, Y. (2007, February 20). *Effect of inoculating microbes in municipal solid waste composting on characteristics of humic acid*. Life Science College, northeast Agriculture University, Harbin 150030, China. Chinese Research Academy of Environmental Science, Beijing 100012, China. College of Food and Science and Nutritional Engineering, China Agricultural University, Beijing 100083.

Melanoidin Induced DNA Damage and Effects on Antioxidative Enzymes in Earthworm (*Eudrilus eugeniae*)

Swabna Vivekanandam, Selvakumar Muniraj and Vasanthy Muthunarayanan

Abstract Toxicology tests on earthworms were conducted in artificial soil, which is composed of 10% Kaolin clay (Fisher Scientific), 20% of coir pith and 70% of sand (grade 70 particle size 0.1–0.3 mm). Melanoidins were applied to artificial soil in different concentrations ranging from 0 (control), 250, 500, 750 and 1000 μg per kg. The toxicological analysis of earthworms treated with synthetic melanoidins and distillery effluent includes antioxidant enzyme analysis (superoxide dismutase, catalase, glutathione peroxidase and glutathione S-transferase), comet assay and histopathological analysis. The investigation facilitates us to conclude that melanoidin stimulates adverse effects on earthworms such as oxidative stress and DNA damage which may be the important mechanisms of melanoidin toxicity to earthworms.

Keywords *Eudrilus eugeniae* · Melanoidin · Superoxide dismutase · Glutathione peroxidase

1 Introduction

Molecular markers or biological markers are referred as the indicators of biological effects of contaminants on organisms which may perhaps be used as diagnostic and extrapolative tests to identify and levy the pollution effects, predominantly to find the effects of low concentrations of composite mixtures of contaminants on the quality of the environment (Livingstone 1993). Biological responses are classified in molecular, subcellular and cellular level. In ecotoxicology, the biomarkers have a major rationale as the conventional tactics have certain limitations about the quantity of chemicals that can cause adverse effects on plants and animals which are considered as the endpoints.

S. Vivekanandam
Department of Biotechnology, St. Joseph's College (Autonomous), Tiruchirappalli, Tamil Nadu 620 001, India

S. Muniraj · V. Muthunarayanan (✉)
Water and Solid Waste Processing Lab, Department of Environmental Biotechnology, Bharathidasan University, Tiruchirappalli, Tamil Nadu 620 024, India
e-mail: vasanthy@bdu.ac.in; drvasanthy02@gmail.com

S. K. Ghosh et al. (eds.), *Bioresource Utilization and Bioprocess*,
https://doi.org/10.1007/978-981-15-1607-8_22

The use of living organisms as biomarkers is a fundamental contemplation which provides the unsurpassed consideration of authentic circumstances of ecosystems. The trials are possible by means of either structural or functional, which are based on the ecosystems. In general, oligochaetes are vastly considered to be used as bioindicators (Peijnenburg et al. 2009). Their significance is copious as they are the ecologically chief invertebrate group. Other reasons to use oligochaetes for ecotoxicological studies are that they are easy to handle and could be cultured under laboratory conditions.

Earthworms prevail in various types of soils ranging from temperate to tropical regions. They are considered as an imperative number in food chains as earthworms are considered as a prime food source for many organisms including birds and mammals which have turned out as the major incentive for the ecologists and ecotoxicologists to study about earthworms (Lokke and Gestel 1998). Taking this into account, earthworm is habitually used as experimental organisms to resolve the deposition of chemicals in the soil (Oste et al. 2001).

As the behavior and morphology of earthworms are created accordingly to cope up with the aqueous and solid phases of the soil, it may be concluded that the earthworms can be used to study about the ill effects caused by both the organic (Vijver et al. 2003) and inorganic (Jager et al. 2003) contaminants added to the soil.

Roughly, the majority of the earthworm species do not react to low concentrations of chemicals, as the chemical constitution of their body takes time for the proper understanding about the chemical or pollutant and the mechanism of its toxicity on their body (Sturzenbaum et al. 1998). As portrayed earlier in the literatures, the cellular organization of the oligochaetes is not convolute, and it also constitutes distinguished organs.

The other purpose to use earthworms as bioindicators is that they have the ability to ensure metals to a great extent. Chlorogocytes play an important role in accumulating metals in the earthworm's body in different patterns resulting in tissue accumulation for different types of metals (Morgan and Morgan 1990).

Hitherto, in the literatures, various effects of biomarkers have been explained. The effects include:

1. Alternations in DNA induced by toxic contaminants which have genotoxic properties.
2. Immunological responses.
3. Stimulation of metal-binding proteins.
4. Hindrance of enzymes.
5. Cohesion of lysosomal membrane.

1.1 Alternations in DNA Induced by Toxic Contaminants Which Have Genotoxic Properties

The conventional reactions include binding of the contaminant or its secondary metabolites to the DNA, DNA strand breakage, exchange of bases, etc.

1.2 Immunological Responses

The immune system crashes in acute conditions which may lead even to mortality. The changes in the immune system namely sublethal are considered as the first level of toxicity induced by the contaminants. The immunological system is compliant and pliable, wherein earthworms, it was observed that the alteration on the immune system restores immediately after the confiscation of earthworms from the contaminant source. Some studies have proved the influence of chemicals in the immune system of the earthworms. Nevertheless, the dosage and the retaliation of the immune system to the contaminants are yet to be unearthed.

1.3 Stimulation of Metal-Binding Proteins

The heavy metals in different concentrations to which the earthworm is exposed may be detoxified by metal-binding proteins such as metallothionein (Huang et al. 1987; Klaassen et al. 1999). The roles of these proteins are not fully understood, but it is thought that these proteins may involve in the regulation of the metal levels in tissues. The use of these proteins as bio indicators may be proficient to quantify the exposure and metal toxicity.

1.4 Hindrance of Enzymes

The inhibition or the hindrance of the enzymes due to the exposure of the contaminants is considered as the conventional biomarker. For example, inhibition of cholinesterases has been studied to understand the exposure of pesticides such as carbamate and organophosphorous using earthworms. The enzymes were responsible for the transmission of nerve signals, and it is reported that the pesticides can trigger a decline of the production and the activity of the specific enzyme (Dikshith and Gupta 1981).

1.5 Cohesion of Lysosomal Membrane

Lysosomes are membrane-bound cell organelles which are structurally and chemically spherical vessels which are capable of breaking down almost all biomolecules including protein and DNA. Any alteration in the membrane of this organelle is considered to be a measure of stress in the organism which is considered to be bioindicator of contaminants in the soil.

2 Materials and Methods

2.1 Earthworm Exposures

Earthworms used in this assay (*Eudrilus eugeniae*) were purchased from an earthworm culturing farm at Periyar Maniyammai University in Thanjavur. Toxicology tests on earthworms were conducted in OECD artificial soil (OECD 2004), which is composed of 10% Kaolin clay (Fisher Scientific), 20% of coir pith and 70% of sand (grade 70 particle size 0.1–0.3 mm). Melanoidins were applied to artificial soil in concentration of 0 (control), 250, 500, 750 and 1000 μg per kg. For each concentration, 20 earthworms were added.

2.2 Enzyme Extraction

Earthworms were killed by introducing into the formaldehyde solution for a minute and were placed in a dissection tray. They were dissected vertically, and its gut was cleared and then placed into a prechilled mortar and pestle and was crushed under ice-cold condition with 0.5 M Phosphate buffer (pH 7.0). The homogenate was centrifuged at 8000 rpm at 4 °C for 30 min. The supernatant was used for the assay enzyme activity and for protein determination (Song et al. 2009).

2.3 Protein Estimation by Bradford Assay

Accurate determination of protein concentration was done by Bradford method (1976). It is widely used and is found to be more reliable for detecting 20–400 μg of protein.

Procedure

To the clean test tubes, about 2.50 ml of Bradford reagent was added. To the reagent, 10–20 μl of earthworm extract is added which produces a blue color by reaction. Its absorbance is measured at 595 nm.

Formula

$$\frac{\text{Test OD}}{\text{Standard OD}} \times \frac{\text{Standard concentration}}{\text{volume of sample}} \times \text{Assay volume} = \text{mg/ml}$$

2.4 *Enzyme Assays*

2.4.1 Superoxide Dismutase Assay

Superoxide dismutase is an enzyme that catalyzes the dismutation of superoxide (O_2^-) into oxygen and hydrogen peroxide. Thus, they are an important antioxidant defense in nearly all the cells exposed to oxygen.

Procedure

100 μl of earthworm extract was taken and mixed with 250 μl of absolute alcohol. Then, 150 μl of chloroform was added, and 1 ml of distilled water was added to the mixture. The reaction mixture was centrifuged at 2500 rpm for 15 min. The whole supernatant was taken and to that 2 ml of Tris buffer was added. OD was measured with time interval of 0, 1, 2 and 3 min (Marklund and Marklund 1974) following the addition of 500 μl of pyrogallol.

Formula

$x = 0\text{ min EDTA OD} - 1\text{ min OD}$

$y = 0\text{ min OD} - 2\text{ min OD/2}$

$z = 0\text{ min OD} - 3\text{ min OD/3}$

$a = \text{Control}$

$b = (x + y + z)/3$

$c = b/a * 100$

$$\frac{c}{50} \times \frac{\text{Assay Volume}}{\text{volume of tissue extract}} \times \text{dilution factor} \times \frac{1}{\text{mg protein}} = \text{unit/mg protein}$$

2.4.2 Catalase Assay

Catalase enzyme is a common enzyme present nearly in all living organisms exposed to oxygen. It catalyzes the decomposition of hydrogen peroxide to water and oxygen. It is a very important enzyme in protecting the cells from oxidative damage by reactive oxygen species. Likewise, catalase has one of the highest turnover numbers of all enzymes; one catalase molecule can convert millions of molecule of hydrogen peroxide to water and oxygen/second.

Procedure

Accurately, 1.5 ml of phosphate buffer was taken, and 0.5 ml of hydrogen peroxide was added to the buffer. To the mixture, 100 μl of earthworm extract is added. At this stage, to the reaction mixture, 250 μl of potassium dichromate was added which when added arrests the reaction at different time intervals. Then, the whole set is kept in boiling water bath for 10 min. After 10 min, the reaction develops a green color for which the absorbance is measured (Sinha 1972).

Formula

$$\text{Test OD} \times \frac{\text{Standard concentration}}{\text{Standard OD}} \times \frac{1}{\text{dilution factor}} \times \frac{1}{\text{mg protein}}$$
$$= \mu\text{gmol/mg protein/min}$$

2.4.3 Glutathione Peroxidase

Glutathione peroxidase is the general name of an enzyme family with peroxidase activity whose main biological role is to protect the organism from oxidative damage. The biochemical function of glutathione peroxidase is to reduce lipid hydro peroxides to their corresponding alcohols and to reduce free hydrogen peroxide to water.

Procedure

200 μl of EDTA was added to the centrifuge tubes. To that, 100 μl of sodium azide was added, and along with that 100 μl of hydrogen peroxide was added. To the reaction mixture, reduced glutathione was mixed. 400 μl of phosphate buffer is added simultaneously. To that, earthworm extract is added followed by incubation of 37 °C for 10 min. Then, the reaction is arrested by adding 500 μl of 10% TCA. Then, the whole mixture is centrifuged at 3000 rpm for 3 min. The supernatant is transferred to a test tube, and 3 ml of disodium hydrogen phosphate is added. One ml of DTNB was further added which develops a yellow color. Absorbance is measured for the reaction mixture at 412 nm (Rotruck et al. 1973).

Formula

$$\text{Test OD} \times \frac{\text{Standard Concentration}}{\text{Standard OD}} \times \frac{1}{\text{Dilution Factor}} \times \frac{1}{\text{mg protein}}$$
$$= \text{nmoles/mg protein/min}$$

2.4.4 Glutathione S-Transferase

Glutathione S-transferases (GSTs), previously known as ligandins, comprise a family of eukaryotic and prokaryotic phase II metabolic isozymes best known for their ability to catalyze the conjugation of the reduced form of glutathione (GSH) to xenobiotic substrates for the purpose of detoxification.

Procedure

To a clean centrifuge tube, 1 ml of phosphate buffer, 100 μl of CDNB and 100 μl of enzyme extract are added. Then, it was made up to the volume of 3 ml and was incubated at 37 °C for 5 min. The reaction gets started by adding 100 μl of reduced glutathione. The absorbance is measured at 340 nm with an interval of 30 s for 5 min. Reaction mixture without enzyme source is used as blank (Mozer et al. 1983).

Formula

$$\text{Conjugate formed/min/mg protein} = \frac{\text{OD} \times 3 \times 1000}{9.6 \times 5 \times \text{mg protein}}$$
$$= \text{nmoles/mg protein/min}$$

2.5 *Comet Assay*

After exposure of the earthworms to the synthetic melanoidins, their coelomocytes were obtained using the non-invasive extrusion method described by Eyambe et al. (1991). Individual earthworms were rinsed in the extrusion medium composed of 5% ethanol, 95% saline, 2.5 mg/ml EDTA and 10 mg/ml guaiacol glyceryl ether (pH 7.3). Coelomocytes were spontaneously secreted in the medium and washed with phosphate-buffered saline (PBS) prior to the comet assay. The cells were collected by centrifugation (3000 μg, 10 min) and placed on ice prior to the comet assay. The comet assay was performed according to Singh et al. (1988), with slight modifications. The cell suspension was mixed with 100 ml of 0.7% low melting agar

(LMA) in PBS at 37 °C and pipetted onto fully frosted slides precoated with a layer of 100 ml 0.8% normal melting agar (NMA). After solidification on ice, another layer of 85 ml LMA was added, and the slides were immersed into a lysis solution (2.5 M NaCl, 10 mM Tris, 100 mM Na_2EDTA (pH 10.0), 1% Na-sarcosinate, 10% dimethyl sulfoxide (DMSO) and 1% Triton X-100). Slides were then incubated in an electrophoresis tank containing 300 mM NaOH with 1 mM Na_2EDTA for 20 min prior to electrophoresis for 15 min at 25 V (300 mA). The slides were then neutralized (0.4 M Tris, pH 7.5) thrice at 5 min intervals and stained with 40 ml ethidium bromide (13 mg/ml) for fluorescence microscopy analysis (a fluorescence microscope) using a digital imaging system. The images of the SCGE were analyzed using CASP (Konca et al. 2003). One hundred cell cores on each slide were counted. The parameter used to quantify the extent of DNA damage was the olive tail moment (OTM). OTM is the product of the distance between the centers of gravity of the head and the center of gravity of the tail and percent tail DNA.

2.6 *Histology of Earthworm*

The histology of gut of earthworm was studied adopting the routine paraffin method (Humason 1979). Gut of earthworm, dissected out from the control and experimental animals, was blotted free of mucus, washed thoroughly in physiological saline, cut into pieces of desired size and fixed in Bouin's fluid fixative immediately after autopsy. Fixation was carried out at room temperature for 24 h, after which the tissues were transferred to 70% alcohol. Several changes of 70% alcohol were given until the yellow color disappeared from the tissues. The tissues were then dehydrated by passing through ascending grades of alcohol, cleared in xylene, infiltrated with molten paraffin and finally embedded in paraffin wax (58 °C MP). Tissue section of 5-μm thick transverse and longitudinal sections were obtained using a rotary microtome (Leica, Germany). The sections, thus obtained, were stained in Harris hematoxylene and eosin, dehydrated using alcohol, cleared in xylene and mounted using dihydroxy phthalate xylol (DPX). The stained slides were observed using a Research Microscope (NikonE400).

3 Results and Discussion

The experimental set up used for the study conducted to understand the effect of differnt concentrations of melanoidin and distillery effluent along with the control is as represented in Plate 1.

Plate 1 Experiments exposing earthworms to synthetic melanoidin and distillery effluent

3.1 *Mortality of Earthworms*

Table 1 denotes the mortality of earthworms during the period of the study. The mortality of the earthworms increased significantly as the concentration of the synthetic melanoidins increased. This signifies that the melanoidins induce mortality in the earthworms.

Initially, 20 earthworms were inoculated in each of the treatment groups.

Table 1 Mortality of earthworms

Concentration/kg	I week	II week	III week	IV week
250 μg	20	15	13	9
500 μg	20	14	9	7
750 μg	20	12	7	4
1000 μg	14	9	6	4
Distillery effluent	10	6	4	1
Control	20	20	20	19

3.2 Estimation of Protein Concentration of Earthworm Sample Treated with Varying Concentrations of Synthetic Melanoidin

In the present study, the total protein content got subsequently increased in the second week of the study period for melanoidins treated group compared to control animals. The total protein concentration in all the four concentrations melanoidins (250, 500, 750, 1000 μg kg^{-1}) and the distillery effluent increased during 14 days and then decreased after 21 days compared to the controls (Table 2 and Fig. 1). Upon all the concentrations, 1000 μg kg^{-1} and distillery effluent showed drastic reduction (0.0512 and 0.0263 mg/ml) in the protein concentration.

The total protein content subsequently increased in the second week of the study period for melanoidins treated group compared to control animals. This could be

Table 2 Estimation of Protein Concentration of Earthworm sample

Concentration/kg	1st week mg/ml	2nd week mg/ml	3rd week mg/ml	4th week mg/ml
250 μg	0.734 ± 0.01	1.77 ± 0.04	1.5756 ± 0.04	0.966 ± 0.01
500 μg	0.732 ± 0.02	1.04 ± 0.02	0.853 ± 0.02	0.313 ± 0.01
750 μg	0.67 ± 0.01	0.922 ± 0.01	0.653 ± 0.01	0.0995 ± 0.09
1000 μg	0.562 ± 0.01	0.776 ± 0.01	0.1714 ± 0.01	0.0512 ± 0.02
Distillery effluent	0.48 ± 0.01	0.347 ± 0.01	0.068 ± 0.01	0.0263 ± 0.01
Control	1.98734 ± 0.06	1.97734 ± 0.01	1.96734 ± 0.06	1.95734 ± 0.06

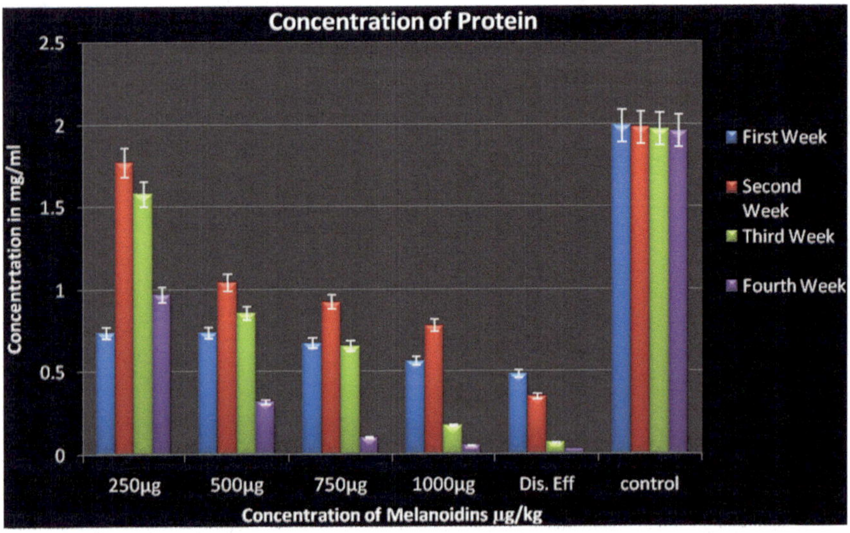

Fig. 1 Graphical representation of Protein Concentration of Earthworm sample

attributed to the fact that melanoidins greatly increase the gene transcription and enhance expression of mRNA (Lescoat et al. 2000).

3.3 *Estimation of Superoxide Dismutase Concentration of Earthworm Sample Treated with Varying Concentrations of Synthetic Melanoidin*

As shown in Table 3 and Fig. 2, the SOD activity in all the four concentrations of melanoidins (250, 500, 750, 1000 μg kg^{-1}) and the distillery effluent increased at the initial stage and then decreased throughout the study period compared to the controls. The SOD activity of the groups treated with distillery effluent was significantly higher on the 7th day (23.492 Unit/mg protein/min) when compared to the other treated groups and control.

Table 3 Estimation of Superoxide Dismutase Concentration of Earthworm sample

Concentration/kg	1st week mg/ml	2nd week mg/ml	3rd week mg/ml	4th week mg/ml
250 μg	2.318 ± 0.01	0.631 ± 0.01	0.401 ± 0.01	0.143 ± 0.01
500 μg	3.3112 ± 0.01	1.652 ± 0.01	0.626 ± 0.01	0.079 ± 0.01
750 μg	5.287 ± 0.01	2.294 ± 0.01	0.988 ± 0.01	0.03 ± 0.01
1000 μg	9.83 ± 0.02	6.64 ± 0.02	1.1912 ± 0.01	0.491 ± 0.01
Distillery effluent	23.492 ± 0.05	10.066 ± 0.03	3.4314 ± 0.02	1.411 ± 0.01
Control	0.286 ± 0.01	0.216 ± 0.01	0.35 ± 0.02	0.238 ± 0.01

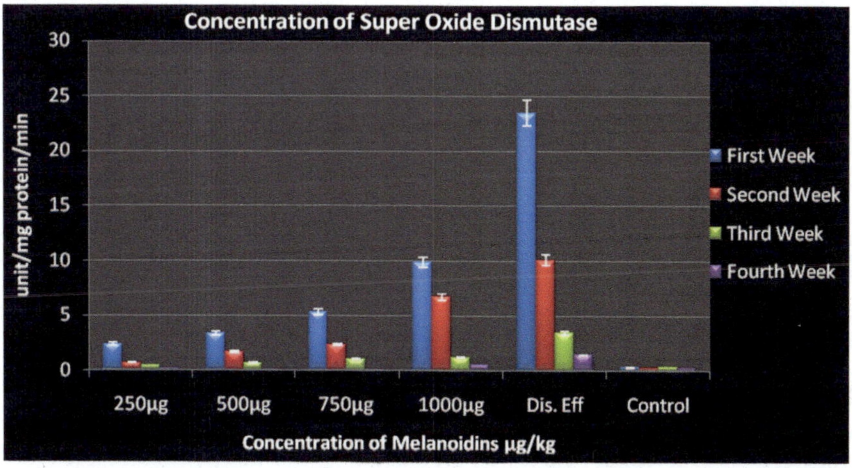

Fig. 2 Graphical representation of Superoxide Dismutase Concentration of Earthworm sample

Table 4 Estimation of Catalase Concentration of Earthworm sample

Concentration/kg	1st week mg/ml	2nd week mg/ml	3rd week mg/ml	4th week mg/ml
250 μg	28.4 ± 0.01	16.3 ± 0.01	15.4 ± 0.01	9.8 ± 0.01
500 μg	23.9 ± 0.01	18.4 ± 0.01	18.2 ± 0.01	9.3 ± 0.01
750 μg	24.8 ± 0.01	29.4 ± 0.01	14.2 ± 0.01	8.8 ± 0.01
1000 μg	43.1 ± 0.1	28.4 ± 0.1	9.4 ± 0.1	4.9 ± 0.1
Distillery effluent	52.5 ± 0.1	38.1 ± 0.1	16.2 ± 0.1	3.6 ± 0.01
Control	48.9 ± 0.1	48.5 ± 0.1	48.2 ± 0.1	47.9 ± 0.1

The decrease in SOD activity signifies an imbalance between the pre-oxidant and antioxidant states in the body leading to an imbalance in systemic redox status (Chakraborty et al. 2007) caused by the melanoidins.

3.4 *Estimation of Catalase Concentration of Earthworm Sample Treated with Varying Concentrations of Synthetic Melanoidin*

Compared to the control, the CAT activity of the earthworm exposed to distillery effluent was markedly higher (52.5 μg mol/mg protein/min) on the 7th day and started to decrease in the following weeks of the study. The CAT activity at the other four concentrations of melanoidins was lower than that of the controls on the 7th day, and statistically, significant reduction was found on the all other days.

The catalase activity may be induced or hindered in the cells that were exposed to Melanoidins. Induction and deterioration of catalase activity are regulated at the mRNA level (Mutoh and Hayashi 1988). The catalase activity tended to increase in the tissues of the earthworm at the beginning treated with effluent and then tend to decrease which shows that the melanoidins affect the catalytic mechanism of the earthworm tissues (Table 4 and Fig. 3).

3.5 *Estimation of Glutathione Peroxidase Concentration of Earthworm Sample Treated with Varying Concentrations of Synthetic Melanoidin*

There was a significant reduction in the GPx activity of earthworms exposed to four concentrations melanoidins (250, 500, 750, 1000 $\mu g\ kg^{-1}$) and the distillery effluent when compared to control except on the 7th day of the study period (Table 5 and Fig. 4).

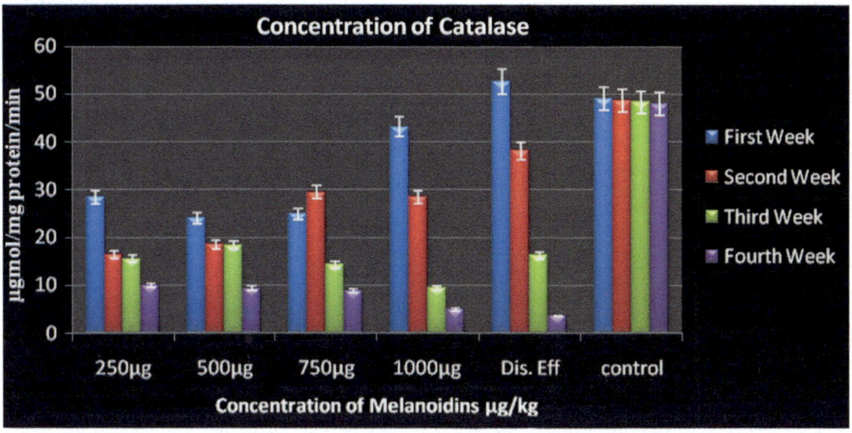

Fig. 3 Graphical representation of Catalase concentration of Earthworm sample

Table 5 Estimation of Glutathione Peroxidase Concentration of Earthworm sample

Concentration/kg	1st week mg/ml	2nd week mg/ml	3rd week mg/ml	4th week mg/ml
250 μg	5.56 ± 0.01	8.26 ± 0.01	4.04 ± 0.01	2.56 ± 0.01
500 μg	8.87 ± 0.01	4.72 ± 0.01	2.94 ± 0.01	1.14 ± 0.01
750 μg	9.54 ± 0.01	5.51 ± 0.01	3.87 ± 0.01	1.07 ± 0.01
1000 μg	16.18 ± 0.1	7.6 ± 0.1	4.08 ± 0.01	1.25 ± 0.01
Distillery effluent	21.48 ± 0.1	18.86 ± 0.1	9.7 ± 0.01	3.86 ± 0.01
Control	4.34 ± 0.01	4.78 ± 0.01	5.13 ± 0.01	5.03 ± 0.01

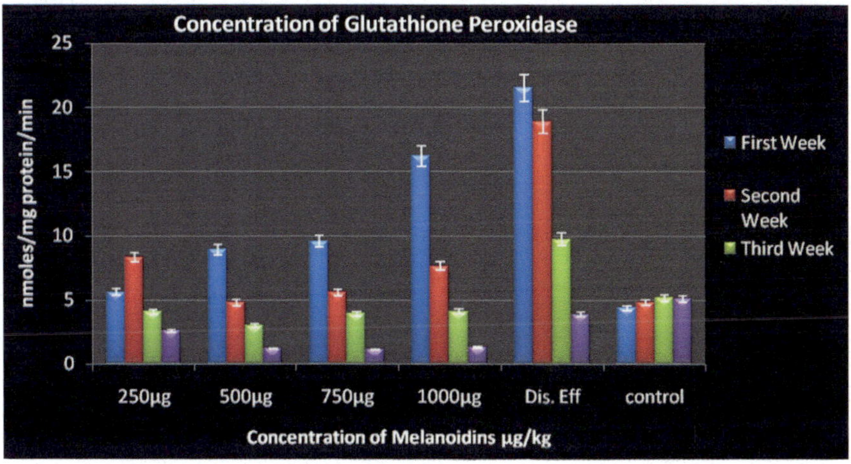

Fig. 4 Graphical representation of Glutathione Peroxidase Concentration of Earthworm sample

Table 6 Estimation of Glutathione S-Transferase Concentration of Earthworm sample

Concentration/kg	1st week mg/ml	2nd week mg/ml	3rd week mg/ml	4th week mg/ml
250 μg	93.167	19.28	7.81	3.722
500 μg	99.31	11.699	9.112	10.216
750 μg	120.283	64.036	10.33	6.716
1000 μg	240.88	197.5	53.23	6.229
Distillery effluent	294.76	167.085	12.192	4.427
control	54.02	53.068	56.83	59.29

The GPx activity decreased from the 14th day throughout the period in all the test groups suggesting that antioxidant defense is overwhelmed by ROS. Mostly, in a system, there will be a balance between the GPx production and lipid peroxide level. Glutathione has ability to control, the levels of lipid peroxides in body. In the animal, the system lost its critical balance of redox states compared to control animals (Song et al. 2009).

3.6 Estimation of Glutathione S-Transferase Concentration of Earthworm Sample Treated with Varying Concentrations of Synthetic Melanoidin

The experimental results obtained from plants suggest that glutathione S-transferase may be closely linked with stress response. Glutathione S-transferase activity seemed to increase at seven days' exposure and then significantly decrease from 14th day. The trend observed after seven days was more convicting as the activity in the control was particularly low at this time. This could explain the lack of a significant effect of melanoidins on GST activity (Table 6 and Fig. 5) (Saint-Denis et al. 2001).

3.7 DNA Damage Evaluation by Comet Assay in Earthworm Coelomocytes

The alkaline comet assay was conducted in earthworm coelomocytes of negative and positive controls exposed to different doses of melanoidins and distillery effluent. The data shown in Table 7 represent the DNA damage. As expected, comparison between positive and negative controls showed that methyl methanesulfonate (MMS) induced a significant increase in DNA migration. In control group, the cells with intact DNA and few comets with very short tail length were seen. The extent of DNA damage along with number of comets was increased with increased doses of melanoidins (250, 500, 750, 1000 $\mu g\ kg^{-1}$) and distillery effluent. As compared to positive control,

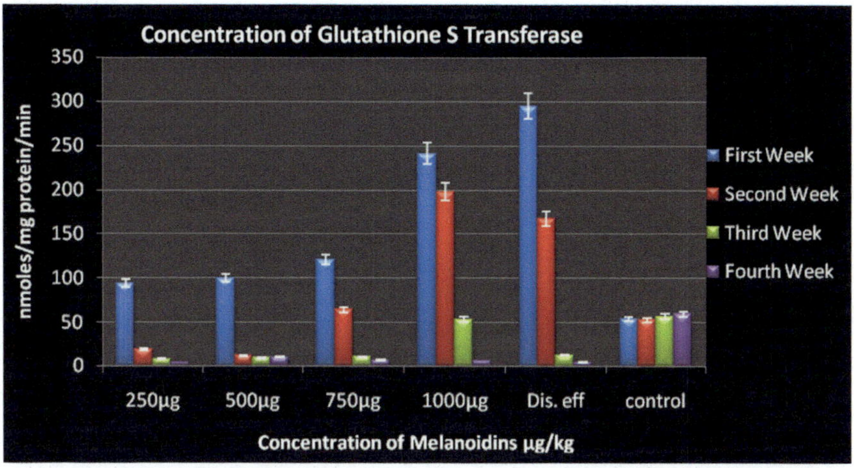

Fig. 5 Graphical representation of Glutathione S-Transferase Concentration of Earthworm sample

Table 7 Comparison of percentage of DNA present in head, tail and tail moment in positive control, negative control and samples—comet assay

S. No.	Dose concentration	% DNA in head (%)	% DNA in tail (%)	Tail moment
1	Negative control	97.00	3.00	0.34
2	Positive control (25 μM MMS)	49.55	50.45	2.53
3	250 μg melanoidins	84.45	15.55	0.68
4	500 μg melanoidins	80.15	19.85	0.94
5	750 μg melanoidins	75.95	24.05	1.23
6	1000 μg melanoidins	69.34	30.66	1.59
7	Distillery effluent	52.36	47.64	2.04

there was a significant decrease in those of 250, 500 and 750 μg/kg melanoidins test groups, whereas no appreciable difference between those of 1000 μg/kg melanoidins, distillery effluent and positive control. The % DNA in comet head decreased in treated groups compared to negative control.

But only, the group of 1000 μg/kg melanoidins and distillery effluent showed significant decrease in % DNA than in positive control. Sharp and significant increase was observed in % DNA in comet tail and tail moment in treated groups compared to negative control. Regarding positive control, there was a notable difference between low dose and positive control, but the other two doses had no significant difference in % DNA in comet tail and tail moment (Table 7, Fig. 6). Correspondingly, Collins and Harrington 2002 have reported the determining of steady-state damage levels following induction of oxidative stress.

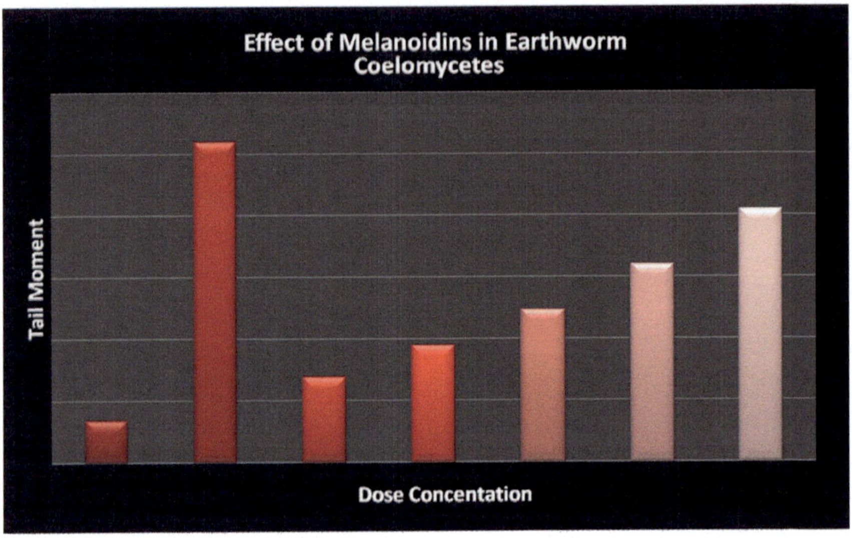

Fig. 6 Graphical representation of Effect of Melanoidins on Earthworm Coelomycetes

3.8 Histopathological Analysis of Earthworms When Treated with Varying Concentrations of Synthetic Melanoidin

There are no visible alterations in the vertical histology of the body wall of *E. eugeniae* in the control (Plate 2). The earthworms exposed to different concentrations of melanoidins and distillery effluent developed varied histological changes (Plate 2). The most prominent changes included degeneration of the circular and longitudinal muscles.

The longitudinal muscle has shown the signs of crakes in major parts of the regions. Also, there was generalized cellular cytolysis, tissue vacuolization and necrosis. Degenerative zones in the longitudinal muscles were noticeable as well as tissue erosion. There were also noticeable cytoplasmic and nuclear alterations in the epidermal cells and those of both circular and longitudinal muscles. Similarly, Gobi and Paramasamy (2010) studied the effect of butachlor herbicide on earthworm *Eisenia fetida* where they have investigated about the histological changes of the earthworm.

In short, the midgut region of distillery effluent exposed earthworms is severely affected as the tissue necrosis is severe. The 750 and 500 μg/kg melanoidins show tissue vacuolization and cellular cytolysis. This shows evidently that the melanoidins and the distillery effluent affect the tissue organization of the earthworms.

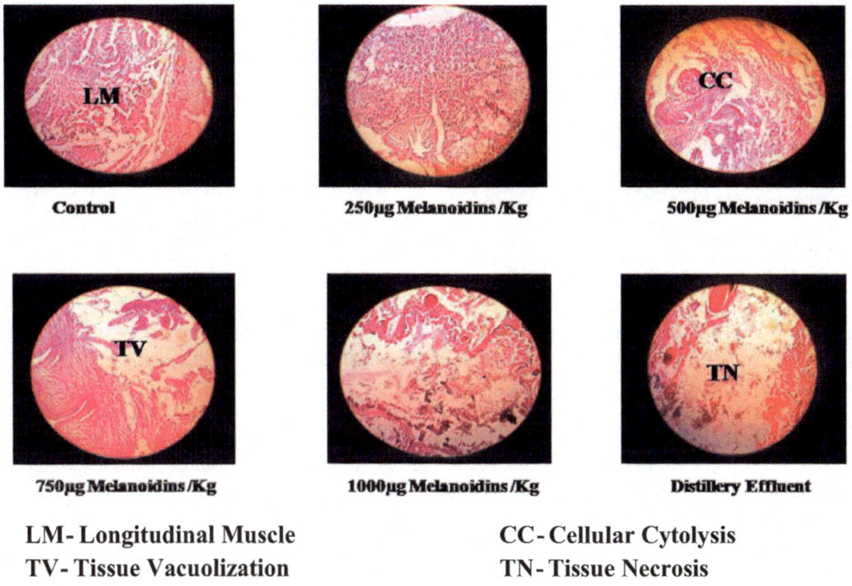

Plate 2 Histopathological analysis of earthworms when treated with varying concentrations of synthetic melanoidin

4 Conclusion

The effect of raw distillery effluent and different concentrations of melanoidin was understood by estimating the antioxidative enzymes. The impact on DNA was made clear following the Comet assay. Further the histological studies makes us understand that the increase in melanoidin increases the damage caused to the earthworms. This shows that melanoidins induce an adverse effect on earthworms which trigger mutilation in the antioxidant enzyme mechanism, DNA damage and cytological damages. And to add a note, this is alleged to be the first investigation to find out the toxicity of melanoidins on earthworms.

Acknowledgements The authors are grateful and would like to thank Department of Science and Technology (DST), MoEF & CC and UGC, Government of India, New Delhi, for the financial support and encouragement. They also acknowledge the DST-FIST, DST-PURSE, UGC-Non-SAP, UGC-SAP and MHRD-RUSA for providing the instrumentation facilities in the department.

References

Bradford, M. M. (1976). Rapid and sensitive method for the quantification of microgram quantities of protein utilizing the principle of protein dye binding. *Analytical Biochemistry, 72,* 248–254.

Chakraborty, I., Mondal, S., Rout, D., Chandra, K., & Islam, S. S. (2007). Structural investigation of a heteroglycan isolated from the fruit bodies of an ectomycorrhizal fungus *Astraeus hygrometricus*. *Carbohydrate Research, 342*(7), 982–987.

Collins, A., & Harrington, V. (2002). Repair of oxidative DNA damage: Assessing its contribution to cancer prevention. *Mutagenesis, 17,* 489–493.

Dikshith, T. S. S., & Gupta, S. K. (1981). Carbaryl induced biochemical changes in earthworm (*Pheretima posthuma*). *Indian Journal of Biochemistry & Biophysics, 18,* 154–169.

Eyambe, S. G., Gaven, A. J., Fitzpatrick, L. C., Venables, B. J., & Cooper, E. L. (1991). A non-invasive technique or sequential collection of earthworm leukocytes during sub chronic immune toxicity study. *Laboratory Animals, 25,* 61–70.

Gobi, M., & Paramasamy, G. (2010). Effect of butachlor herbicide on earthworm *Eisenia fetida*—Its histological perspicuity. *Applied and Environmental Soil Science,* 758–850.

Huang, P. C., Morris, S., Dinman, J., Pine, R., & Smith, B. (1987). Role of metallothionein in detoxification and tolerance to transition metals. *Experientia Suppl, 52,* 439–446.

Humason, G. L. (1979). *Animal tissue techniques* (4th ed.). San Francisco: W. H. Freeman and Co.

Jager, T., Fleuren, R. H. L. J., Hogendoorn, E. A., & De Korte, G. (2003). Elucidating the routes of exposurefor organic chemicals in the earthworm, *Eisenia andrei* (Oligochaeta). *Environmental Science and Technology, 37,* 3399–3404.

Klaassen, C. D., Liu, J., & Choudhuri, S. (1999). Metallothionein: an intracellular protein to protect against cadmium toxicity. *Annual Review of Pharmacology and Toxicology, 39,* 267–294.

Konca, K., Lankoff, A., Banasik, A., Lisowska, H., Kuszewski, T., Gozdz, S., et al. (2003). A cross-platform public domain PC image-analysis program for the comet assay. *Mutation Research, 534,* 15–20.

Lescoat, P., Ali-Haimoud-Lekhal, D., & Bayourthe, C. (2000). Effets de *Saccharomyces cerevisiae* et *Aspergillus oryzae* sur la digestion et le fonctionnement ruminal: Rencontres autour des Recherches sur les Ruminants, Paris, France.

Livingstone, D. M. (1993). Temporal structure in the deep-water temperature of four Swiss lakes: A short-term climatic change indicator. *Verhandlungen des Internationalen Verein Limnologie, 25,* 75–81.

Lokke, H., & Gestel, C. A. M. (1998). *Handbook of soil invertebrate toxicity tests*. Chichester: Wiley.

Marklund, S., & Markulnd, G. (1974). Involvement of the superoxide anion radical in the autooxidation of pyrogallol and a convenient assay for superoxide dismutase. *European Journal of Biochemistry, 47,* 469–474.

Morgan, J. E., & Morgan, A. J. (1990). The distribution of cadmium, copper, lead, zinc and calcium in the tissues of the earthworm *Lumbricus rubellus* sampled from one uncontaminated and four polluted soils. *Oecologia, 84,* 559–566.

Mozer, T. J., Tiemeier, D. C., & Jaworski, E. G. (1983). Purification and characterization of corn glutathione S-transferase. *Biochemistry, 22,* 1068–1072.

Mutoh, N., & Hayashi, Y. (1988). Isolation of mutants of *Schizosaccharomyces pombe* unable to synthesize cadystin, small cadmiumbinding peptides. *Biochemical and Biophysical Research Communications, 151,* 32–39.

OECD. (2004). Guidelines for Testing of Chemicals No. 222: Earthworm Reproduction Tests (Eisenia foetida/Eisenia andrei). Organisation for Economic Co-operation and Development (OECD), Paris.

Oste, L. A., Dolfing, J. K., Ma, W. C., & Lexmond, T. M. (2001). Cadmium uptake by earthworms as related to the availability in the soil and the intestine. *Environmental Toxicology and Chemistry, 20,* 1785–1791.

Peijnenburg, Vijiver, M. G., Martina, G. V. (2009). Monitoring metals in terrestrial environments within a bioavailability framework and a focus on soil extraction. *Ecotoxicology and Environmental Safety, 67*, 163–179.

Rotruck, J. T., Pope, A. L., Ganther, H. E., Swanson, A. B., Hafeman, D. G., & Hoekstra, W. G. (1973). Selenium: Biological role as a component of glutathione peroxidase. *Science, 179,* 588–590.

Saint-Denis, M., Narbonne, J. F., Arnaud, C., & Ribera, D. (2001). Biochemical responses of the earthworm *Eisenia fetida andrei* exposed to contaminated artificial soil: Effects of lead acetate. *Soil Biology & Biochemistry, 33,* 395–404.

Singh, N. P., McCoy, M. T., Tice, R. R., & Schneider, E. L. (1988). A simple technique for quantitation of low levels of DNA damage in individual cells. *Experimental Cell Research, 175,* 184–191.

Sinha, K. A. (1972). Colorimetric assay of catalase. *Analytical Biochemistry, 47,* 389–394.

Song, Y., Zhu, L. S., Wang, J., Wang, J. H., Liu, W., & Xie, H. (2009). DNA damage and effects on antioxidative enzymes in earthworm (*Eisenia foetida*) induced by atrazine. *Soil Biology & Biochemistry, 41,* 905–909.

Sturzenbaum, S. R., Kille, P., & Morgan, A. J. (1998). Heavy metal-induced molecular responses in the earthworm, *Lumbricus rubellus* genetic fingerprinting by directed differential display. *Applied Soil Ecology, 9,* 495–500.

Vijver, M. G., Vink, J. P. M., Miermans, C. J. H., & Van Gestel, C. A. M. (2003). Oral sealing using glue: A new method to distinguish between intestinal and dermal uptake of metals in earthworms. *Soil Biology & Biochemistry, 35,* 125–132.

Co-composting of Cattle Dung and Pigeon Pea Stalk in Large-Scale Pit Systems Using Passive Aeration

S. Modi, S. M. Mahajani, V. Honkalaskar and A. Garg

Abstract The growing requirement of synthetic inputs, including fertilizers, has been one of the reasons for the increasing investment in farming and depletion of soil fertility, leading to the agrarian crisis in a set of regions in India. In order to mitigate the crisis, it is imperative to develop effective manure fertilizers and supplant the synthetic fertilizers. In this regard, the present study attempts to find out the most suitable ratio of cattle dung to pigeon pea stalk for composting. The results from pit composting of different ratios of cattle dung to the pigeon pea stalk (4:0, 4:1 and 4:2 by volume) are compared and the spatial variations in temperature were recorded. The highest temperature, as well as sanitation requirement (temperature ≥ 55 °C for at least three days), could only be achieved with the mixture containing cattle dung to pigeon pea stalk in the ratio of 4:1. Thus, this proportion is suggested for the efficient composting process.

Keywords Co-composting · Cattle dung · Pigeon pea stalk

1 Introduction

Presently, the agrarian crisis is one of the major burgeoning problems of India, where around 58% population depends on agriculture and agro-based businesses to earn their primary livelihood (National Sampling Survey Organization 2014).

S. Modi · S. M. Mahajani
Department of Chemical Engineering, IIT Bombay, Mumbai 400076, India
e-mail: modisujit1995@gmail.com

S. M. Mahajani
e-mail: sanjaym@iitb.ac.in

V. Honkalaskar
Sulokchetana Pratishthan, Himalay Bungalow, Adarsha Colony, Tamjai Nagar, Satara 415002, India
e-mail: honkalaskar@gmail.com

A. Garg (✉)
Environmental Science and Engineering Department, IIT Bombay, Mumbai 400076, India
e-mail: a.garg@iitb.ac.in

S. K. Ghosh et al. (eds.), *Bioresource Utilization and Bioprocess*,
https://doi.org/10.1007/978-981-15-1607-8_23

The agrarian crisis is a multidimensional and complex phenomenon that is observed through a set of symptoms including a constantly rising input intensiveness of the agricultural practices, depleting agro-ecological indicators, and increasing vulnerability of the agrarian system to various exogenous factors such as climatic conditions, support prices of the yield, and labour availability (Honkalaskar et al. 2018; Reddy and Mishra 2009; Karmarkar 2007). Against this background, inorganic fertilizers account for 40–50% of the total synthetic inputs (others being pesticides, herbicides, and growth hormones) for the cultivation of cotton (with pigeon pea as an inter-crop) in Maharashtra (Honkalaskar et al. 2018). The synthetic inputs also have been found to be one of the reasons for the depletion of the physical, chemical, and biological structure of the soil (Wilson and Tisdell 2001; Milosevic and Govedarica 2002; Hussain et al. 2009), thereby depleting the land fertility.

If composts supplant inorganic fertilizers, a substantial amount of input cost can be reduced. However, traditional composting practices are often associated with inefficient composting leading to incomplete decomposition, presence of weed seeds, and growth of pathogens. A set of the reasons for this include inappropriate feedstock proportion (*C/N* ratio), unsuitable moisture conditions, and uneven aeration. Different feedstocks are generally mixed in proper proportions to achieve a desired C/N ratio of ~30 in order to trace an optimal path of composting. There is a substantial literature describing optimal proportions of different residues and cattle dung. For instance, Sharma et al. (2018) used a mixture of cattle manure, sawdust, and flower waste, whereas Muscolo et al. (2018) composted a mixture of olive oil, straw, manure, and broadleaf vegetables. In other studies, a mixture of greenhouse manure and animal manure was also tried for composting process (Külcü and Yaldiz 2014) while Varma et al. (2017) used vegetable waste, cattle manure, sawdust, and dry leaves. However, composting of pigeon pea stalk with cattle dung is yet to be explored. Pigeon pea is second-largest pulse crop of India producing around 12–15 million tons of pigeon pea stalk in a year (Directorate of Pulses Development 2017).

The two main objectives of the current work are (i) to determine the ratio of cattle dung to pigeon pea stalk for efficient composting process, and (ii) to investigate spatial variations of the rate of composting in large-scale pit systems (where the temperature is used as a response parameter).

Three different ratios of the stalk and dung were composted in pit systems. Temporal and spatial variations in temperature along the length, width, and height within pits were recorded. Depending on the temperature profile, the optimum ratio of the raw feed is suggested.

2 Materials and Methods

2.1 Experiment Set-up

Three pits having dimensions of 3 ft (depth) × 10 ft (length) × 5 ft (width) were used for the composting runs that were relied on passive aerations (Fig. 1a). The substrate compositions in the pits are presented in Table 1. The particle size of the pigeon pea stalk was in the range of 0.75–1.5 in. The feedstock was well mixed before addition to the pits. Water was added intermittently to maintain the optimum moisture conditions (50–60%) (Haug 1993).

During the experiments, temperatures at 0.5 ft depth from the point of intersection of top surface diagonals were recorded daily using a thermometer (Testo 174),

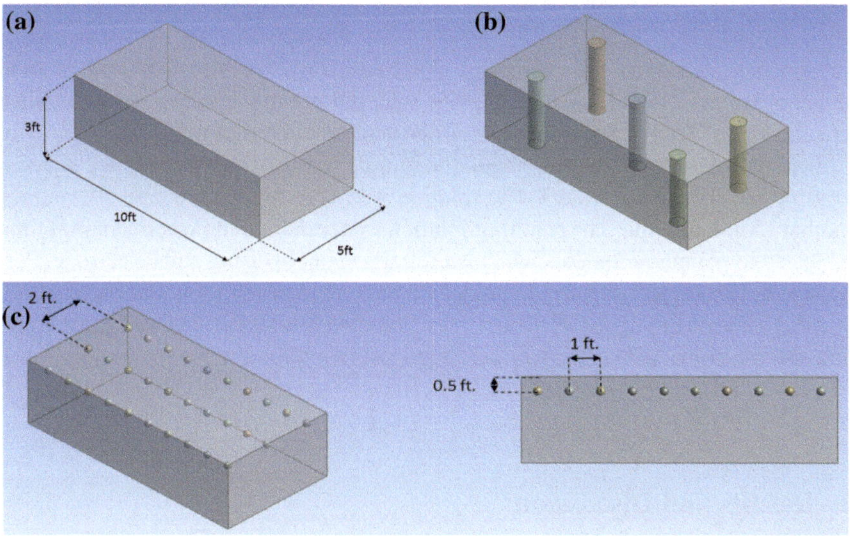

Fig. 1 Schematic diagram of **a** pit dimensions, **b** locations of sampling and spatial variation measurements along the depth, and **c** locations of spatial variation measurement along the length and width

Table 1 Substrate composition in different pits

Parameters	Pit number		
	1	2	3
Cattle dung: Pigeon pea stalk (by volume)	4:0	4:1	4:2
Pigeon pea stalk (kg)	–	95	190
C:N ratio	13.4	16.1	18.5

whereas the temperatures at a depth of 1.5 ft were recorded after every 30 min interval using a temperature data logger (Testo 175 T3). At each of the five locations (shown in Fig. 1b), 6 points were located at 0.5, 1, 1.5, 2, 2.5, and 3 ft depth for spatial variation measurement along the depth. A total thirty points were located for the spatial variation measurement along the length and width (Fig. 1c).

To collect the representative samples from the whole pit, five locations (including four locations equidistant from the four corners and one at the central location of the cross-section of the pit), as shown in Fig. 1b, were selected for sampling. At each of these locations, a through hole of diameter 10 cm was dug. The material extracted from all the holes was properly mixed and 0.5 kg material from the resulting mixture was used as the representative sample.

2.2 Analytical Methods

The dry bulk density of the substrate was measured by weighing a definite amount of sample in a 15 l calibrated vessel. The material was not compressed once filled into the vessel for the measurement. Dried and powdered samples were used for further analysis. For pH measurement, the sample was diluted with distilled water at a ratio of 1:20 (sample: water). The solution was shaken for 5 min and allowed for another 15 min to settle. The resulting solution was then filtered (Whatman no. 4) and subjected to pH measurements which were carried out by using a pH electrode. Percentage of carbon and nitrogen in cattle dung and pigeon pea stalk were determined by using Thermo Finnigan CHNS analyser. Inductively coupled plasma-atomic emission spectroscopy (ICP-AES) was used to measure micronutrients (including B, Ca, Fe, K, Mg, Zn, and P) (Manu et al. 2013).

3 Results and Discussion

3.1 Characteristics of Substrates

The characteristics of cattle dung and pigeon pea stalk are presented in Table 2. *C*/*N* ratios of cattle dung and pigeon pea stalk were 13.4 and 57, respectively. The bulk density of bulking agent (pigeon pea stalk) was around 4 times of the cattle dung. Among the various micronutrients, Ca, Fe, K, and Mg were found in major concentrations.

Table 2 Feed characteristics

Property	*C/N*	Density (kg/m^3)	B	Ca	Cu	Fe	K	Mg	Zn	P
Cattle dung	13.4	160.4	40	1085	ND	866	1131	920	36	ND
Pigeon pea stalk	57	656	ND	811	35	305	436	180	86	190

Note Concentration of micronutrients is in ppm and "ND" indicates non-detectable (i.e., the value <0.05 ppm)

3.2 Overall Temperature Profile

The complete temperature profiles for all the pits are shown in Fig. 2. The maximum temperatures attained were 50.4 °C (pit 1, 68th day), 56.9 °C (pit 2, 67th day), and 51.3 °C (pit 3, 68th day), whereas the average temperatures were 44.3 °C (pit 1), 49.3 °C (pit 2), and 45.2 °C (pit 3). The time duration required for the transition from the mesophilic phase to the thermophilic phase, which indicates rates of decomposition, was in the following order:

$$\text{pit } 3(10\text{th day}) > \text{pit } 1(13\text{th day}) > \text{pit } 2(18\text{th day})$$

The following major points were noted from the above study:

i. Waste composition in the considered range had no significant effect on the time required for the completion of active decomposition period, i.e., drop in the temperature to ambient condition.
ii. Highest peak temperature and waste decomposition were attained with the cattle dung and pigeon pea stalk ratio of 4:1 (by volume).
iii. Sanitation requirement for composting (temperatures greater than or equal to 55 °C for at least three days) was fulfilled only for the mixture of cattle dung with pigeon pea stalk in the ratio of 4:1 (by volume).

The temperature in all pits reduced to ambient condition on approximately 112th day (at the end of the active decomposition period). Possible reasons for the longer active decomposition period are as follows:

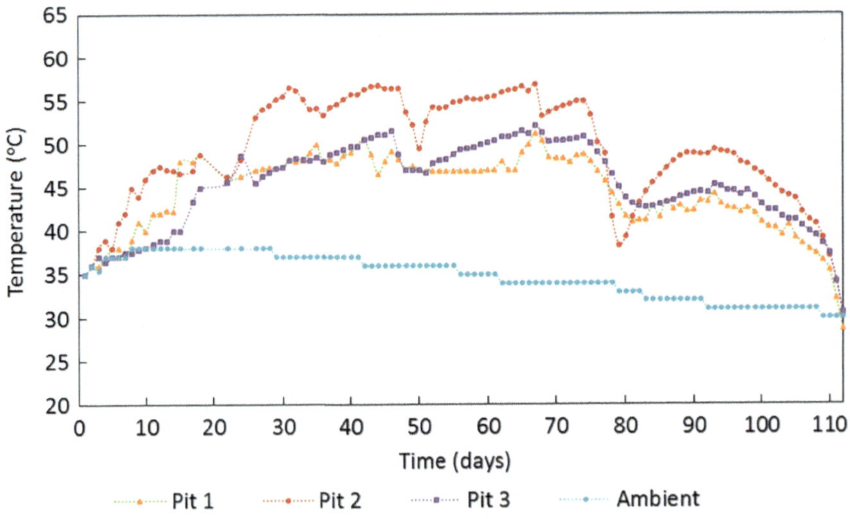

Fig. 2 Illustration of temperature variation (at 1.5 ft depth) with time

i. Initial low values of *C/N* ratio may limit the growth of microbes (Kong et al. 2018).
ii. Lack of microbial population in the beginning which was also indicated from longer time requirement (10–18 days) for getting into the thermophilic phase.
iii. Creation of anaerobic zones in the pits may adversely affect the rate of reaction.
iv. Pigeon pea straw contains lignin up to 21% and high lignin content substrate degrades slowly (Tuomela et al. 2000; Solomon et al. 2016).

In spite of the substrate of pit 3 having higher *C/N* ratio, i.e., more closer towards the optimum compared to other two pits, the temperatures attained were lesser compared to pit 2 (Haug 1993). This supports the fact that C/N is not a complete representation of the degradability (Wang and Witarsa 2016). The lower temperature in pit 3 is attributed to its higher lignin contents, which is due to the larger proportion of pigeon pea stalk in it.

3.3 Spatial Variation of Temperature

Temperature variation along the length and depth are illustrated in Fig. 3. The temperatures near the boundaries were lesser compared to those recorded at the centre of the pit. Along the length and width, the temperature increased from the boundary towards the centre, i.e., up to 5 ft and 2.5 ft in length and width, respectively, after which it was decreased continuously as moved towards opposite boundary (Fig. 3a). The value of temperature at a particular depth is the average of values at that depth at five different locations (Fig. 1b). Similarly, the temperature at a particular distance from the boundary is the average of values at 3 locations along the width at the same distance from the boundary (Fig. 1c). Among all pits, the maximum temperature variation along the length (9 °C) and depth (6 °C) was recorded in pit 3. The spatial variation measurements were carried out twice a week. The measurements shown in Fig. 3 were recorded on 66th day.

3.4 pH

pH trends in all pits is shown in Fig. 4a. It decreased initially up to 20th day which may be due to the formation of organic acids, and then the increase up to 66th day can be attributed to the generation of alkaline salts such as of ammonia. After 66th day, the pH variation was negligible in all the pits.

Fig. 3 Temperature variation profiles along the **a** length and **b** depth of three pits

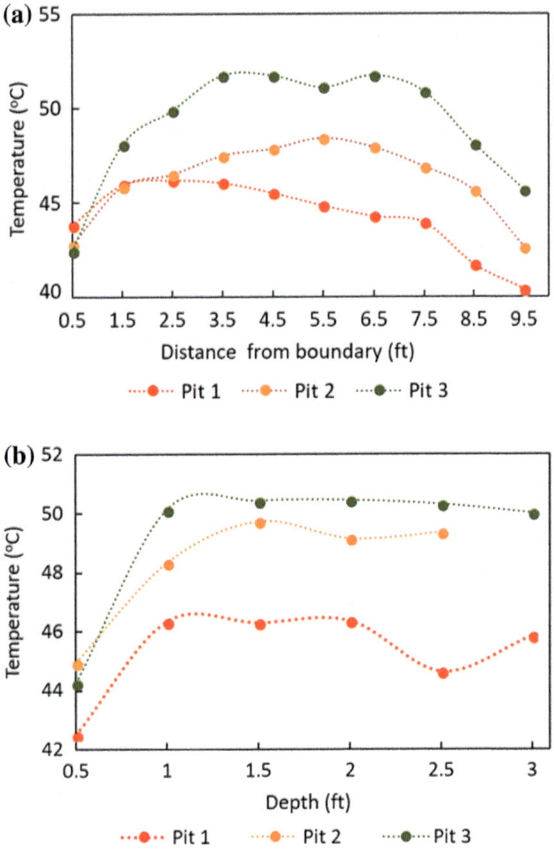

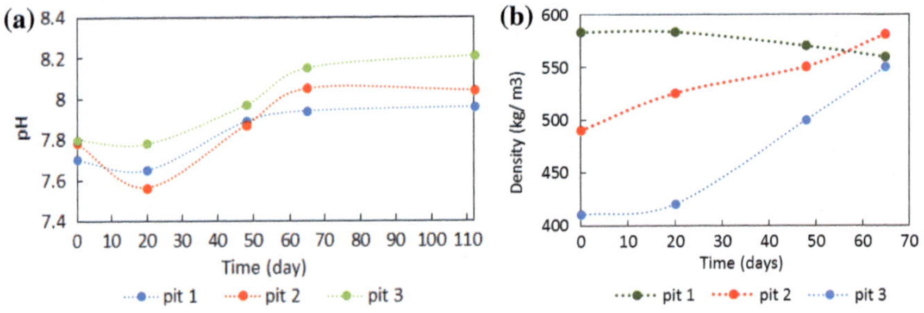

Fig. 4 Plots showing **a** pH variation with time and **b** density variation with time

3.5 *Bulk Density*

Changes in dry bulk density of waste material are shown in Fig. 4b. Dry bulk density is mainly the function of particle size and shape. Increase in its value indicates the reduction in particle size. The increase in the bulk density of pit 3 was higher (410–550 kg/m^3) than pit 2 (490–580 kg/m^3). No significant changes were observed with pit 1, as it no stalk was added into this pit (coarse particles).

4 Conclusions

Considering the availability of both the substrates, this study used different ratios of cattle manure and pigeon pea stalk for finding optimum composition for the composting process. The scarcity of microbial source at the starting and/or generation of anaerobic zones during composting can be the possible reasons for the extended active decomposition period. Based on maximum temperatures attained during composting, density changes, and sanitation requirement condition, the optimum composting ratio of cattle dung to pigeon pea stalk can be suggested as 4:1 (by volume). The spatial variations of temperature were recorded as high as 9 °C along the length in pit 3. The generated data can be used as a basis for designing further experiments as well as for the validation of mathematical models. However, chemical characterization of the substrate other than *C*/*N* ratio such as fibre analysis (soluble substrate, cellulose, hemicellulose, and lignin) and physical characterization can be effectively used to decide optimal composting conditions.

Acknowledgements This work was financially supported by Tata Centre for Technology and Development, Institute of Technology Bombay, Mumbai, India and for the experimental studies space and feedstock was provided by the organization "Deendayal Bahuuddeshiya Prasarak Mandal", Yavatmal, Maharashtra, India. Mr. Minnath Ghode and other co-workers from the same organization helped in carrying out experimental studies.

References

Directorate of Pulses Development. (2017). Annual Report 2016-17. Ministry of Agriculture & Farmers Welfare (Department of Agriculture, Cooperation & Farmers Welfare), Government of India (33–35).

Fortuna, H. (2015). *Climate change may be causing Indian farmers to take their own lives*. https://qz.com/india/1042708/farmer-suicides-in-india-could-be-connected-to-climate-change-according-to-a-new-study/. Accessed on 23-09-18.

FAO. (2018). *Organic agriculture: can organic farmers produce food for everybody*. http://www.fao.org/organicag/oa-faq/oa-faq7/en/. Accessed: 22-06-2018.

Haug, R. T. (1993). *The practical handbook of compost engineering*. Boca Raton, Florida: CRC Press.

Honkalaskar, V. H., Bagde, N. D., & Kedare, S. B. (2018). Understanding Agrarian crisis: A participatory and qualitative systemic analysis. *Journal of Agricultural Studies, 6*(1), 260–297.
Hussain, S., Siddique, T., Saleem, M., Arshad, M., & Khalid, A. (2009). Impact of pesticides on soil microbial diversity, enzymes, and biochemical reactions. *Advances in Agronomy, 102,* 159–200.
Karmarkar, N. (2007). *Agrarian crisis in India: An overview*. Hyderabad, India: ICFAI University Press.
Kong, Z., Wang, X., Liu, Q., Li, T., Chen, X., Chai, L., et al. (2018). Evolution of various fractions during the windrow composting of chicken manure with rice chaff. *Journal of Environmental Management, 207,* 366–377.
Külcü, R., & Yaldiz, O. (2014). The composting of agricultural wastes and the new parameter for the assessment of the process. *Ecological Engineering, 69,* 220–225.
Manu, M., Kumar, R., & Garg, A. (2013). Physical and chemical characterization of yard waste. *International Journal of Applied Engineering Research, 8,* 1891–1896.
Milosevic, N. A., & Govedarica, M. M. (2002). Effect of herbicides on microbiological properties of soil. *Matice Srpske Journal of Natural Sciences, 102,* 5–21.
Misra, R., Roy, R., & Hiraoka, H. (2003). *On-farm composting methods*. Rome, Italy: Food and Agricultural Organization of the United Nation.
Muscolo, A., Papalia, T., Settineri, G., Mallamaci, C., & Jeske-Kaczanowska, A. (2018). Are raw materials or composting conditions and time that most influence the maturity and/or quality of composts? Comparison of obtained composts on soil properties. *Journal of Cleaner Production, 95,* 93–101.
NSSO. (2014). *Statistics on economy and society*. http://www.indianstatistics.org/nssoreports.html. Accessed: 10-02-2018.
Reddy, D. N., & Mishra, S. (2009). Agriculture in the reforms regime. *Agrarian crisis in India* (1–19).
Sharma, D., Yadav, K. D., & Kumar, S. (2018). Role of sawdust and cow dung on compost maturity during rotary drum composting of flower waste. *Bioresource Technology, 264,* 285–289.
Solomon, T., Wondimagegne, B., Desalegn, A., Asress, Z., Niguse, B., & Misganawu, W. (2016). Evaluation of the effect of pigeon pea dried leave supplementations on feed intake and live weight change on local sheep breed in eastern Amhara, Ethiopia. *International Journal of Advanced Research in Biological Sciences, 3,* 81–85.
Tiwary, D. (2016). *Farmer suicides up 40 percent in a year, Karnataka shows sharpest spike*. https://indianexpress.com/article/india/india-news-india/farmer-suicidecase-in-india-crop-failure-drought-dry-zones-indian-monsoon-2984125/. Accessed: 02-7-18.
Tuomela, M., Vikman, M., Hatakka, A., & Itävaara, M. (2000). Biodegradation of lignin in a compost environment: a review. *Bioresource Technology, 72,* 169–183.
Varma, V. S., Das, S., Sastri, C. V., & Kalamdhad, A. S. (2017). Microbial degradation of lignocellulosic fractions during drum composting of mixed organic waste. *Sustainable Environment Research, 27*(6), 265–272.
Wang, Y., & Witarsa, F. (2016). Application of Contois, Tessier, and first-order kinetics for modelling and simulation of a composting decomposition process. *Bioresource Technology, 220,* 384–393.
Wilson, C., & Tisdell, C. (2001). Why farmers continue to use pesticides despite environmental, health and sustainability costs. *Ecological Economics, 39*(3), 449–462.

Effective Recycling of Flower Waste as Organic Manure

P. Swapna and V. V. Lakshmi

Abstract Flower waste is being one of the major organic solid wastes generated in temples, functions, festivals and ceremonies. With the emerging techniques in horticultural practices and with the increase of usage, flowers are cultivated in large scale, and hence waste is also released in an equal proportion. The flower wastes are discarded besides as dumps attracting flies and pathogenic organisms. Hence, the rotten and desiccated flowers are amounting to major health hazard and serious environmental pollution. So, safe and effective degradation to manage the flower waste dumps to value-added products is highly desirable. The development of bacterial consortium is one of the innovative and effective methods to degrade the flower waste. The current study is taken up to recycle the flower waste into organic manure which is most promising and potent alternative to chemical fertilizers for the sustainable development in agriculture.

Keywords Flower waste degradation · Environmental pollution · Bacterial consortium

1 Introduction

The solid wastes are released as a result of human activities, and the waste generation increases drastically with the industrialization and urbanization. The management of the solid waste has become a big challenge to society as well as to the government too, due to the sheer volume as well as the diversity of the source. The variation in solid organic matter content depends on source of waste generation as well as the economic conditions, literacy social and cultural customs of the place (Bandara et al. 2007). Thus, the management of solid wastes can be complex, and economic issue and numerous other factors add constraints to the process (The Expert Committee 2010). On average cities in India, the per capita solid waste generated is 0.8–1 kg

P. Swapna · V. V. Lakshmi (✉)
Sri Padmavati Mahila Visva Vidyalayam, Tirupati, Andhra Pradesh, India
e-mail: lakshmivedula28@gmail.com

P. Swapna
e-mail: swapnamsc9@gmail.com

S. K. Ghosh et al. (eds.), *Bioresource Utilization and Bioprocess*,
https://doi.org/10.1007/978-981-15-1607-8_24

and is steadily increasing year after year. The waste collected is mainly dumped in landfill or finds their way to water bodies thereby becoming major causing pollution (Mishra 2013; Singh and Singh 2007). Separation of solid waste from agricultural and animal waste origin has a great potential of being bioprocessed to organic fertilizers, biofuels and other potentially value-added products generating wealth from waste and reducing load of pollution on the environment (Mohanty et al. 2009).

Flowers play an important role in society and are used extensively in religious, cultural, social events and functions for decoration as well as all ceremonies from birth, marriage to death (Jadhav et al. 2013). In Indian culture, worship is an integral part of human life, and flowers are offered at places of worship like temples, mosques and gurudwaras throughout the year without distinction of religion. Huge quantities of flowers are used in pilgrim centers like Tirumala and Tirupati, where nearly 1500 kg of flowers are used for worship at the Temple of Lord Venkateswara alone daily. This usage rockets significantly to several folds during festival season and annual celebrations like Brahmotsavams, Vasanthotsavams, Pushpa Yagam and Vaikunta Ekadasi. In respect and due to sanctity, the temple wastes are released into water bodies or dumped in the not reachable places of land. The flower dumps attract insects and eels as they rot causing serious health hazards and leading to degradation of environmental quality (Gurav and Pathade 2011). When discharged into water, they choke the rivers and water bodies by depleting the oxygen while they rot and adding the chemical fertilizers and pesticides used for their growth. These toxic changes adversely affect the friable aquatic ecosystem as well as lives of animals inhabiting and lead to frequent mass deaths of fish. It is estimated that more than 80,000 tons of floral wastes find way into river bodies each year without proper treatment, and this has a great impact on society. Thus, maintaining quality environment requires huge expenditure for the massive clean and rejuvenates water bodies locally and nationally. The scenario is visible in even major holy rivers like Ganga which has several important pilgrim centers on its banks leading to environmental degradation. In addition, 40% of the flowers produced are not sold and due to their shorter shelf life to form an important fraction of solid waste (Masure and Patil 2014). Therefore, there is an urgent need to develop suitable, cost-effective and eco-friendly process for bioconversion of flower waste (Fig. 1).

2 Materials and Methods

2.1 Sample Collection and Isolation

One hundred and fifty soil samples were collected from the temples and flower dumped areas of Tirupati in a sterile sampling covers and transferred to the laboratory aseptically. One gram of soil sample was weighed, and a tenfold serial dilution was performed in a sterile saline. 0.1 ml of the 10^{-5} and 10^{-6} dilution was spread plated on the minimal medium containing 1% flower waste. The plates were incubated at

Fig. 1 Flower waste dumps near temples and besides roads of Tirupati

37 °C for 24 h and screened for prominent growth and degradation on the plates were selected. Five bacterial isolates showing significant degradation were selected and maintained as pure culture and subjected to secondary screening. In the secondary screening, the isolates were inoculated in minimal broth containing 1% flower waste and incubated in a shaking incubator at 37 °C and 150 rpm speed. Three isolates showing best degradation visually were selected for further study.

2.2 *Identification of the Bacterial Isolates*

Morphological, cultural characterization of the isolates was carried by performing gram staining and spore staining. Growth parameters were studied at selected temperatures (27, 37 and 45 °C) and pH (4.5, 7.5 and 8.5). Biochemical characterization of the isolates was carried out by IMVic, sugar fermentation tests using standard protocols (Cappuccino and Sherman 2009). Production of oxidase, catalase and urease amylase enzymes was tested following standard protocols of Mackie and McCartney (1989).

2.3 *Preparation of Mixed Culture*

The mixed culture was prepared by inoculating 100 μl of each in combination with two isolates each as well as all the three isolates together in nutrient broth and incubated at 37 °C for 24 h and checked for compatibility and coexistence microscopically after gram staining. The antagonism if any between the organisms was tested by plate co-culture method (Sarkar and Chourasia 2017).

2.4 Degradation of Flower Waste Using Mixed Culture Consortium

Hundred ml of minimal medium containing 1% flower waste was prepared in seven 250 ml conical flasks. The media were inoculated with 10%V/V of B1, B2, B3 isolates (10^6 CFU/ml) individually in first three flasks, respectively. The next three flasks were inoculated with the combination of two cultures each; whereas, the seventh flask was inoculated with mixed bacterial culture of three isolates (10^6 CFU/ml) to a final volume of 10%V/V. The flasks were incubated in a shaking incubator at 37 °C and 150 rpm speed (Mirdamadian et al. 2011). Sample was withdrawn at regular intervals of 24 h up to 4 days and centrifuged at 5000 rpm for 10 min to get the clear supernatant. The degradation of the floral substrate was determined by the estimation of reducing sugars present in the supernatant adopting DNS method. The absorbance was recorded at 540 nm, and glucose (1 mg/ml) was taken as standard.

2.5 Optimization Parameters of Degradation

Optimization of flower waste degradation parameters was carried out by varying the substrate concentration in the range of 1–15%, and the temperature of degradation includes the effect of substrate concentration and varied temperatures such as room temperature, 37 and 45 °C using the mixed culture consortium. The degradation efficiency was determined by the amount of reducing sugars produced.

2.6 Preparation of Organic Manure

The flower degraded product (FDP) was obtained by semisolid state fermentation adopting the method of Grazziotin et al. (2006) and Kumari Chittturi and Lakshmi (2015). The flower waste was crushed into small pieces to produce FDP in 1500 g quantity batches, 50 g of crushed flower wastes were taken in each 2 l conical flasks, and these flasks were supplemented with the 100 ml minimal broth and sterilized in autoclave at 121 °C and 15 lb/in.^{2} broth pressure for 10 min. The flasks were cooled and inoculated with 15% of mixed culture (10^9 Cfu/ml) and incubated in a shaking incubator at 37 °C and 150 rpm speed up to 5 days to achieve complete degradation. The product after degradation was treated as FDP which is used as organic manure.

2.7 *Effect of FDP Amendment to Soil*

FDP amendment to the soil influences the nutrient uptake, moisture retention capacity, biomass composition, etc., were determined by adopting Hadas and Kautsky (1994). The various amendments performed in the current study included T_1: Soil sample (1500 g), T_2 (20%VC): Soil sample (1200 g) + 300 g vermicompost, T_3 (20%FDP): Soil sample (1200 g) + 300 g FDP. The garden soil was mixed with the amendments at 20% in 2 l size containers covered with perforated covers. The moisture content was maintained during the period of experimentation. Duplicates were maintained and incubated at 37 °C for 60 days. The parameters analyzed were soil moisture retention capacity (AOAC 2000), soil pH and electrical conductivity (Jackson 1973), CO_2 evolution (Anderson Anderson and Domsch 1989), biomass—bacterial, fungal and actinomycetes count (Allen and Nelson 1910).

3 Results and Discussion

3.1 *Isolation and Screening of Bacteria*

More than 50 isolates from the different colonies show prominent growth on the minimal agar with 1% flower waste were selected in the primary screening. Among them, five isolates show degradation of flower waste visually (Table 1). In the secondary screening, three best bacterial isolates (B1, B2, B3) effectively degraded the flower waste to slimy slurry in 4 days in a shaking condition at 37 °C and 150 rpm speed. The color of the slurry turned to dark brown is shown in Fig. 2.

Table 1 Visual degradation of the isolates

S. No.	Isolates	Degradation (visually)
1	B1	+++
2	B2	+++
3	B3	+++
4	B4	++
5	B5	++

+++ Complete degradation, ++ Partial degradation

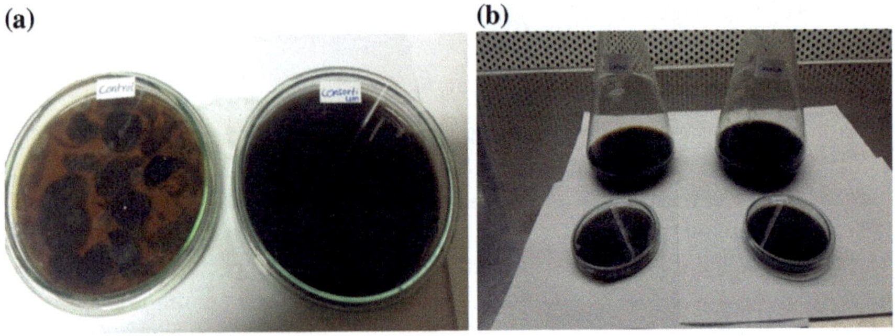

Fig. 2 Degradation of flower in **a** plate culture, **b** liquid culture

3.2 Identification of Bacterial Isolates and Development of Mixed Culture

The identification of the three isolates was carried out, and the results of morphological cultural and biochemical tests are summarized in Table 2. The preliminary identification showed that all the isolates belong to the Bacillus sp. Molecular characterization of the isolates to identify them at species level is in progress.

Bacterial mixed cultures were prepared using the three potent isolates B1, B2 and B3 grown together, and the compatibility was checked by gram staining. The slides were observed under 100× oil immersion, and the gram positive rods were observed. Three different morphologies of short rod for B1, long rods for B2 and bacilli in chains were observed to be in almost equi-proportion. Antagonism assay results of cross streaking method (Fig. 3) showed that no inhibition was observed between the organisms when incubated even up to 40–72 h. Each indicates that the mixed isolate culture was highly stable and could be used as mixed culture for flower degradation.

3.3 Degradation of Flower Waste Using Mixed Bacterial Culture

The flower waste degradation was achieved within 4 days in shaking incubator at 150 rpm speed and 37 °C. The flower waste in the broth becomes dark slimy slurry after degradation. The samples (fermented broth) were collected regularly up to 4 days and reducing sugars in the supernatant were estimated by DNS method. The reducing sugars were found increasing from the first day to second day and then start decreasing from the third day onwards. The better degradation was achieved in mixed culture of the isolates compared with any of the single isolate based on the reducing sugars produced in the fermented broth Table 3.

Table 2 Identification of bacterial isolates

Test	Bacterial isolates			
	Condition	B1	B2	B3
Growth at different temperature (°C)	4	–	–	–
	27	+	+	+
	37	+	+	+
	45	+	+	+
Growth at different pH	4.5	–	–	–
	7.5	+	+	+
	8.5	+	+	+
Oxygen requirement		Aerobic	Aerobic	Aerobic
Morphological identification	Colony morphology	Dull white colonies	Milky white colonies	White raised colonies
	Microscopic morphology	Gram positive short rods	Gram positive long rods	Gram positive rods in chains
	Endospore staining	Spore forming	Spore forming	Spore forming
Sugar fermentation tests	Glucose	+	+	+
	Galactose	–	+	–
	Maltose	+	–	+
	Sucrose	+	+	+
	Starch	+	–	+
Biochemical tests	Indole	–	–	–
	MR	–	+	–
	VP	+	+	–
	Citrate	–	+	+
Enzymatic tests	Urease	+	+	+
	Oxidase	+	+	+
	Catalase	+	+	+
	Starch hydrolysis	+	–	+

3.4 *Optimization Parameters*

To increase the degradation of flower waste, substrate concentration of substrate and temperature of degradation were studied and the results are shown in Fig. 4. There was a study increase in the production of reducing sugar up to 8% substrate concentration beyond which the efficiency reduced significantly. Maximum degradation of flower waste was obtained at substrate concentration of 8% with production of 1445.5 μg/ml

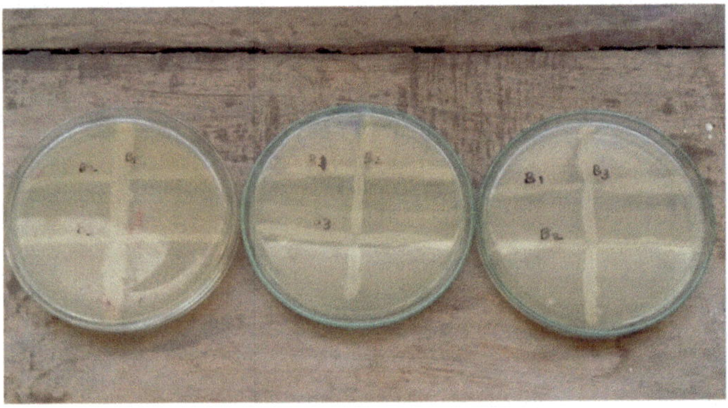

Fig. 3 Antagonism assay results of the isolates

Table 3 Flower waste degradation with single and mixed cultures

S. No.	Isolates	Final pH	TCD[a] (in hours)	Reducing sugars in µg/ml			
				Days			
				1st	2nd	3rd	4th
1	Control	7.10	–	300.0	310.0	290.0	300.0
2	B1	7.15	60	367.4	633.0	542.1	495.8
3	B2	7.21	72	350.4	520.6	439.2	320.0
4	B3	7.36	72	561.9	733.3	622.2	607.4
5	B1 + B2	7.48	72	540.4	636.8	566.4	380.2
6	B2 + B3	7.56	72	478.9	620.4	550.8	382.4
7	B3 + B1	7.80	72	550.6	735.8	575.6	490.6
8	B1 + B2 + B3	7.90	48	607.4	882.9	600.7	592.5

[a]Time required for complete degradation (in hours)

reducing sugars. The rate of degradation was found to be high at temperature 37 °C (1385.6 µg/ml) which was the optimum temperature for the degradation flower waste.

3.5 *Effect of FDP Amendment on Soil*

Amendment of soil with FDP resulted in good of the moisture retention capacity of the soil up to 45 days as compared to controls. Maximum moisture content of 57% was observed in soil amended with 20% FDP followed by 20% VC showing 55%. The moisture content of control sample was 50%. The moisture range of 50–75% is known to facilitate the growth of bacteria like *Bacillus* sp., *Pseudomonas* sp. *and Clostridium* sp., in soil and increase soil activity as well as biomass (Alexander

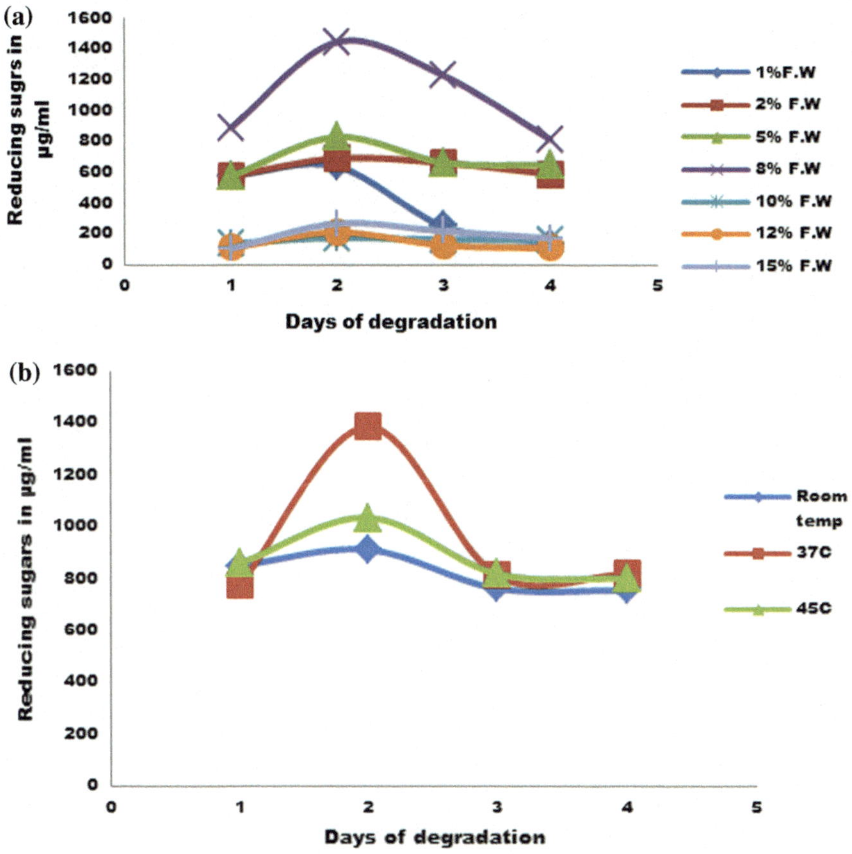

Fig. 4 Effect of **a** substrate concentration, **b** temperature on flower degradation

1977). The pH of the control soil was slightly alkaline at 7.35. There was a significant increase in soil pH on the amendment of floral amendment to pH 8.24 which reduced to 7.5 by 15 days and continued the trend of maintaining slightly alkaline pH. Similar observations were also made in earlier studies (Shouche et al. 2011; Jain 2016). Soil EC values were in normal range though gradually increased with incubation time, and the magnitude of increase was higher in 20% FDP amended soils than the control, followed and VC. The increase in the EC indicates effective availability of nutrients to the plant.

This was also supported by the results of rate of CO_2 evolution due to decomposition of soil organic matter which increase steadily up to 45 days followed by beginning to tapper by 60th day. Maximum CO_2 evolution as observed in FDP amendment and VC amendments was comparable suggesting that FDP can also have good potential organic manure, and the results were shown in Fig. 5.

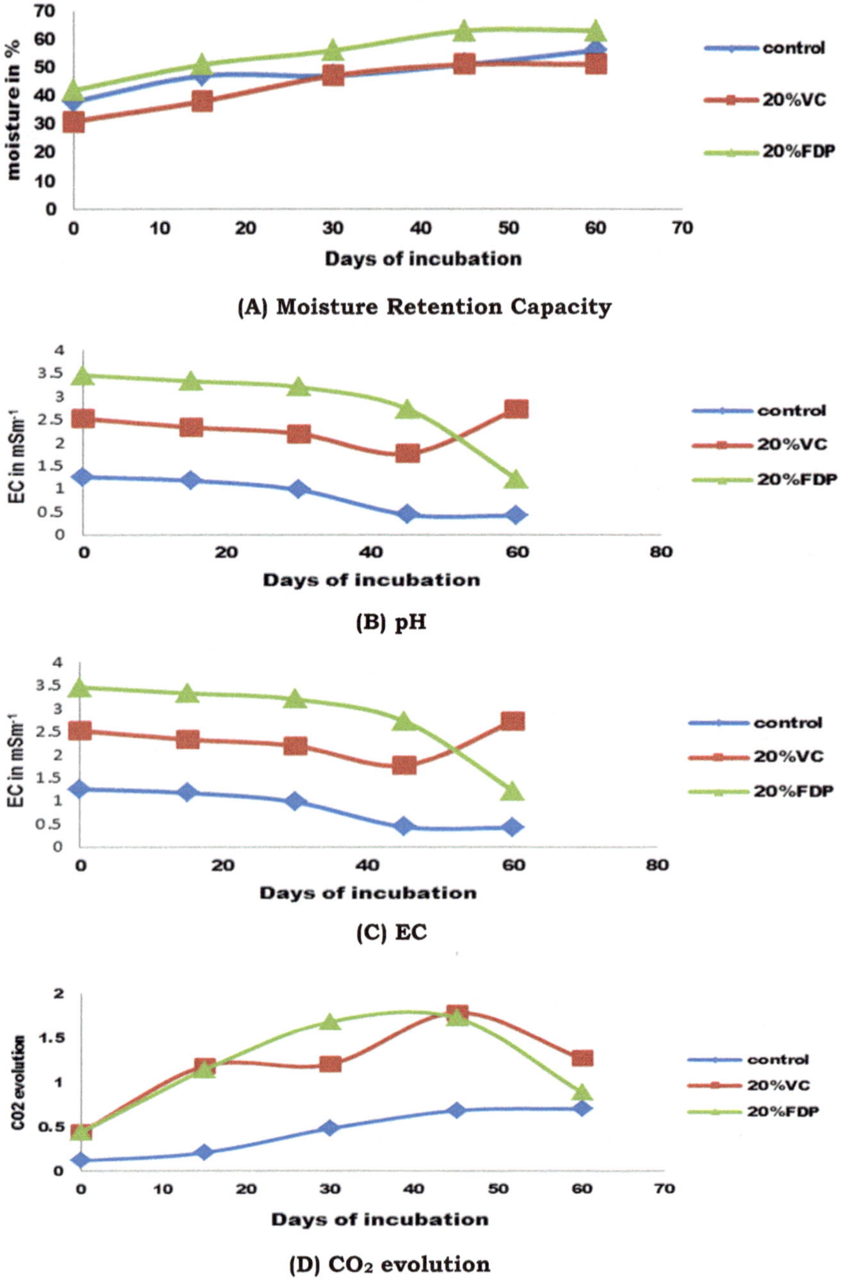

Fig. 5 Comparison of soil parameters in amended soils

Table 4 Microbial counts of the various amendments

S. No.	Soil amendments	Days of incubation				
		0	15	30	45	60
(a) Bacterial count			*CFU/ml*			
1	Control	6×10^8	8×10^8	10×10^8	11×10^8	8×10^8
2	20% VC	8×10^8	12×10^8	15×10^8	17×10^8	13×10^8
3	20% FDP	13×10^8	16×10^8	18×10^8	21×10^8	18×10^8
(b) Fungal count						
1	Control	10×10^4	13×10^4	15×10^4	20×10^4	17×10^4
2	20% VC	14×10^4	16×10^4	20×10^4	22×10^4	22×10^4
3	20% FDP	20×10^4	25×10^4	33×10^4	39×10^4	35×10^4
(c) Actinomycetes count						
1	Control	3×10^3	4×10^3	6×10^3	8×10^3	3×10^3
2	20% VC	4×10^3	5×10^3	6×10^3	10×10^3	8×10^3
3	20% FDP	5×10^3	7×10^3	9×10^3	10×10^3	8×10^3

4 Soil Biomass

The bacterial, fungal and actinomycetes count were recorded by colony counter, the CFU per ml was calculated based on the number of colonies, volume of the media plated and dilution factor, and the results were tabulated in Table 4. The highest bacterial biomass was seen in 20% FDP by 45th day followed by 20% VC. Similar trend was also observed in actinomycetes as well as fungal counts. Organic amendments stimulate both intra and extra cellular enzymes that stimulate microbiological activity (Pascual et al. 1999; Mclatchey and Reddy 1998).

5 Conclusion

A mixed isolate culture was found to have high efficacy in degrading flower waste with reduction in the time required compared to degradation by a single isolate. The mixed isolate culture exhibited compatible and no antagonistic activity was observed among the isolates indicating stability and effectiveness in degradation of the flower waste. The optimum substrate concentration was 8% and with efficient flower waste degradation achieved at 37 °C. The parameters of pH and EC observed with soil amendment with FDP were comparable to vermicompost amendments in similar conditions. Thus, eco-friendly organic manure having high nutritive values can be prepared out of flower degraded product. This highly desirable and cost-effective organic manure has the potential to emerge as an alternative to the chemical fertilizers.

References

Alexander, M. (1977). *Introduction to soil microbiology* (2nd ed.). New York: Wiley.

Allen, E. J., & Nelson, E. W. (1910). On the artificial culture of marine plankton organisms. *Journal of the Marine Biological Association of the United Kingdom, 8*, 421–474.

Anderson, T. H., & Domsch, K. H. (1989). Ratios of microbial biomass carbon to total organic carbon in arable soils. *Soil Biology & Biochemistry, 21*, 471–479.

AOAC. (2000). *Official methods of analysis of the AOAC* (18th ed.). Washington, DC.

Bandara, N. J., Hettiaratchi, J. P., Wirasinghe, S. C., & Pilapiiya, S. (2007). Relation of waste generation and composition to socio-economic factors: A case study. *Environmental Monitoring and Assessment, 135*(1), 31–39.

Cappuccino, J. G., & Sherman, N. (2009). *Microbiology; A laboratory manual* (pp. 161–165). Harlow: Benjamin.

Kumari Chittturi, Ch M, & Lakshmi, V. V. (2015). Fermentative production of keratinase using solid agricultural wastes. *International Journal of Scientific & Engineering Research, 6*(2), 50–55.

Grazziotin, A., Pimentel, F. A., de Jong, E. V., & Brandelli, A. (2006). Nutritional improvement of feather protein by treatment with microbial keratinase. *Animal Feed Science and Technology, 126*(1–2), 135–144.

Gurav, M. V., & Pathade, G. R. (2011). Production of vermicompost from temple waste (Nirmalya): A case study. *Universal Journal of Environmental Research and Technology, 1*(2), 182–192.

Hadas, A., & Kautsky, L. (1994). Feather meal, a semi-slow-release nitrogen fertilizer for organic farming. *Fertilizer Research, 38*, 165–170.

Jackson, M. L. (1973). *Soil chemical analysis* (1st ed.). New Delhi, India: Prentice Hall of India Pvt. Ltd.

Jadhav, A. R., Chitanand, M. P., & Shete, H. G. (2013). Flower waste degradation using microbial consortium. *Journal of Agriculture & Veterinary Science, 3*(5), 01–04.

Mackie & McCartney (1989) *Practical medical microbiology* (13th ed.) Vol. 2 (pp. 43–160). London: Churchill Livingstone.

McLatchey, G. P., & Reddy, K. R. (1998). Regulation of organic matter decomposition and nutrient release in a wetland soil. *Journal of Environmental Quality, 27*(5), 1268–1274.

Mirdamadian, S. H., Khayam-Nekoui, S. M., & Ghanavati, H. (2011). Reduce of fermentation time in composting process by using a special microbial consortium. *World Academy of Science, Engineering and Technology, 52*, 475–479.

Mishra, N. (2013). Temple waste, a concern. *Times of India*. Retrieved from http://www.timesofindia.indiatimes.com.

Jain, N. (2016). Waste management of temple floral offering by vermicomposting and its effect on soil and plant growth. *International Journal of Environmental & Agriculture Research., 2*(7), 89–94.

Masure, P. S., & Patil, B. M. (2014). Conceive of waste flowers. *International Journal of Engineering Research &Technology, 3*(10), 1072–1075.

Mohanty, S. K., Behera, S., Swamin, M. R., & Ray, R. C. (2009). Bio ethanol production from mahula (Madhuca latifolia L.) flowers by solid-state fermentation. *Applied Energy*, 640–644.

Pascual, J. A., García, C., & Hernandez, T. (1999). Lasting microbiological and biochemical effects of the addition of municipal solid waste to an arid soil. *Biology and Fertility of Soils, 30*(1–2), 1–6.

Sarkar, Payel, & Chourasia, Rounak. (2017). Bioconversion of organic solid wastes into biofortified compost using a microbial consortium. *International Journal of Recycling of Organic Waste in Agriculture, 6*, 321–334.

Shouche, S., Pandey, A., & Bhati, P. (2011). Study about the changes in physical parameter during vermicomposting of floral waste. *Journal of environmental research and development., 6*(1), 63–68.

Singh, S. K., & Singh, R. S. (2007). Study on municipal solid waste and its management practices in Dhanbad-Jharia coalifield. *Indian Journal of Environmental Protection, 18*(11), 850–852.
The Expert Committee. (2010). Manual on municipal solid waste management, The Ministry of Urban Development. *The Government of India, 1*(2), 295–789.

Food Waste, a Good Option for Biodiesel Production

Nibedita Sarkar, Byong-Hun Jeon, Pradip Kumar Chatterjee and Amit Ganguly

Abstract Biodiesel is a type of renewable energy source. Researchers have considered it as a substitute for fossil fuel. Combustion of biodiesel generates fewer pollutants such as CO_2, CO, particulate matter, except NO_x. Fatty acid methyl esters (FAMEs) are the key components of biodiesel which can be synthesized through transesterification of lipid in the presence of alcohol, base, acid, enzyme or solid catalyst. Traditional biodiesel from rapeseed, palm, sunflower, jatropha and soya bean has been strongly criticized as they require long extension of lands for cultivation. In that context, food waste is better choice for biodiesel production. The reports of FAO of the United Nations say that approximately 1.3 billion tons of food waste is disposed through worldwide. Exponential growth of population will cause a continuous increase of food waste generation in developing countries in Asia. Usually, food waste is disposed in landfills. This is causing world's mounting food waste disposal problem. This practice of disposing food waste disposal in landfills has harmful effect in human life such as bad odour, air pollution and leaching. Carbon dioxide, methane and other toxic gaseous substances are emitted from landfills. Therefore, food waste can be utilized as non-edible resources for biodiesel production which is a better option for not to use lands limited for food crop. The cost of traditional biodiesel production is relatively high based on high cost of feedstock as well as biodiesel production technologies. This article reviews the aspects of biodiesel production from food waste as well as potential of biodiesel production from food waste.

Keywords Renewable energy · Biodiesel · Food waste · Greenhouse gas emission

N. Sarkar · P. K. Chatterjee · A. Ganguly (✉)
CSIR Central Mechanical Engineering Research Institute Durgapur, Durgapur, India
e-mail: amitganguly022@gmail.com

N. Sarkar
e-mail: nibedita_sarkar@rediffmail.com

P. K. Chatterjee
e-mail: pradipcmeri@gmail.com

B.-H. Jeon
Earth Resources and Environmental Engineering, Hanyang University, 222-Wangsimni-Ro, Seongdong-Gu, Seoul 04763, Republic of Korea
e-mail: bhjeon@hanyang.ac.kr

S. K. Ghosh et al. (eds.), *Bioresource Utilization and Bioprocess*,
https://doi.org/10.1007/978-981-15-1607-8_25

1 Introduction

Researchers have taken several initiatives for an alternative fuel to beat the challenge regarding limited source of fossil fuels. In that context, biodiesel can be the most efficient fuel as well as replacement of highly demanding petro-diesel in transportation sector. Fuel properties of biodiesel resemble with petro-diesel. Higher flash point and lubricating quality improve the efficiency of biodiesel as high flash point makes biodiesel safer during storage as well as transportation. Biodiesel implies a good ignition quality compared to traditional diesel due to its high cetane number. Additionally, emissions of particulate matter, CO, are less in case of biodiesel compared to fossil diesel (Verma et al. 2016). However, disadvantages of biodiesel are included as high viscosity, lower energy content, lower engine speed and power, high cloud and pour point, high nitrogen oxide emission, high price and engine erosion (Demirbas 2009). High viscosity is the drawback of biodiesel due to harmful effect on fuel injection equipment (Balat and Balat 2008). At low temperature region, biodiesel gets cold and starts to gel up, chocking of fuel filters due to high cloud and pour point.

Edible, non-edible and waste cooking oil or animal oil is the source for biodiesel synthesis. Total world energy consumption report says that diesel is the main fuel in industrial, transport as well as agricultural sectors in developed and developing countries. According to ASTM standards, B100 is the pure form of biodiesel which is mostly acceptable in transport sector; otherwise, it can be used as blended form with a range of 5–20%, i.e. B20. Chemically, biodiesel is monoalkyl ester and is produced by means of transesterification process. Transesterification is a reaction between lipid specifically triacylglycerol and an alcohol to produce esters and glycerol as by-product in the presence of catalyst. Carbonyl carbon of the fatty acid ester receives nucleophilic attack of the incoming alkoxide (R2O−) which synthesizes a tetrahedral intermediate that can go through the reverse reaction or proceeds to the forward reaction. Transesterification reaction is a combination of three successive reversible reactions. At first, triglycerides are converted to diglycerides, simultaneously diglycerides are converted to monoglycerides, and finally monoglycerides into glycerol, with a yield of one ester molecule from each glyceride at each step.

2 Food Waste

Food waste can be defined as the food material that is generally uneaten and simply thrown away. Food waste is a result of food loss. However, it consists of enough nutritious component for human consumption. Food waste or loss generally occurs at various stages such as production, processing, retailing and consumption. Reports by Food and Agricultural Organization of the United Nations say that approximately 1.3 billion tons of food waste is discarded globally without any further use (Gustafsson et al. 2011; Melikoglu et al. 2013). India ranks 7th in terms of food waste generation where Russian Federation is at the top of the list. In India, food waste consists of 4% meat and 70% vegetable and fruit wastes. However, economic cost of the meat is more compared to vegetable and fruit waste. Report says that Indians waste an amount of food equal to the total food consumption in UK. Exponential growth of population will cause an continuous increase of food waste generation in developing countries in Asia. In recent years, food waste has become a significant concern in India. A large amount of food is discarded in weddings, canteens, hotels, social and family functions which results in garbage bins, landfills to be overloaded. Ministry of Agriculture report says that each year India generates a huge amount of food waste with a total cost of Rs. 50,000 crores. Additionally, food waste has great social and environmental impact. A large proportion of municipal solid waste is composed of food waste. Food waste generally is divided into household food waste, food-processing waste, canteen and restaurant waste. Streets scattered with food waste have become a part of life in several places of India. Inadequate collection as well as improper disposal issues of food waste is creating a serious problem in present India. Dumping of food waste in landfills has been considered as the most easiest and economic way of disposal. Approximately, 90% of the food waste is directly disposed as landfills which results in various health and aesthetic issues (Baskarn and Aiswarya 2016; Das et al. 1998). Additionally, rising cost of waste disposal and the lack of land space in metropolitan cities have become major concerns in present days. Moreover, landfills emit methane which is a potent global warming gas (GHGs). The decomposition of organic matter in dumped manner creates a severe health problem like bad odour, air pollution and leaching. Approximately, 282.6 million tons of methane emission was reported by United States Environmental Protection Agency (USEPA) in 2000, wherein 36.7 million tons was due to landfill emissions (Rena et al. 2018).

However, food waste production as well as disposal issues can be resolved by better management system at food industries as well as level of eating. Currently, several initiatives have been taken to utilize food wastes into valuable products. Sustainable valorization may be a great choice to solve the waste disposal problem to reduce its environmental impact. Present valorization methods include incineration, anaerobic digestion and processing which can be used to recycle food wastes (A food waste & yard waste plan for Hong Kong 2014; Monitoring of solid waste in Hong Kong 2015). Additionally, conversion of food waste to efficient liquid fuel is a major concern in recent days as it allows complete utilization of food waste. In that context, biodiesel production from food waste may be a great idea to produce some worth from waste with zero cost. To accelerate this process, chemical, chemo-enzymatic and multi-enzymatic routes can be used.

Generally, food waste is composed of carbohydrate, protein, lipid and phosphates where lipid can be extracted and used for biodiesel production. Edible oil has been a traditional choice for biodiesel production though it increases the production cost due to the high cost of the feedstock (OECD 2007). Alternatively, non-edible oil is extensively examined in academic and industry as well. However, severe controversy has been started between civil societies and stakeholders arguing that increasing use of arable land for non-food biofuel crops will create problem for growing food crops due to scarcity of land, water and other resources (Fargione et al. 2008; Tilman et al.

2009). Jatropha is the primary choice for the biodiesel industries in India for a long period, and jatropha is generally cultivated in waste degraded land with low nutrition level. Gui et al. reported that approximately 1.72 million hectares of land is reserved for jatropha cultivation in maximum states in India for biodiesel production (Pham et al. 2014). But emission of toxic substance after jatropha cultivation makes it unfavourable for biodiesel production (Pham et al. 2014). In that context, food waste can be a better choice for biodiesel production. Food waste is considered as efficient feedstock for biodiesel production since it is disposed without further use. Food waste generally includes (i) rotten fruits and vegetables, (ii) fish and poultry organs, intestine, meat trimmings and other residues, (iii) fruits and vegetable peelings, (iv) meat, fish, shellfish shells, bones, (v) food fats, sauces, condiments, (vi) soup pulp and herbal medicinal pulp, (vii) eggshells, cheeses, ice cream, yogurts, (viii) tea leaves, tea bags, coffee grounds, (ix) bread, cakes, biscuits, desserts, jam, (x) cereals of all types, e.g. rice, noodles, oats, (xi) plate scrapings and leftover of cooked food, (xii) BBQ raw or cooked leftovers and (xiii) different pet foods (Karmee 2016).

According to various reports, different kinds of food wastes are the promising feedstock for liquid biofuels as they contain nutritious components (Pham et al. 2014; Karmee and Lin 2014b). Food wastes are totally unhygienic though it contains a large amount of carbohydrate, amino acid, lipid and phosphates. Carbohydrate and amino acid hydrolysate can be utilized for bioethanol and other valuable products, whereas lipid portion can be extracted for biodiesel production (Karmee and Lin 2014a). Furthermore, bio-oil can be synthesized by direct conversion of food waste through pyrolysis.

3 Biodiesel from Food Waste

Biodiesel is the major concern of the transport sector of the present world. Biodiesel consists of either saturated or unsaturated fatty acid methyl ester. Saturation or unsaturation of fatty acid depends on the types of feedstock. Food waste and waste cooking oil are low-cost feedstock and better resource for biodiesel production compared to edible as well as non-edible vegetable oils (Kiran et al. 2014; Yaakob et al. 2013).

Lipid extraction from food waste is required for the production of biodiesel. To meet the aim, generally food waste is mixed with water with ratio of 1:100 (w/v) and vortexed to make slurry. Further, the slurry is mixed with non-aqueous solvents like n-hexane, chloroform, diethyl ether, etc. Lipid will be extracted by the action of organic solvent. Afterwards, the obtained mixture is transferred into a separating funnel. The organic layer is separated and evaporated under reduced pressure to obtain the organic solvent-free lipid. Oil-producing micro-organisms such as *Aspergillus awamori* or *Aspergillus oryzae* also can be used for lipid extraction from the unpalatable form of food waste. Micro-organisms are mixed with the slurry of food waste and incubated for 5–7 days at 30 °C to accumulate proteolytic, amylolytic enzymes. After the incubation period, the liquid hydrolysate fraction and lipid-rich fungal biomass were

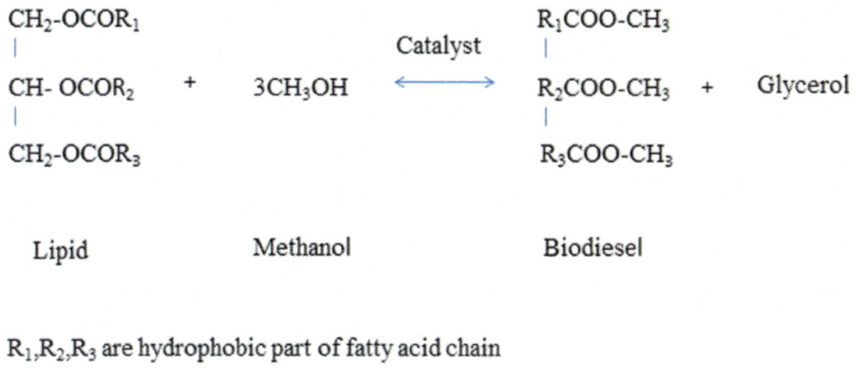

Fig. 1 Transesterification process

obtained. Simultaneously, fungal biomass is separated out and heated to 100 °C to extract the lipid simply by decanting (Karmee 2016).

Transesterification process is generally employed for biodiesel production from the extracted lipid. The process can be carried out in several ways such as using an alkali catalyst, acid catalyst, biocatalyst, heterogeneous catalyst or alcohols in their supercritical state. Transesterification reaction can be summarized as (Fig. 1).

Though base catalyst is commercially used for biodiesel production due to its low cost and high rate of reaction, it leads to saponification rather than transesterification in the presence of water. Moreover, extra downstream processing steps have to be included for neutralization, separation of the glycerol as well as washing of biodiesel product with water to remove impure products. Additionally, both acid and base catalysed reactions are not efficient due to their corrosiveness and toxic nature towards the environment as well as requirement of high reaction temperature. Problems regarding base and acid catalysed reaction can be nullified by using enzymatic transesterification as enzyme action is not hindered by the presence of free fatty acids and water content. However, biocatalyst is substantially more expensive, and the reaction rate is much slower compared to base catalysts (Pham et al. 2014).

4 Potential of Biodiesel Production from Food Waste

Due to limited fossil fuel source, the economic as well as social growth of the country will be controlled by biofuel in the coming years. Additionally, greenhouse gas emission will be lowered by the use of biofuels. Also, a gradual decrease on the dependence on politically unstable fossil fuel-rich nations will have economically positive impact on various countries. Cost of the feedstock is the main bottleneck for biodiesel production. In that context, food waste is a great choice for biodiesel production as it is no cost resource and disposed without further use. However, classification, transportation and pre-treatment of food waste include the main costs.

Hence to establish a commercial plant for biodiesel production, a prior analysis of techno-economics is required counting (i) design and cost estimation of biofuel plant, (ii) methodology improvement, (iii) real market data, (iv) techno-economics of the production facility and (v) cost of the biofuel (Economics of Biofuels 2015). The future policy should be focused on using this no-value resource to produce high-value products. Metropolis in India is facing air pollution with current food waste disposal techniques due to limited space. Hence, utilization of food waste can nullify the problems and can build a sustainable bio-based economy.

Nowadays, several companies are budding throughout the world to convert waste into energy. RWL Water Group generally collects wastes from slaughterhouses, breweries, dairy farms and coffee shops and uses it for energy conversion (2015). Enerkem Alberta Biofuels LP is situated in Edmonton, Canada, which targets to transfer approximately 100,000 t of municipal waste into 38 million litres of biofuel and valuable products per year (Waste-to-biofuels and chemicals facility 2015). M/S Asia Bio-energy Pvt. Ltd (ABIL) situated in Chennai, India, converts 5.1 MW MSW to energy using technology "biogas-induced mixing arrangement (BIMA)".

5 Conclusion

Food waste is a zero cost and non-edible feedstock. Food waste can prove to be an efficient resource for high-quality biodiesel production as it is more available compared to virgin vegetable oils. Proper and useful valorization of food waste is the major concern in academic as well as industrial research field. The future scope should aim the efficient and cost-effective valorization of food waste into biofuels and value-added products at commercial scale by chemical and biochemical methods.

Acknowledgements The authors are grateful to CSIR-CMERI for providing all the facilities to conduct the research work.

Funding This work was financially supported by the Science and Engineering Research Board (SERB), Department of Science and Technology, Government of India, under National Post-Doctoral Fellowship Scheme [Grant Number: PDF/2017/001189/ES].

References

A food waste & yard waste plan for Hong Kong. (http://www.enb.gov.hk/en/files/FoodWastePolicyEng.pdf); 2014–2022. Accessed 11 Feb 2015.

Balat, M., & Balat, H. (2008). A critical review of biodiesel as a vehicular fuel. *Energy Conversion and Management, 49,* 2727–2741.

Baskarn, G., & Aiswarya, R. (2016). Trends in catalytic production of biodiesel from various feedstocks. *Renewable and Sustainable Energy Reviews, 57,* 496–504.

Das, D., Srinivasu, M., & Bandyopadhyay, M. (1998). Solid state acidification of vegetable waste. *Indian Journal of Environmental Health, 40*(4), 333–342.

Demirbas, A. (2009). Progress and recent trends in biodiesel fuels. *Energy Conversion and Management, 50,* 14–34.

Economics of Biofuels. (2015). (http://yosemite.epa.gov/EE%5Cepa%5Ceed.nsf/webpages/Biofuels.html). Accessed 25 May 2015.

Fargione, J., Hill, J., Tilman, D., & Polasky, S. (2008). Hawthorne P Land clearing and the biofuel carbon debt. *Science, 319,* 1235–1238.

Gustafsson, U., Wills, W., & Draper, A. (2011). Food and public health: Contemporary issues and future directions. *Critic. Public Health, 21*(4), 385–93.

Karmee, S. K. (2016). Liquid biofuels from food waste: Current trends, prospect and limitation. *Renewable and Sustainable Energy Reviews, 53,* 945–953.

Karmee, S. K., & Lin, C. S. K. (2014a). Lipids from food waste as feedstock for biodiesel production: Case Hong Kong. *Lipid Technology, 26,* 206–209.

Karmee, S. K., & Lin, C. S. K. (2014b). Valorisation of food waste to biofuel: Current trends and technological challenges. *Sustainable Chemical Processes, 2,* 22. https://doi.org/10.1186/s40508-014-0022-1.

Kiran, E. U., Trzcinski, A. P., Ng, W. J., & Liu, Y. (2014). Bioconversion of food waste to energy: a review. *Fuel, 134,* 389–399.

Melikoglu, M., Lin, C. S. K., & Webb, C. (2013). Analysing global food waste problem: Pinpointing the facts and estimating the energy content. *Cent Eur J Eng, 3*(2), 157–164.

Monitoring of solid waste in Hong Kong. (2015). (https://www.wastereduction.gov.hk/en/assistancewizard/waste_red_sat.htm). Accessed 11 Feb 2015.

OECD. (2007). Municipal waste generation. In OECD (Ed.), *Municipal waste generation.* Tokyo: OECD Publishing.

Pham, T. P. T., Kaushik, R., Parshetti, G. K., Russell, M., & Balasubramanian, R. (2014). Food waste-to-energy conversion technologies: Current status and future directions. *Waste Management.*

Rena, Y., Yua, M., Wua, C., Wanga, Q., Gaoa, M., Huanga, Q., et al. (2018). A comprehensive review on food waste anaerobic digestion: Research updates and tendencies. *Bioresource Technology, 247,* 1069–1076.

Tilman, D., Socolow, R., Foley, J. A., Hill, J., Larson, E., Lynd, L., et al. (2009). Beneficial biofuels–the food, energy, and environment trilemma. *Science, 325,* 270–271.

Verma, D., Raj, J., Pal, A., & Jain, M. (2016). A critical review on production of biodiesel from various feedstocks. *Journal of Scientific and Innovative Research, 5*(2), 51–58.

Wasted energy: Converting discarded food into biofuels promises global energy boon. (2015). http://www.scientificamerican.com/article/food-waste-toenergy/. Accessed 25 May 2015.

Waste-to-biofuels and chemicals facility. (2015). (http://www.edmonton.ca/programs_services/garbage_waste/biofuels-facility.aspx). Accessed 25 May 2015.

Yaakob, Z., Mohammad, M., Alherbawi, M., Alam, Z., & Sopian, K. (2013). Overview of the production of biodiesel from waste cooking oil. *Renewable and Sustainable Energy Reviews, 18,* 184–193.

Growth and Enzyme Activity of Lipolytic Bacteria Isolated from Degrading Oil Cakes

Sutripta Sarkar and Subhasree Banerjee

Abstract Lipases are water soluble and industrially important enzymes and find application in the dairy industry, manufacturing detergents, pharmaceutical industries, etc. Lipases also have a potential role in waste management and biofuel production. They are acyl hydrolases and play a role in fat digestion by cleaving long-chain triglyceride into polar lipids. Due to the opposite polarity between the enzyme (hydrophilic) and their substrate (lipophilic), lipase reaction occurs at an interface between the aqueous and oil phase. In our previous study, we had isolated a few lipolytic strains from degrading oil cakes. They were microbiologically and biochemically characterized. Thirteen strains showed good lipolytic activity (Sarkar and Chatterji 2018). In this study, we further characterize these strains by growing them in different oil medium. Growth, cell count, and enzyme activity were assessed in nine different oil mediums (tributyrin, Tween 20, Tween 80, castor oil, coconut oil, olive oil, mustard oil, and used oil). Used oil was collected from the wok of a fried food street-side vendor. For growth and enzyme activity studies, a well-known lipase producing bacteria of the *Pseudomonas* species was taken as reference standard. While most of the thirteen strains showed reasonably good cell count in all the mediums used, isolated strain LC showed maximum activity in castor oil medium. In our previous study, stain LJ had shown maximum activity in standard growth medium (starch oil medium) for lipolytic bacteria. Though many other bacterial strains have reported higher activities, the lipase activity of strain LC was five times higher than the *Pseudomonas* species. Other stains like LA, LK, LG, LJ, and LH showed good activity in different oil mediums. This work helps in assessing the capacity of different strains in breaking down the fatty acids of different sizes. Thus, these lipolytic strains can find application in waste treatment and other relevant industries.

Keywords Lipases · Degrading oil cakes · Waste treatment · Biofuel product

---

S. Sarkar (✉)
Post Graduate, Department of Food and Nutrition, Barrackpore Rastraguru Surendranath College, 85, Middle Road, Barrackpore 700120, India
e-mail: sarkar.sutripta@gmail.com

S. Banerjee
Post Graduate, Department of Microbiology, Barrackpore Rastraguru Surendranath College, 85, Middle Road, Barrackpore 700120, India

S. K. Ghosh et al. (eds.), *Bioresource Utilization and Bioprocess*,
https://doi.org/10.1007/978-981-15-1607-8_26

1 Introduction

Lipases are serine hydrolases that catalyze the hydrolysis and synthesis of esters formed from glycerol and long-chain fatty acids and act at the oil–water interface (Kanmani et al. 2015). Lipases catalyze esterification and transesterification reactions, which are used for biodiesel formation (Boonmahome and Mongoltharnuk 2013). They are used for the conversion of glycerides and fatty acids to fatty acid alkyl ester, i.e., biodiesel (Ribeiro et al. 2011). Several researchers have used both extracellular and intracellular lipases and different varieties of raw materials like animal fats, vegetable oils, used cooking oil, acidic waste from oil mills, etc. for biodiesel production (Shimada et al. 1999; Ban et al. 2001; Kose et al. 2002; Correa et al. 2011). Esters of short and medium-chain carboxylic acids and alcohol moieties synthesized by lipase are used in the food industries to enhance aroma and flavor (Aravindan et al. 2007). Lipases are extensively used in waste management (Okino-Delgado et al. 2017). They are used in the aerobic treatment of activated sludge, where thin oil layers are removed from the surface of the aerated tanks by lipases in order to facilitate proper aeration (Hou 2002). They are also used in the treatment of fat-laden wastewater generated from dairies, slaughterhouses, and fish processing units. Lipases help in enhancing the biodegradability of fatty wastewaters (Cammarota and Friere 2006). An enzyme hydrolytic step preceding the biological degradation helps in reduction of the fat particle size making it easier for microbes to assimilate it (Valladao et al. 2011). Therefore, lipases can be substituted for the highly alkaline processes used conventionally which generate alkaline waste (Gog et al. 2012). With the use of modern techniques like site-specific mutagenesis and protein engineering, it is now possible to customize enzymes for specific targets (Stemmer 1994; Lutz 2010; Akubulut et al. 2013).

Literature is available on several bacterial and fungal lipases which have found application in industries (Kanmani et al. 2015). Several types of lipid-rich wastes can be utilized as a growth medium for lipase producing bacteria. Groundnut and mustard oil cakes (Joseph et al. 2011), grease waste (Kumar et al. 2011), lignocellulosic agricultural waste such as rice husk and wheat bran, etc. (Mala et al. 2007; Colla et al. 2010) have been studied extensively for their potential as substrate for lipase production.

In this study, thirteen lipase producing strains which were previously isolated from degrading oil cakes (Sarkar and Chatterji 2018) are characterized for their growth and enzyme activities in different oil mediums. The current study attempts to explore new microbial strains or consortium which has a future prospect in industries, waste management, and biofuel production.

2 Materials and Methods

Isolation and biochemical characterization methods of the isolated strains have been detailed in Sarkar and Chatterjee 2018. Following medium were used for growth of lipolytic bacteria: LB broth/agar (0.5% Yeast extract, 1% Tryptone, 1% NaCl, 2% Agar, pH 7). Olive oil (1%) (V/V), Agar—2%, pH was maintained at 7.5. Tributyrin medium was prepared by adding Peptone—5 gm, Yeast Extract—3 gm. Tributyrin—10 ml per 1000 ml of distilled water (pH 7.0). For Tween 20 medium, Tween 80 medium, olive oil medium, coconut oil medium, and castor oil medium, 4.7 ml of Tween 20, Tween 80, olive oil, coconut oil, and castor oil were added to 1000 ml of distilled water along with 10 gm peptone and 2 gm $CaCl_2$. pH of the medium maintained at 7. 2% agar was added to the broths for making plates. The 13 strains and a lipase producing standard strain of *Pseudomonas sp.* were inoculated in these different oil mediums and incubated overnight at 37 °C. Microbial cell cultures were then diluted in sterile 9% NaCl solution and plated in the respective agar mediums. After overnight incubation at 37 °C, colonies formed were counted. Microbial cells were plated in triplicate.

For lipase assay, the method described by Kanwar et al. (2005) and as reported by Sarkar and Chatterjee (2018) was followed. Para nitrophenol (PNP) stock of 60 μg/ml concentration was prepared for the standard curve. 20 mm stock solution of para nitrophenol palmitate (PNPP) was prepared in isopropanol and 0.05 M Tris-HCl buffer of pH 8.5. Twenty-four hours old cultures of different isolated strains and standard strain, *Pseudomonas aeruginosa*, grown in different oil medium were centrifuged at 10,000 rpm for 20 min at 4 °C and the supernatant was collected. PNPP and the buffer were added to the supernatant and incubated at 37 °C for 10 min.

Ethanol: acetone (1:1) mixture was added to stop the reaction. Optical density was measured at 410 nm. Blank for each set of assay tubes was made by adding heat-inactivated (boiling in water bath for 10 min.) crude enzyme and everything else was remained same. The absorbance of these enzme blank tubes were subtracted from the respective test sample tubes and actual absorbance was recorded for each strain.

One unit (1U) was defined as that amount of enzyme that liberated 1 μmol of pNPP per minute per ml under the test conditions (Karadzic et al. 2006).

3 Result and Discussion

Thirteen strains were isolated from degrading mustard oil cake by dilution method except *Pseudomonas sp.* All thirteen strains were characterized morphologically and biochemically and their gram characters were also noted. Most of the cells were coccus baring strain L_E, which was rod shaped. All the strains were Gram +ve and also gave positive catalase test. Strains L_E and L_F gave positive results for indole test.

Strains L_A, L_B, L_C, L_D, L_E, L_F showed positive methyl red (MR) test and Voges–Proskauer (VP) test was positive in case of L_E, L_F, L_H, L_I, L_J (Sarkar and Chatterji 2018).

These strains (LA, LB, LC, LD, LE, LF, LG, LH, LI, LJ, LK, LL, LM, and Pseudomonas sp.) show lipolytic activity, i.e., lipase degrading activity by producing clear zone around their colony (data not shown). Eight different oil mediums were used in this experiment.

All cell count values given are mean of triplicate values (Figs. 1, 2, 3, 4, 5, 6, 7 and 8).

Most of the isolates showed a higher cell count in the oil mediums when compared to the standard strain. Figures 9 and 10 show the enzyme activity of the isolates in different oil medium.

Most of the strains showed some activity in all the oil mediums but strains LA, LK, LG, LJ, LH showed much higher activity when compared to the standard pseudomonas strain. Many previously isolated strains have reported much higher activity; these isolates still have a potential of being used in waste treatments. There impact when used as consortium needs to be explored.

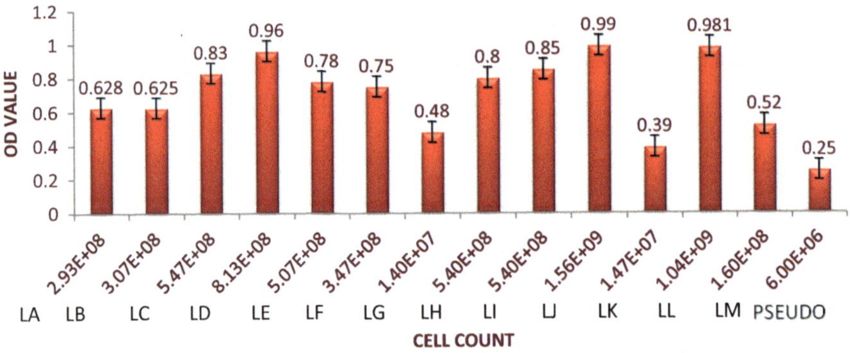

Fig. 1 Cell count in tributyrin medium

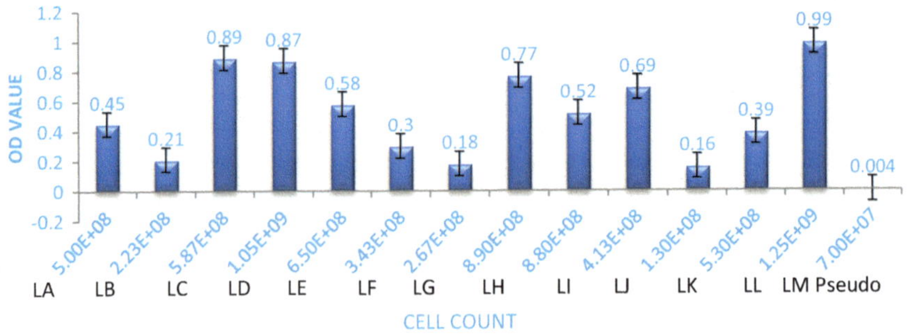

Fig. 2 Cell count in tween 80 medium

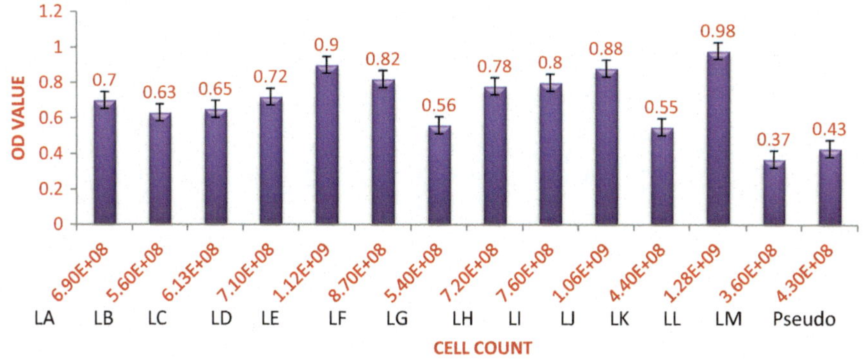

Fig. 3 Cell count in tween 20 medium

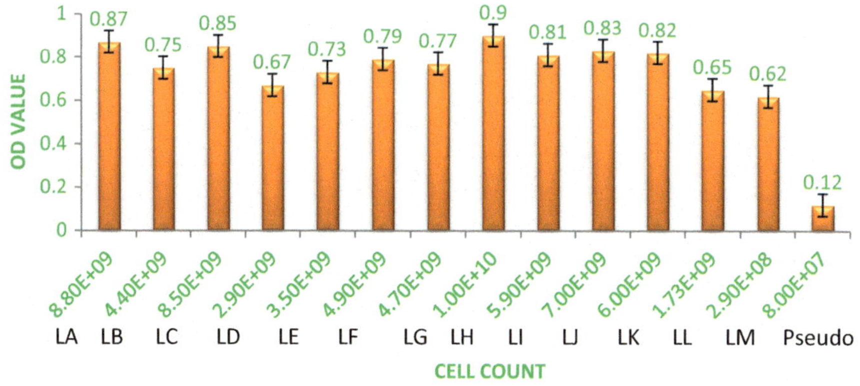

Fig. 4 Cell count in olive oil medium

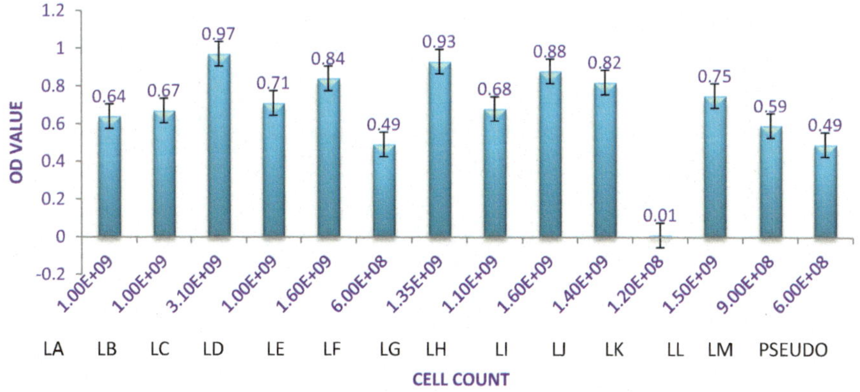

Fig. 5 Cell count in castor oil medium

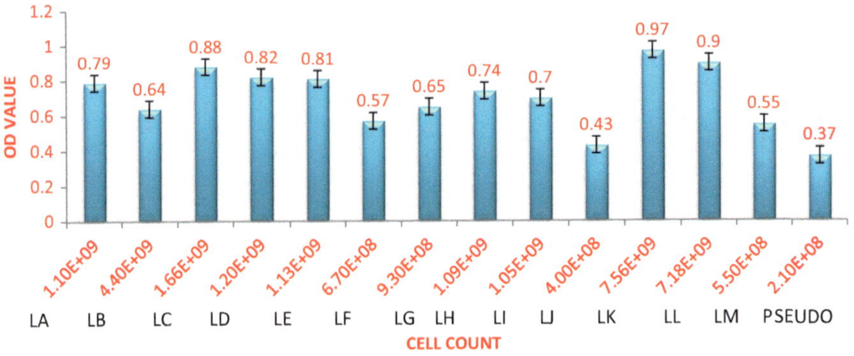

Fig. 6 Cell count in coconut oil medium

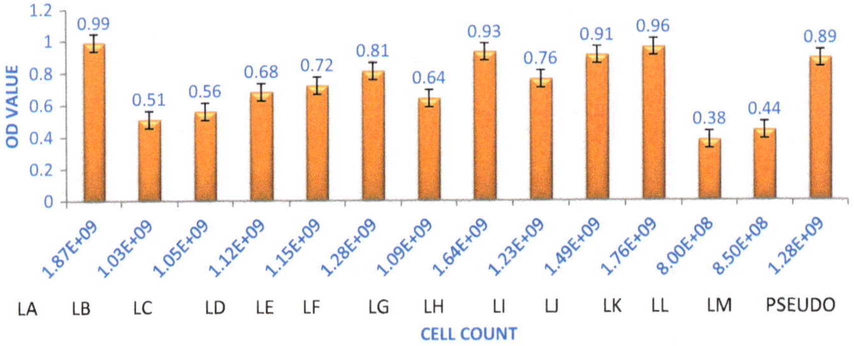

Fig. 7 Cell count in mustard oil medium

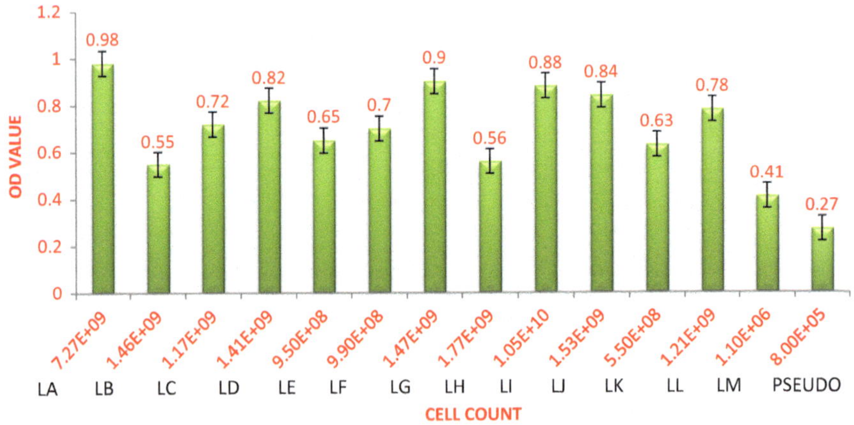

Fig. 8 Cell count in used oil medium

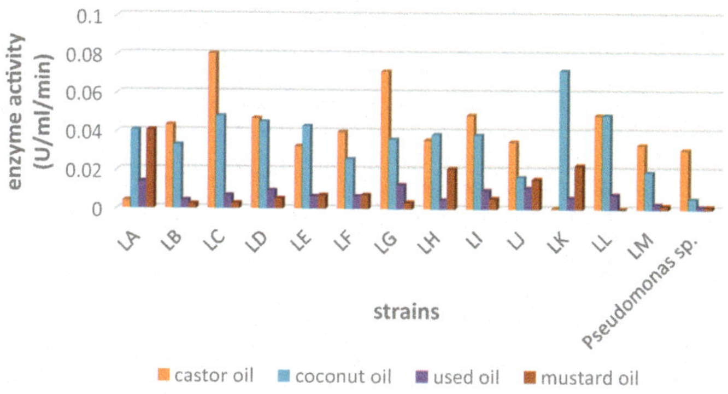

Fig. 9 Enzyme activity of different strains in castor oil, coconut oil, used oil and mustard oil medium

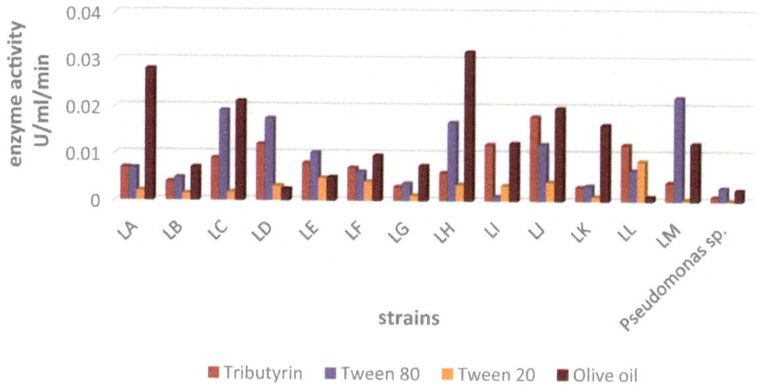

Fig. 10 Enzyme activity of different strains in Tributyrin, Tween 80, Tween 20 and olive oil medium

4 Conclusions

Lipase is a commercially important enzyme but very few bacterial lipases are actually used in industries. The present study was an attempt to isolate and identify a few lipase producing bacterial strains from oil industry waste which can be commercially viable and can be effectively used for waste management and bioremediation. Highest activity was shown by the isolates LA, LC, LK, LG, LJ, LH but it was low compared to other fungal lipase activities. However, that the isolates could utilize a variety of oils and fats was as a unique feature of this study.

Acknowledgements Authors wish to thank Principal, Barrackpore Rastraguru Surendranath College for Encouragement and Support.

References

Akbulut, N., Ozturk, M. T., Pijning, T., Ozturk, S. I., & Gumusel, F. (2013). Improved activity and thermostability of Bacillus pumilus lipase by directed evolution. *Journal of Biotechnology, 164*, 123–129.

Aravindan, R., Anbumathi, P., & Viruthagiri, T. (2007). Lipase applications in food industry. *Indian Journal of Biotechnology, 6*, 141–158.

Ban, K., Kaieda, M., Matsumoto, T., Kondo, A., & Fukuda, H. (2001). Whole cell biocatalyst for biodiesel fuel production utilizing Rhizopus oryzae cells immobilized within biomass support particles. *Biochemical Engineering Journal, 8*(1), 39–43.

Boonmahome, P., & Mongkolthanaruk, W. (2013). Lipase-producing bacterium and its enzyme characterization. *Journal of Life Science and Technologies, 1*(4), 196–200.

Cammarota, M. C., & Freire, D. M. G. (2006). A review on hydrolytic enzymes in the treatment of wastewater with high oil and grease content. *Bioresource Technology, 97*(17), 2195–2210.

Correa, I. N. S., Souza, S. L., Catran, M., et al. (2011). Enzymatic biodiesel synthesis using a byproduct obtained from palm oil refining. *Enzyme Research, 2011*, 8. Article ID 814507.

Colla, L. M., Rizzardi, J., Pinto, M. H., Reinehr, C. O., Bertolin, T. E., & Costa, J. A. (2010). Simultaneous production of lipases and biosurfactants by submerged and solid-state bioprocesses. *Bioresource Technology, 101*, 8308–8314.

Gog, A., Roman, M., Tos, M., Paizs, A. C., & Irimie, F. D. (2012). Biodiesel production using enzymatic transesterification: Current state and perspectives. *Renewable Energy, 39*(1), 10–16.

Hou, C. T. (2002). Industrial uses of Lipases. In T. M. Kuo & H. Gardner (Eds.), *Lipid Biotechnology* (pp 387–395). USA: Marcel Dekker Inc.

Joseph, B., Upadhyaya, S., & Ramteke, P. (2011). Production of cold-active bacterial lipases through semisolid state fermentation using oil cakes. *Enzyme Research, 2011*. Article ID 796407.

Kanmani, P., Arvind, J., & Kumaresan, K. (2015). An insight into microbial lipases and their environmental facet. *International Journal of Environmental Science and Technology, 12*, 1147–1152.

Kanwar, S., Kaushal, R., et al. (2005). Methods for inhibition of residual lipase activity in colorimetric assay: A comparative study. *Indian Journal of Biochemistry & Biophysics, 42*, 233–237.

Karadzic, I., Masui, A., Zivkovic, L. I., & Fujiwara, N. (2006). Purification and characterization of an alkaline lipase from *Pseudomonas aeruginosa* isolated from putrid mineral cutting oil as component of metal working fluid. *Journal of Bioscience and Bioengineering, 102*, 82–89.

Kose, O., Tuter, M., & Aksoy, H. A. (2002). Immobilized *Candida antarctica* lipase-catalyzed alcoholysis of cotton seed oil in a solvent-free medium. *Bioresource Technology, 83*(2), 125–129.

Kumar, S., Katiyar, N., Ingle, P., & Negi, S. (2011). Use of evolutionary operation (EVOP) factorial design technique to develop a bioprocess using grease waste as a substrate for lipase production. *Bioresource technology, 102*, 4909–4912.

Lutz, S. (2010). Beyond directed evolution-semi-rational protein engineering and design. *Current Opinion in Biotechnology, 21*(6), 734–743.

Mala, J. G. S., Edwinoliver, N. G., Kamini, N. R., & Puvanakrishnan, R. (2007). Mixed substrate solid state fermentation for production and extraction of lipase from Aspergillus niger MTCC 2594. *Journal of General and Applied Microbiology, 53*(4), 247–253.

Okino-Delgado, C. H., Do Prado, D. Z., Facanali, R., Marques, M. M., Nascimento, A. S., Fernandes, C. J., et al. (2017). Bioremediation of cooking oil waste using lipases from waste water. *PLOS One, 12*(10), e0186246.

Ribeiro, B. D., Machado de Castro, A., Coelho, M. A. Z., & Freire, D. M. G. (2011). Production and use of lipases in bioenergy: A review from the feedstocks to biodiesel production. *Enzyme Research*. https://doi.org/10.4061/2011/615803.

Sarkar, S., & Chatterji, A. (2018). Characterization of lipase-producing bacteria from degrading oil cakes. In S. K. Ghosh (Ed.), *Utilization and management of bioresources* (pp. 253–260). Singapore: Springer Nature.

Shimada, Y., Watanabe, Y., Samukawa, T., et al. (1999). Conversion of vegetable oil to biodiesel using immobilized *Candida Antarctica* lipase. *Journal of the American Oil Chemists' Society, 76*(7), 789–793.

Stemmer, W. P. C. (1994). DNA shuffling by random fragmentation and reassembly: In vitro recombination for molecular evolution. *Proceedings of the National Academy of Sciences of the United States of America, 91*(22), 10747–10751.

Valladao, A. B. G., Cammarota, M. C., & Freire, D. M. G. (2011). Performance of an anaerobic reactor treating poultry abattoir wastewater with high fat content after enzymatic hydrolysis. *Environmental Engineering Science, 28*(4), 299–307.

Printed by Printforce, the Netherlands